International Conference on Informatics, Technology and Engineering (InCITE 2017)

IOP Conference Series: Materials Science and Engineering
Volume 273

Bali, Indonesia
24 – 25 August 2017

ISBN: 978-1-5108-5269-3
ISSN: 1757-8981

Printed from e-media with permission by:

Curran Associates, Inc.
57 Morehouse Lane
Red Hook, NY 12571

Some format issues inherent in the e-media version may also appear in this print version.

Printed by Curran Associates, Inc. (2017)

For permission requests, please contact the Institute of Physics
at the address below.

Institute of Physics
Dirac House, Temple Back
Bristol BS1 6BE UK

Phone: 44 1 17 929 7481
Fax: 44 1 17 920 0979

techtracking@iop.org

Additional copies of this publication are available from:

Curran Associates, Inc.
57 Morehouse Lane
Red Hook, NY 12571 USA
Phone: 845-758-0400
Fax: 845-758-2633
Email: curran@proceedings.com
Web: www.proceedings.com

Table of contents

Volume 273

International Conference on Informatics, Technology and Engineering 2017 (InCITE 2017)

24–25 August 2017, Bali, Indonesia

Accepted papers received: 15 November 2017
Published online: 28 November 2017

Preface

Papers

The Role of IT in Sustainable Enterprise

Sustainable Energy & Earth Resources

IOP Conf. Series: Materials Science and Engineering **273** (2017) 011001 doi:10.1088/1757-899X/273/1/011001

International Conference on Informatics, Technology and Engineering 2017 (InCITE 2017)

Sustainable Technology and Innovation: Opportunities and Challenges

24-25 August 2017: Bali, Indonesia

http://incite.ubaya.ac.id

A collaborative activity jointly organized by:

Universitas Surabaya
INDONESIA

Suranaree University of Technology
THAILAND

National Taiwan University of Technology
TAIWAN

Solar Energy Research Center, Dalarna University
SWEDEN

University of Wollongong
AUSTRALIA

Supported by:

Discovery Kartika Plaza Hotel
Bali, INDONESIA

IOP Conf. Series: Materials Science and Engineering **273** (2017) 011001 doi:10.1088/1757-899X/273/1/011001

Scientific Committee (International Reviewers)

- Assoc. Prof. Azharul Karim, Ph.D. (Queensland University of Technology, AUSTRALIA)
- Prof. Dinesh Kant Kumar, Ph.D. (Royal Melbourne Institute of Technology, AUSTRALIA)
- Prof. Willy Susilo, Ph.D. (University of Wollongong, AUSTRALIA)
- Assoc. Prof. Yassierli, Ph.D. (Institut Teknologi Bandung, INDONESIA)
- Prof. Ali Altway, Ph.D. (Institut Teknologi Sepuluh Nopember, INDONESIA)
- Prof. Dr-Ing. I Made Londen Batan (Institut Teknologi Sepuluh Nopember, INDONESIA)
- Assoc. Prof. Setiyo Gunawan, Ph.D. (Institut Teknologi Sepuluh Nopember, INDONESIA)
- Prof. Renanto Handogo, Ph.D. (Institut Teknologi Sepuluh Nopember, INDONESIA)
- Prof. Mauridhi Hery Purnomo, Ph.D. (Institut Teknologi Sepuluh Nopember, INDONESIA)
- Prof. Nur Iriawan, Ph.D. (Institut Teknologi Sepuluh Nopember, INDONESIA)
- Prof. I Nyoman Pujawan, Ph.D. (Institut Teknologi Sepuluh Nopember, INDONESIA)
- Asst. Prof. Budi Hartono, Ph.D. (Universitas Gadjah Mada, INDONESIA)
- Asst. Prof. Hanung Adi Nugroho, Ph.D. (Universitas Gadjah Mada, INDONESIA)
- Asst. Prof. Dr.rer.nat. Lanny Sapei (Universitas Surabaya, INDONESIA)
- Asst. Prof. Nemuel Daniel Pah, Ph.D. (Universitas Surabaya, INDONESIA)
- Prof. Joniarto Parung, Ph.D. (Universitas Surabaya, INDONESIA)
- Prof. Lieke Riadi, Ph.D. (Universitas Surabaya, INDONESIA)
- Prof. Katsuhiko Takahashi, Ph.D. (Hiroshima University, JAPAN)
- Asst. Prof. Dr.Eng. Wahyudiono (Nagoya University, JAPAN)
- Prof. Anton Satria Prabuwono, Ph.D. (King Abdulaziz University, KINGDOM OF SAUDI ARABIA)
- Assoc. Prof. Oki Muraza, Ph.D. (King Fahd University of Petroleum & Minerals, KINGDOM OF SAUDI ARABIA)
- Assoc. Prof. Azizi Abdullah, Ph.D. (Universiti Kebangsaan Malaysia, MALAYSIA)
- Assoc. Prof. Siti Norul Huda Sheikh Abdullah, Ph.D. (Universiti Kebangsaan Malaysia, MALAYSIA)
- Assoc. Prof. Md. Jan Nordin, Ph.D. (Universiti Kebangsaan Malaysia, MALAYSIA)
- Assoc. Prof. Mohammad Faidzul Nasrudin, Ph.D. (Universiti Kebangsaan Malaysia, MALAYSIA)
- Assoc. Prof. Rosmadi Fauzi, Ph.D. (University of Malaya, MALAYSIA)
- Assoc. Prof. Md. Nasir Sulaiman, Ph.D. (Universiti Putra Malaysia, MALAYSIA)
- Prof. Ravindra S. Goonetilleke, Ph.D. (Hong Kong University of Science & Technology, PRC)
- Assoc. Prof. Tan Kay Chuan, Ph.D. (National University of Singapore, SINGAPORE)
- Asst. Prof. Aldy Gunawan, Ph.D. (Singapore Management University, SINGAPORE)
- Asst. Prof. Hendry Raharjo, Ph.D. (Chalmers University of Technology, SWEDEN)
- Assoc. Prof. Waree Kongprawechnon, Ph.D. (Sirindhorn International Institute of Technology, THAILAND)
- Asst. Prof. Itthisek Nilkhamhang, Ph.D. (Sirindhorn International Institute of Technology, THAILAND)
- Assoc. Prof. Vatanavongs Ratanavaraha, Ph.D. (Suranaree University of Technology, THAILAND)
- Assoc. Prof. Yupaporn Ruksakulpiwat, Ph.D. (Suranaree University of Technology, THAILAND)
- Assoc. Prof. Peerapong Uthansakul, Ph.D. (Suranaree University of Technology, THAILAND)

IOP Conf. Series: Materials Science and Engineering **273** (2017) 011001 doi:10.1088/1757-899X/273/1/011001

Steering Committee

Chair:	Assoc. Prof. Ir. Markus Hartono, Ph.D., CHFP, IPM
Honorary Members:	Prof. Joniarto Parung, Ph.D.
	Prof. Suksun Horpibulsuk, Ph.D.
	Prof. Nai-Wei Lo, Ph.D.
	Prof. Mats Rönnelid, Ph.D.
	Prof. Willy Susilo, Ph.D.
Members:	Assoc. Prof. Amelia Santoso, Ph.D.
	Asst. Prof. Djuwari, Ph.D.
	Mr. Agung Prayitno
	Assoc. Prof. Emma Savitri, Ph.D.
	Assoc. Prof. Budi Hartanto, Ph.D.
	Mr. Sunardi Tjandra
	Asst. Prof. Nemuel Daniel Pah, Ph.D.
	Assoc. Prof. Elieser Tarigan, Ph.D.
	Assoc. Prof. Jaya Suteja, Ph.D.
	Asst. Prof. Dr.rer.nat. Lanny Sapei
	Assoc. Prof. Hudiyo Firmanto, Ph.D.
	Assoc. Prof. Restu Kartiko Widi, Ph.D.

Organizing Committee

Chair:	Assoc. Prof. Eric Wibisono, Ph.D.
Secretary:	Assoc. Prof. Rudy Agustriyanto, Ph.D.
Treasurers:	Ms. Dhiani Tresna Absari
	Ms. Arum Soesanti
Secretariat:	Mr. Rahman Dwi Wahyudi (Coordinator)
	Ms. Yuana Elly Agustin
	Ms. Akbarningrum Fatmawati
	Ms. Yenny Sari
	Mr. Njoto Benarkah
Program:	Ms. Melissa Angga (Coordinator)
	Assoc. Prof. Susila Chandra, Ph.D.
	Mr. Susilo Wibowo
	Mr. Yunus Fransiscus
	Mr. Moh. Arbi Hadiyat
	Ms. Indri Hapsari
	Ms. Susana Limanto
	Mr. I Made Ronyastra
Website:	Mr. Daniel Soesanto
	Mr. Marcellinus Ferdinand Suciadi
Design:	Ms. Tyrza Adelia
Documentary:	Mr. Henry Hermawan

Invited Speakers

Prof. Suksun Horpibulsuk, Ph.D.
Director
Center of Excellence in Innovation for Sustainable Infrastructure Development
Suranaree University of Technology
THAILAND

Prof. Nai-Wei Lo, Ph.D.
Director
Taiwan Information Security Center
National Taiwan University of Science and Technology
TAIWAN

Prof. Mats Rönnelid, Ph.D.
Energy and Environmental Technology
Solar Energy Research Center
Dalarna University
SWEDEN

Prof. Willy Susilo, Ph.D.
Director
Center for Computer and Information Security Research
University of Wollongong
AUSTRALIA

IOP Conf. Series: Materials Science and Engineering **273** (2017) 011001 doi:10.1088/1757-899X/273/1/011001

Welcome Remarks, Chair of the Steering Committee

Welcome to Bali! Welcome to our very first International Conference on Informatics, Technology and Engineering (InCITE) 2017 held by Faculty of Engineering Universitas Surabaya (Ubaya), in collaboration with University of Wollongong (Australia), Solar Energy Research Center – Dalarna University (Sweden), Suranaree University of Technology (Thailand), and National Taiwan University of Science and Technology (Taiwan).

This international conference will be enlivened by a series of keynote speeches and parallel sessions delivered by scholars, researchers, practitioners and academicians who are coming from 5 different countries and more. More specifically, it is hoped that InCITE may link researchers and practitioners from various branches of engineering disciplines from around the world. This year's conference theme of Sustainable Technology and Innovation – Opportunities and Challenges will bring you to the critical awareness of what we have done and what we should contribute to our sustainable environment, society and economy, through the applied technology and innovation. All participants will disseminate information on the relevant and recent research and practice in engineering-based sustainability.

We would like to say thank you to all keynote speakers, presenters, and reviewers/scientific committees for the generous supports. In addition, our thank you to the Ubaya Foundation, Rector of Ubaya, Dean of Faculty of Engineering Ubaya, OC Chairman and members, SC members, and all InCITE committees.

We wish you a very pleasant and memorable stay and research networking event in InCITE 2017 Bali. We are looking forward to seeing you again at the 2nd InCITE 2019! Thank you very much. *Matur nuwun sanget.*

Assoc. Prof. Ir. Markus Hartono, Ph.D., CHFP, IPM

IOP Conf. Series: Materials Science and Engineering **273** (2017) 011001 doi:10.1088/1757-899X/273/1/011001

Welcome Remarks, Chair of the Organizing Committee

Rector of University of Surabaya: Prof. Dr. Joniarto Parung,
Dean of Faculty of Engineering, University of Surabaya: Dr. Amelia Santoso,
Honorary Keynote Speakers: Prof. Dr. Suksun Horpibulsuk, Prof. Dr. Nai-Wei Lo, Prof. Dr. Mats Rönnelid, and Prof. Dr. Willy Susilo,
Fellow Participants, Distinguished Guests, Ladies and Gentlemen:

First of all, welcome to Bali, Indonesia, and welcome to the first International Conference on Informatics, Technology and Engineering (InCITE) 2017!

It is still vivid in my memory, one and a half year ago, when some colleagues and officials of our Faculty of Engineering discussed the possibility of organizing an international event, to substitute national seminars that some of our study programs held annually or bi-annually. The call for an international event is a necessity given 30 years of Faculty of Engineering's existence, and the dawn of University of Surabaya's Silver Anniversary next year. Such a level of maturity prompts us to contribute more to a larger scale. An international event will have greater exposure to international community, and consequently greater impact to us all.

The following process, however, was far from easy. We were inexperienced, but we were faithful to our mission. It took us some time until we were able to formulate the conference theme, found prominent scholars in the selected theme, and negotiated with them. We are very grateful that all four speakers whom we approached are here with us today, to deliver their insights on opportunities and challenges in sustainable technology and innovation. Let's give our big hands to them!

Sessions beyond those with our invited speakers will deliver four sub-themes, namely: *sustainable design & innovation*, *sustainable manufacturing & processes*, *sustainable energy & earth resources*, and *the role of IT in sustainable enterprise*. We are glad to inform you that our conference has attracted 67 papers from the first round of acceptance. After careful selection by a panel that consists of high-profile international reviewers around the world, we passed 50 papers. We are thankful to our international reviewers who worked very hard providing feedback to the submitted papers. We are indebted to such great service that they have given.

I sincerely hope that the exchange of knowledge throughout this event, be it from within the substance of academic papers or during the conference time, will enhance our professional network and benefit us in the long run. Thank you to all our speakers, reviewers, participants, and most of all my committee members who have been hand-in-hand with me in this long journey! You all have made our dream come true!

We hope you will have a wonderful conference and memorable stay in Bali this week. We are looking forward to seeing you again in the next two years!

Assoc. Prof. Eric Wibisono, Ph.D.

IOP Conf. Series: Materials Science and Engineering **273** (2017) 011001 doi:10.1088/1757-899X/273/1/011001

Conference Photographs

2017

InCITE
2017

IOP Conf. Series: Materials Science and Engineering **273** (2017) 011002 doi:10.1088/1757-899X/273/1/011002

Peer review statement

All papers published in this volume of *IOP Conference Series: Materials Science and Engineering* have been peer reviewed through processes administered by the proceedings Editors. Reviews were conducted by expert referees to the professional and scientific standards expected of a proceedings journal published by IOP Publishing.

IOP Conf. Series: Materials Science and Engineering **273** (2017) 012001 doi:10.1088/1757-899X/273/1/012001

Text-based CAPTCHAs over the years

Y W Chow and W Susilo

Institute of Cybersecurity and Cryptology, School of Computing and Information Technology, University of Wollongong, Australia

E-mail: wsusilo@uow.edu.au

Abstract. The notion of CAPTCHAs has been around for more than two decades. Since its introduction, CAPTCHAs have now become a ubiquitous part of the Internet. Over the years, research on various aspects of CAPTCHAs has evolved and different design principles have emerged. This article discusses text-based CAPTCHAs in terms of their fundamental requirements, namely, security and usability. Practicality necessitates that humans must be able to correctly solve CAPTCHA challenges, while at the same time automated computer programs should have difficulty solving the challenges. This article also presents alternative paradigms to text-based CAPTCHA design that have been examined in previous work. With the advances in techniques to defeat CAPTCHAs, the future of auto- mated Turing tests is an open question.

Keywords: CAPTCHA, security, Turing test, usability

1. Introduction

The term CAPTCHA is an acronym for "Completely Automated Public Turing test to tell Computers and Humans Apart" and was coined by von Ahn et al. [1] in their seminal paper on proposing the challenge of using hard Artificial Intelligence (AI) problems for security. CAPTCHAs refer to automated challenge-response tests that humans can pass, but current computer programs cannot pass [1]. Before the introduction of the term CAPTCHA, the notion of "automated Turing tests" was previously proposed by Naor [2] in an unpublished manuscript. Furthermore, CAPTCHAs have also been referred to in other work as "Human Interaction Proofs" (HIPs) [3, 4].

These days, CAPTCHAs are a common part of the Internet and are used on various websites to distinguish whether an online service is legitimately being used by a human, or being accessed by a computer program (otherwise known as a "bot"). Its use has been effective, to a certain extent, in deterring the automated abuse of online services intended for humans [5]. As examples, CAPTCHAs have been effective in preventing malicious bot programs from signing up for thousands of free email accounts and in deterring bots from sending out thousands of spam messages every minute [6].

Numerous CAPTCHA schemes have emerged over the years. The reason for this is because many schemes have been found to be vulnerable to automated attacks, and practitioners have attempted to design more robust schemes. In addition, there has been much research on examining various CAPTCHA schemes for design flaws and weakness that can be exploited to defeat the overall scheme [6–12]. This highlights the challenging and daunting task of developing a good scheme that is robust against auto- mated attacks. To further complicate matters, not only must a good CAPTCHA be secure against attacks, it must also be easily usable by humans. It would defeat the underlying purpose of a CAPTCHA if a human had difficulty in correctly solving a challenge generated from such a scheme.

To date, there are a variety of different CAPTCHA categories and approaches. To illustrate its diversity, different approaches in CAPTCHA schemes include text-based, image-based, audio-based, containing animation, game-based requiring interaction, and so on. This article mainly focuses on text-based CAPTCHAs, which have been the most predominant form since its inception. Text-based CAPTCHAs generally consist of a sequence of alphanumeric characters, which can easily be entered via a keyboard, that are presented in the form of an image. The image may contain some sort of visual noise (e.g., a textured background) and the characters are typically distorted to impede recognition by automated computer attacks. The widespread popularity of the text-based approach is due to its intuitiveness from a human usage point of view, as well as its low implementation cost in terms of being able to automatically generated different challenges easily and present this to the user without requiring much computational resources [13].

This article discusses research on text-based CAPTCHAs, and how this has progressed and evolved over the years. It presents the fundamental design issues, namely, security and usability, along with the various forms of text-based CAPTCHAs that have been proposed by researchers and practitioners as alternatives to the traditional form of CAPTCHAs. In addition, this article also presents research on methods for defeating such CAPTCHAs.

2. Design issues

The security and usability of a resulting CAPTCHA are two of the main issues that must be considered in the design of a scheme. Security refers to the robustness of a scheme against automated attacks by a computer program, whereas usability refers to the ease at which a human can correctly solve the challenges.

The tradeoff between CAPTCHA security and usability is a difficult act to balance. This is because any practical scheme necessitates that humans must be able to correctly solve the challenges with a high success rate, while at the same time the probability that a computer program can correctly solve it must be very small. It has been stated that the design of a robust CAPTCHA must somehow capitalize on the difference in ability between humans and current computer programs [5]. With advances in research areas like computer vision and machine learning, it is debatable whether or not it is possible to design a CAPTCHA that is easy for humans but difficult for computer programs, and as such this remains an open question.

2.1. Security

Over the years, CAPTCHA security has been the topic of much research. Early work on the security of CAPTCHAs by Mori and Malik [7] showed that automated computer programs were capable of solving text-based CAPTCHAs at high success rates. In their work, they demonstrated a method of solving CAPTCHAs based on matching shape contexts of characters based on an image database of known objects, even in the midst of severe visual clutter. Their work highlighted the fact that with knowledge of a scheme's font and lexicon, CAPTCHAs based on language models are susceptible to dictionary attacks. Research also showed that distorted objects could be recognized using distortion estimation and removal techniques [8].

In other pioneering work, Chellapilla et al. [4] showed that machine learning algorithms were capable of breaking a variety of CAPTCHA schemes, without having to rely on language models. In their work, they proposed that solving a text-based CAPTCHA challenge consists of two main stages: segmentation and recognition. Segmentation refers to finding the locations of characters, whereas recognition is the identification of the characters. Their work demonstrated that the recognition problem is a trivial task for machine learning algorithms. This in turn led to the widely accepted segmentation- resistant principle for text-based CAPTCHA design, which states that if a CAPTCHA can be segmented into its constituting characters, it is essentially broken.

In later work examining the strengths and weaknesses of text-based CAPTCHAs, it was observed that the segmentation-resistant principle alone is not enough to guarantee that a CAPTCHA scheme is secure against automated attacks [10]. In fact, Google's popular "reCAPTCHA", which was text-based at the time and designed to be segmentation- resistant, was successfully broken using a holistic approach of recognizing shape con- texts of entire words [14] and also in an approach using heuristic character segmentation and recognition [15].

2.2. Usability

Practicality dictates that ease of use for a human is another fundamental requirement of CAPTCHAs. However, this requirement often conflicts with the security requirement. For example, as previously discussed, CAPTCHAs based on language models are susceptible to attacks due to a readily available dictionary of words. Nevertheless, research has shown that humans find familiar text, like dictionary words, easier to read as opposed to unfamiliar text [16]. As such, a potential compromise is to adopt the use of random "language-like" strings, like phonetic text or Markov dictionary strings, which are pronounceable but not actual words. Although, unlike completely random strings, characters like vowels will appear at higher frequencies when compared with other characters in pronounceable strings.

To impede automated attacks, many CAPTCHAs employ the use of visual clutter, e.g., in the form of background noise. From a usability perspective, humans must first be able to distinguish the text from any background clutter. Hence, too much clutter will make a scheme difficult to use. On the other hand, the use of color or background textures can make the task of perceiving the text from the background easier. However, it has been shown that the inappropriate use of color and background textures can be problematic in terms of both usability and security [17]. In general, if the color or background texture can easily be separated from the text using an automated program, it does not contribute to the security of the CAPTCHA. If so, it may be better not to use it, as it may harm usability by making it difficult to see the actual text or be distracting for humans.

3. Design variants

To overcome the limitations of traditional text-based CAPTCHAs, several different paradigms of CAPTCHAs have emerged. Some of these design variants are discussed in this section.

3.1. 3D-based

There have been a number of attempts at designing 3D-based CAPTCHAs. These approaches are typically implemented by rendering 3D models of text-objects, or of other objects, when generating the CAPTCHA challenges. The assumption of such CAPTCHAs is that the addition of the third dimension increases the difficulty for computers to identify 3D content, while being easy for humans as the perception of 3D is a natural part of the human visual system. Nevertheless, successful techniques for extracting textual content from 3D text-based CAPTCHAs to defeat them have previously been examined [12].

An example of a 3D CAPTCHA scheme, which is based on the rendering of 3D text-objects, was presented in Imsamai and Phimoltares [18]. However, it can be seen that in their approach that characters can easily be extracted, because no distortion was applied and the entire front face of characters were rendered using the same shade. Ross et al. [19] discuss previous approaches of using 3D models in CAPTCHAs and have pointed out problems due to the scalability of some approaches in terms of the amount of manual effort required to generate the challenges, and also observed that computer vision techniques can identify 3D objects in scenes. They proposed a prototype implementation called "Sketcha", which is based on oriented line drawings of 3D models. The user's task is to correctly orient images containing these 3D model line drawings. In other work, Susilo et al.

[20] examined an approach of generating 3D CAPTCHAs using stereoscopic images.

3.2. Animated

Animated CAPTCHAs are another design variant in which a time dimension is incorporated into the scheme. The main purpose of the time dimension is to distribute information required to solve a CAPTCHA challenge over multiple animation frames. Humans can easily process the information presented in the animated frames, whereas the assumption is that computers will have difficulty extracting this information. This has been dubbed the "zero knowledge per frame principle", because information required to solve an animated CAPTCHA is not completely contained within a single image [21]. However, it has been shown that information across multiple animation frames can be extracted to defeat such CAPTCHAs [11].

Examples of animated CAPTCHAs include a method based on moving letters amid a noisy background [21], the use of animation technology to provide better security over static images with distorted text [22], and an approach based on the idea of presenting distorted text on the face of a deforming surface [23]. Chow and Susilo [24] devised an animated 3D CAPTCHA named AniCAP, which was designed with the segmentation- resistant principle in mind. Their approach is based on motion parallax, the perception of depth through motion, where humans are supposed to distinguish the main characters located in the foreground from the background characters.

3.3. Other work

Other text-based CAPTCHA design variants include schemes where the challenge is to identify the location of characters, rather than to solely to identify the characters them- selves [25]. In addition, other than the traditional use of CAPTCHAs, it has been pro- posed that they can be used as graphical passwords [26].

Using machine learning, Bursztein et al. [27] presented a generic approach of using computers to solve text-based CAPTCHAs. They concluded that traditional purely text- based CAPTCHAs are likely to decline in utility, and that significant effort is required to rethink the way in which automated Turing tests are performed.

4. Conclusion

The research and design of CAPTCHAs have evolved over the years. Nowadays, CAPTCHAs remain a ubiquitous part of the Internet for deterring the automated abuse of online services intended for humans, and there are a variety of diverse schemes that have emerged over time. This article discusses text-based schemes in terms of their security and usability issues. In addition, this article also presents different design paradigms for text-based CAPTCHAs that have been proposed as alternatives to their traditional counterparts. With the debate surrounding the continual usage of CAPTCHAs and the various techniques for defeating CAPTCHAs that have been developed, the future ofautomated Turing tests is an open question.

References

[1] von Ahn L, Blum M, Hopper N J and Langford J 2003 CAPTCHA: Using hard AI problems for security *Lect. Notes Comp. Sc.* **2656** pp 294–311

[2] Naor M 1996 *Verification of a human in the loop or Identification via the Turing Test* http://www.wisdom.weizmann.ac.il/~naor/PAPERS/human.pdf

[3] Chew M and Baird H S 2003 BaffleText: A human interactive proof *SPIE Proc.* **5010** pp 305–316

[4] Chellapilla K and Simard P Y 2004 Using machine learning to break visual human interaction proofs (HIPs) *NIPS*

[5] Chellapilla K, Larson K, Simard P Y and Czerwinski M 2005 Building segmentation based human-friendly human interaction proofs (HIPs) *Lect. Notes Comp. Sc.* **3517** pp 1–26
[6] Yan J and Ahmad A S E 2007 Breaking visual CAPTCHAs with naive pattern recognition algorithms *ACSAC* pp 279–291
[7] Mori G and Malik J 2003 Recognizing objects in adversarial clutter: Breaking a visual CAPTCHA *CVPR* **1** pp 134–144
[8] Moy G, Jones N, Harkless C and Potter R 2004 Distortion estimation techniques in solving visual CAPTCHAs *CVPR* **2** pp 23–28
[9] Chellapilla K, Larson K, Simard P Y and Czerwinski M 2005 Computers beat humans at single character recognition in reading based human interaction proofs (HIPs) *CEAS*
[10] Bursztein E, Martin M and Mitchell J C 2005 Text-based CAPTCHA strengths and weaknesses *ACM Conf. on Computer and Communications Security* pp 125–138
[11] Nguyen V D, Chow Y-W and Susilo W 2012 Breaking an animated CAPTCHA scheme *Lect. Notes Comp. Sc. Applied Cryptography and Network Security - 10th Int. Conf. Proc.* **7341** pp 12–29
[12] Nguyen V D, Chow Y-W and Susilo W 2014 Security of text-based 3D CAPTCHAs *Computers & Security* **45** pp 84–99
[13] Chellapilla K, Larson K, Simard P Y and Czerwinski M 2005 Designing Human Friendly Human Inter-action Proofs (HIPs) *CHI ACM* pp 711–720
[14] Baecher P, Buscher N, Fischlin M and Milde B 2011 Breaking reCAPTCHA: A holistic approach via shape recognition *Future Challenges in Security and Privacy for Academia and Industry* **354** pp 56–67
[15] Cruz-Perez C, Starostenko O, Uceda-Ponga F, Aquino V A and Reyes-Cabrera L 2012 Breaking re-CAPTCHAs with unpredictable collapse: Heuristic character segmentation and recognition *Lect. Notes Comp. Sc.* **7329** pp 155–165
[16] Wang S-Y, Baird H S and Bentley J L 2006 CAPTCHA challenge tradeoffs: Familiarity of strings versus degradation of images *ICPR IEEE Computer Society* **3** pp164–167
[17] Yan J and Ahmad A S E 2008 Usability of CAPTCHAs or Usability Issues in CAPTCHA Design *SOUPS, ACM Int. Conf. Proceeding Series* pp 44–52
[18] Imsamai M and Phimoltares S 2010 3D CAPTCHA: A next generation of the CAPTCHA *Proc. of the Int. Conf. on Information Science and Applications (ICISA)* pp 1–8
[19] Ross S A, Halderman J A and Finkelstein A 2010 Sketcha: A CAPTCHA based on line drawings of 3D models *WWW ACM* pp 821–830
[20] Susilo W, Chow Y-W and Zhou H-Y 2010 STE3D-CAP: Stereoscopic 3D CAPTCHA *Lect. Notes Comp. Sc. Cryptology and Network Security* **6467** pp 221–240
[21] Cui J-S, Mei J-T, Zhang W-Z, Wang X and Zhang D 2010 A CAPTCHA implementation based on moving objects recognition problem *ICEE* pp 1277–80
[22] Athanasopoulos E and Antonatos S 2006 Enhanced CAPTCHAs: Using animation to tell humans and computers apart *Communications and Multimedia Security* **4237** pp 97–108
[23] Fischer I and Herfet T 2006 Visual CAPTCHAs for document authentication *8th IEEE Int. Workshop on Multimedia Signal Processing (MMSP 2006)* pp 471–474
[24] Chow Y-W and Susilo W 2011 AniCAP: An animated 3D CAPTCHA scheme based on motion parallax *Lect. Notes Comp. Sc. Cryptology and Network Security - 10th Int. Conf. Proc.* **7092** pp 255–271
[25] Nguyen V D, Chow Y-W and Susilo W 2014 A CAPTCHA scheme based on the identification of character locations *Lect. Notes Comp. Sc. Information Security Practice and Experience - 10th Int. Conf. Proc.* **8434** pp 60–74
[26] Zhu B B, Yan J, Bao G, Yang M and Xu N 2014 Captcha as graphical passwords - a new security primitive based on hard AI problems *IEEE T. Inf. Foren. Sec.* **9** pp 891–904
[27] Bursztein E, Aigrain J, Moscicki A and Mitchell J C 2014 The end is nigh: Generic solving of

text-based captchas *8th USENIX Workshop on Offensive Technologies*

IOP Conf. Series: Materials Science and Engineering **273** (2017) 012002 doi:10.1088/1757-899X/273/1/012002

Computer vision system for egg volume prediction using backpropagation neural network

J Siswantoro, M Y Hilman and M Widiasri

Department of Informatics Engineering, University of Surabaya, Raya Kalirungkut, Surabaya 60293, Indonesia

E-mail: joko_siswantoro@staff.ubaya.ac.id

Abstract. Volume is one of considered aspects in egg sorting process. A rapid and accurate volume measurement method is needed to develop an egg sorting system. Computer vision system (CVS) provides a promising solution for volume measurement problem. Artificial neural network (ANN) has been used to predict the volume of egg in several CVSs. However, volume prediction from ANN could have less accuracy due to inappropriate input features or inappropriate ANN structure. This paper proposes a CVS for predicting the volume of egg using ANN. The CVS acquired an image of egg from top view and then processed the image to extract its 1D and 2 D size features. The features were used as input for ANN in predicting the volume of egg. The experiment results show that the proposed CSV can predict the volume of egg with a good accuracy and less computation time.

1. Introduction

Indonesia is one of the top egg producing countries in the world. According to the data from Directorate General of Livestock, Ministry of Agriculture, Republic of Indonesia, Indonesia produced egg at about 1.43 billion kilograms in 2016 (http://www.pertanian.go.id/ap_pages/mod/datanak). The huge egg production should be followed by a rapid sorting system. Egg sorting is a process to classify egg based on internal and external qualities. There are several aspects used for assessing the quality of egg. Amongst the aspects is volume [1].

The use of computer vision system (CVS) is an appropriate solution for measuring the volume of egg. Several CVSs have been proposed for measuring the volume of egg, including 2D CVS and 3D CVS [2]. Although 3D CVS has a high accuracy, as proposed by Siswantoro et al. [3], this system requires a high computational cost, due to the number of processed image. Soltani et al. [1] have proposed a 2D CVS to predict the volume of egg based on pappus theorem and artificial neural network (ANN). The best prediction was achieved by using ANN with 28 neurons in hidden layer. In predicting the volume of egg using ANN, Soltani et al. [1] used two features, the major and minor diameters of egg, as the input of ANN. By using only this two features, ANN may produce inaccurate prediction. This is because two eggs, with the same major and minor diameters but different cross section area, may have different volume. Gonzalez et al. [4] have proposed a 2D CVS to estimate the mass and volume of passion fruit using ANN combined with Principal Component Analysis (PCA) and Linear Discriminant Analysis (LDA). Color, texture, size, and shape were used to define a model for estimating mass and volume. Although the system used many features and used combination of LDA, PCA, and ANN to estimate volume, the system achieved only 73%in terms of the correlation coefficient, with a typical error of 31.58% for testing data. This result could be caused by the fact that the proposed CVS used inappropriate input features or inappropriate ANN structure in volume estimation.

ANN is a nonlinear model that mimics the biological nervous system. It has been widely used to solve various classification and prediction problems [5]. The performance of ANN is strongly related with input features and its structure [6]. One dimensional and 2D sizes of egg, including length, width, area, and perimeter, are strongly related to the volume of egg. In addition, 1D and 2D sizes of egg can be easily extracted from the image of egg. Therefore, there is a need to investigate the use of 1D and 2D sizes as input features for ANN in predicting the volume of egg. Furthermore the appropriate structure of ANN in predicting the volume of egg also needs to be investigated. This paper aims to propose a 2D CVS for predicting the volume of egg using ANN based on 1D and 2D sizes extracted from the image of egg. The rest of the paper is organized as follow. Section 2 explains the detail of proposed CVS. Section 3 describes samples and method used in validation. The experimental results and discussion are provided in Section 4. Conclusion is drawn in Section 5.

2. Proposed computer vision system

The proposed CVS acquired the image of measured egg one by one. The image was then processed to produce a binary image. Several features were extracted from the binary image, including length, width, area, and perimeter. The features were used as the input of ANN to predict the volume of egg. The system consisted of hardware and software, designed to facilitate all steps in egg volume prediction. The following subsections describe the detail of hardware and software for the proposed CVS.

2.1. Hardware

The hardware consisted of a camera, illumination sources, a computer, a multi USB adapter, a USB extender cable and a container box. Figure 1shows the configuration of the proposed CVS hardware. The camera was used to acquire the image of measured egg. The proposed CVS employed Logitech® HD Webcam c270h. Two LED lamps were used as illumination sources. The camera and the lamps were connected to the multi USB adapter which was connected to the computer using the USB extender cable. An Intel® Core™ i3 3217U Processor with Windows 7 Ultimate 64 bit operating system and 4 GB RAM was used in the proposed CVS. The camera, the lamps, and the measured eggs were contained in the container box.

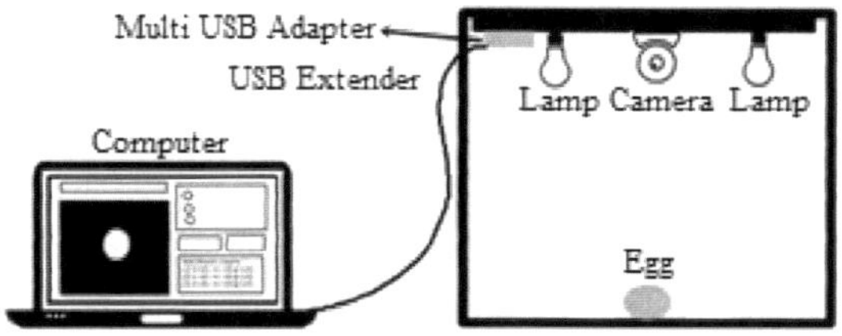

Figure 1.The configuration the proposed CVS hardware.

2.2. Software

The software was developed to control the camera, to process the acquired image, and to predict the volume of measured eggs. It consisted of several steps including image acquisition, pre-processing, segmentation, features extraction, and volume prediction. The software was implemented using C# in Visual Studio 2010 IDE and library Emgu CV 2.3.10, a cross platform .Net wrapper to the Open CV image processing library. The following subsections explain the steps of the software.

2.2.1. Image acquisition. To predict the volume of measured egg, the proposed CVS used an image as input either from the camera or from a file. The camera captured the image of measured egg from top view in a black background. The background was chosen to make the segmentation step able to be easily performed. The image was acquired in RGB (Red Green Blue) color space. The dimension of the image was 640×480 pixels with resolution 96 dpi in both vertical and horizontal directions. During image acquisition, the measured egg was located in the bottom of container box, in any position and any orientation as long as the whole of egg can be captured. The system also had facility for saving the image to a JPEG file. The file would be used for training neural network or could be used to predict the volume of egg containing in the image. The examples of acquired image can be seen in Figure 2 (a).

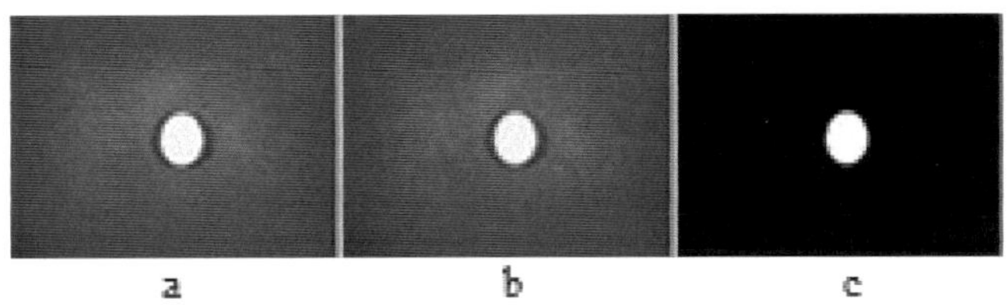

Figure 2. The examples of (a) acquired image, (b) pre-processing result, and (c) segmentation result.

2.2.2. Pre-processing .In this step, the acquired image was firstly converted to a grayscale image. The intensity of a pixel in the grayscale image was obtained from the R, G, and B intensities of the pixel in the RGB image [7]. A 5×5 Gaussian filter was then applied to the grayscale image to increase the quality of image by means of noise reducing [8]. The examples of pre-processing result can be seen in Figure 2 (b).

2.2.3. Segmentation. Segmentation aims to separate an object from its background in the grayscale image. The proposed CVS employed automatic thresholding to perform segmentation. A threshold value T was automatically determined using a simple iteration, as explained in Gonzalez and Woods [8]. The result of this step is a binary image. Any pixel in the grayscale image with grey level value greater than T was transformed into white pixel in the binary image, and any pixel in the grayscale image with grey level value less than or equal T was transformed into black pixel in the binary image. The examples of segmentation result can be seen in Figure 2 (c).

2.2.4. Features extraction. Several features were extracted from the binary image. The proposed CVS used 1D and 2D sizes of egg which are strongly related to the volume of egg, to predict volume. The features consisted of length, width, area, and perimeter. Length was defined as the longest distance between two pixels in the boundary of egg. The line between these pixels was considered as major axis. Width is defined as the longest distance between two pixels in the boundary of egg in perpendicular direction to major axis [9]. Area and perimeter are defined as area and perimeter of egg in binary image. By this definition, area and perimeter were measured by counting the number of white pixels in binary image and by counting the number of white pixel in the boundary of egg, respectively [10]. Figure 3 shows the illustration of length, width, area, and perimeter measurements.

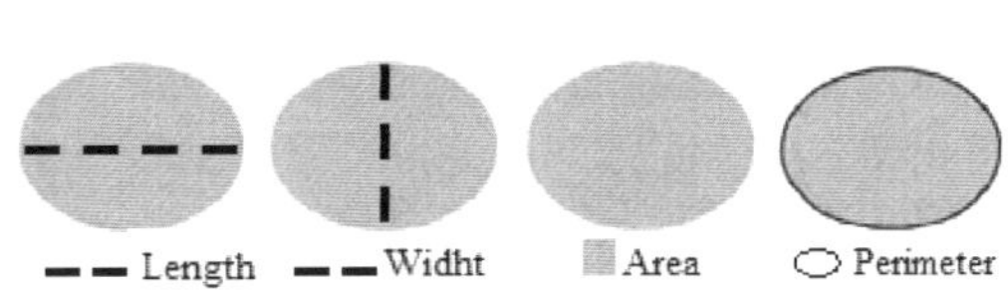

Figure 3. The illustration of length, width, area, and perimeter measurements

2.2.5. Volume prediction. The proposed CVS employed an ANN to predict the volume of egg. The structure of ANN consisted of three layers: an input layer, a hidden layer, and an output layer. The input layer had four neurons, which corresponded to four features extracted from egg image. The output layer had a neuron, which corresponded to the volume of egg. The number of neuron in the hidden layer was determined empirically from two neurons until seven neurons in an experiment, such that the best structure of neural network was obtained. Mean square error and correlation coefficient between the predicted output and the actual output of ANN were used as criteria to choose the best structure. Figure 4 shows the structure of ANN used to predict the volume of egg.

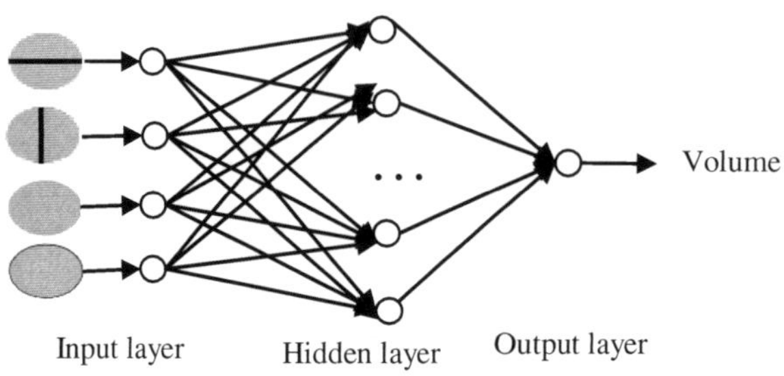

Figure 4. The structure of ANN used to predict the volume of egg.

Sigmoid functions were used as transfer functions both from the input layer to the hidden layer and from the hidden layer to the output layer. The ANN was trained using back propagation algorithm with momentum [11]. All input and output variables were normalized to interval [-1, 1] before training and testing phases. This normalization aimed to reduce the bias in the model and to speed up the training process [12]. To predict the volume of egg, the predicted output of ANN was transformed to original scale using the inverse of transformation used in normalization.

3. Validation

To validate the proposed CVS, an experiment had been performed using 80 samples of egg. The samples were chosen randomly from traditional market in Surabaya Indonesia. To train the ANN, the samples were separated into two sets, one for training set and the rest for testing set. The ratio between training set and testing set was 70:30. The actual volume for every egg was measured using water displacement method for training the ANN and validation process. Once ANN was trained, the volume of each sample was predicted using the proposed CVS. To assess the accuracy of the proposed CVS, the predicted volume and the actual volume of every sample were compared using absolute relative error (ARE). Let V_A and V_P be the actual and predicted volumes of every sample, respectively, then ARE was calculated using equation (1).

$$ARE = \frac{|V_P - V_A|}{V_A} \times 100\% \quad (1)$$

4. Results and discussion

The experiment result show that the best structure of neural network was achieved using three neurons in the hidden layer. By using this structure, the mean square error and correlation coefficient of neural network were 2.3338 and 0.9738 for all samples, respectively. This result shows that the predicted volume of egg is very close and has a strong linear correlation to the actual volume. Therefore, the proposed CVS used the neural network with four neurons in input layer, three neurons in hidden layer, and one neuron in output layer to predict the volume of egg. The correlation between the predicted volume and the actual volume of egg is depicted in Figure 5.

The predicted and actual volumes for all samples are summarized in Table 1. As can be seen in Table 1, mean, standard deviation, minimum, and maximum of predicted volume were close to mean, standard deviation, minimum, and maximum of actual volume. The proposed CVS produced mean ARE of 2.2078 %, for all samples. This result shows that the proposed CVS has a good accuracy in predicting the volume of egg. There were six samples (7.5%) with ARE more than 5%. This could be because the samples had darker shell color than the other samples. The darker shell color would be caused by some pixels near the boundary of egg recognized as background in segmentation process. To overcome this drawback, a local segmentation, as proposed by [13], can be applied to replace automatic thresholding. For comparison, the volume of each sample was also measured using volume prediction method proposed by [1] and obtained mean square error of 26.1226, correlation coefficient of 0.7281, and mean ARE of 3.5016%. This comparison results show that the proposed method archives better performance than volume prediction method proposed by [1].

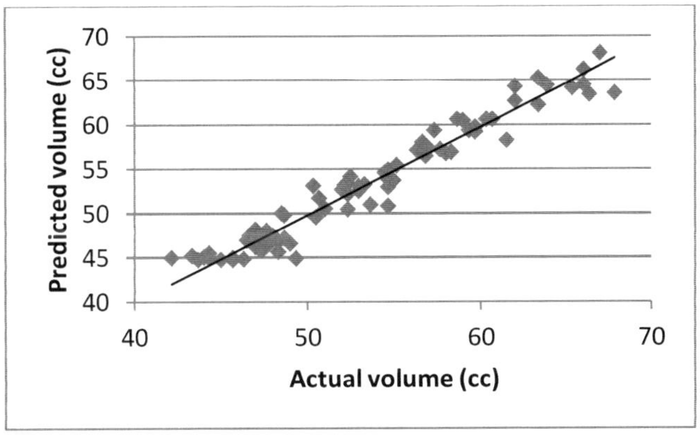

Figure 5. The correlation between the actual volume and the predicted volume of egg for all samples.

Table 1. The summary of volume prediction for all samples.

	Num. of samples	Mean	Std. dev.	Min	Max
Predicted volume (cc)	80	53.0833	6.6110	44. 6720	68.0608
Actual volume (cc)	80	53.2979	6.6672	42.1667	67.8333
ARE (%)	80	2.2078	1.8381	0.0013	9.1478

Furthermore, the paired *t*-test was performed to show that the mean of predicted volume and actual volume are not significantly different. The test was used because the two volume data come from same sample. The test was performed using a significance level of 0.05 with the hypothesis $H_0 : \mu_1 = \mu_2$ vs. $H_1 : \mu_1 \neq \mu_2$, where μ_1 and μ_2 are the mean of predicted volume and actual volume, respectively. The result of the paired *t*-test is shown in Table 2. It can be seen in Table 2 that the paired *t*-test resulted *p* value of 0.2109. This *p* value was less than the significant level 0.05. Therefore, H_0 would be rejected at the level of 0.05 and it can be inferred that the mean of predicted volume and actual volume is not significantly different.

From computation time point of view, the proposed CVS needed 2.636 s for training phase with 56 (70%) training samples. To predict the volume of egg, the computing time of the system was only 0.0156 s per sample on average. This result shows that the proposed CVS can be applied to perform volume measurement for egg sorting in real time production quality inspection.

Table 2. The result of paired *t*-test.

Paired differences		T	do	p value
Mean	Std. dev.			
-0.2147	1.5221	-1.2614	79	0.2109

5. Conclusion

In this paper, a computer vision system for egg volume prediction using artificial neural network is proposed. The system consisted of hardware and software used for image acquisition, image processing, and volume prediction. The image of measured egg was captured from top view and was then processed to extract the features of measured egg, including length, width, area, and perimeter. The features were inputted to neural network for predicting the volume. The experiment result shows that the proposed system achieved absolute relative error of 2.2078% on average compared to the actual volume. Furthermore, the predicted volume had a good linear correlation with the actual volume with correlation coefficient of 0.9738. The result of statistical analysis shows that the means of the predicted volume and the actual volume are not significantly different. For feature research, the proposed system can be extended to measure the volume of other food stuffs such as lemon, orange, and tomato. However, additional features used for predicting the volume might be needed to increase the accuracy.

References

[1] Soltani M, Omid M and Alimardani R 2015 Egg volume prediction using machine vision technique based on pappus theorem and artificial neural network *J. Food Sc. Technol.* **52** pp 3065–71

[2] Moreda G P, Ortiz-Cañavate J, García-Ramos F J and Ruiz-Altisent M 2009 Non-destructive technologies for fruit and vegetable size determination – A review *J. Food Eng.* **92** pp 119–136

[3] Siswantoro J, Prabuwono A S and Abdullah A 2014 Volume measurement algorithm for food product with irregular shape using computer vision based on monte carlo method *J. ICT Res. Appl.* **8** pp 1–17

[4] Gonzalez J P B, Ortiz F A P and Rodriguez C P P 2017 Mass and volume estimation of passion fruit using digital images *IEEE Lat. Am. T.* **15** pp 275–282

[5] Siswantoro J, Prabuwono A S, Abdullah A and Idrus B 2016 A linear model based on Kalman filter for improving neural network classification performance *Expert Syst. Appl.* **49** pp 112–122

[6] Cui A, Xu H and Jia P 2009 Anti-germ performance prediction for detergents based on elman network on small data sets *Advanced Data Mining and Applications: 5th International Conference, ADMA, Beijing, China Proc.* eds R Huang, Q Yang, J Pei, J Gama, X Meng and X Li (Berlin, Heidelberg: Springer) pp 79–90

[7] Bradski G and Kaehler A 2008 *Learning OpenCV: Computer Vision with the OpenCV Library* (Sebastopol, CA: O'Reilly Media, Inc.)

[8] Gonzalez R C and Woods R E 2002 *Digital Image Processing* (New Jersey: Prentice Hall)

[9] Yimyam P, Chalidabhongse T, Sirisomboon P and Boonmung S 2005 Physical properties analysis of mango using computer vision *Proc. ICCAS*

[10] Zheng C and Sun D-W 2008 Object measurement methods *Computer Vision Technology for Food Quality Evaluation* ed Sun D-W (Amsterdam: Academic Press) chapter 3 pp 57–80

[11] Alpaydin E 2010 *Introduction to machine learning* (Cambridge, Massachusetts: MIT Press)

[12] Priddy K L and Keller P E 2005 *Artificial Neural Networks: An Introduction* (Bellingham, Washington: SPIE Press)

[13] Siswantoro J, Prabuwono A S, Abdullah A and Idrus B 2015Automatic image segmentation using sobel operator and k-means clustering: A case study in volume measurement system for food products *Proc. Int. Conf. on Science in Information Technology (ICSITech)* p 13

IOP Conf. Series: Materials Science and Engineering **273** (2017) 012003 doi:10.1088/1757-899X/273/1/012003

A multi-hop relay path selection algorithm considering path channel quality and coordinating with bandwidth allocation

Y C Lai[1], R Jayadi[1] and J N Lai[2]

[1] Department of Information Management
National Taiwan University of Science and Technology, Taipei, Taiwan
[2] Department of Information Technology
Hsing Wu Institute of Technology, New Taipei, Taiwan

E-mail: laiyc@cs.ntust.edu.tw

Abstract. Many multi-hop relay path selection algorithms were proposed. However, these algorithms do not consider the channel condition of the overall path and coordinate with the bandwidth allocation algorithm. In this paper, we proposed a greedy based multi-hop relay path selection algorithm considering channel quality and available bandwidth of overall path to provide high throughput in varying channel conditions. From the simulation results, our proposed algorithm actually provides higher throughput and outperforms the previous works.

1. Introduction

The wireless network is generally constructed by a base station (BS) and some subscriber stations (SSs), where the former is responsible for managing bandwidth resource and the latter support the connectivity between subscribers and the BS [1]. In order to fill the BS coverage gap shadowed by buildings, enable rapid network deployment, and reduce the construction cost, the multi-hop technology and a new station, called the relay stations (RS) are introduced [2]. RS adopts the multi-hop relaying technique to extend the radio coverage of the network, increase the signal-to-noise-ratio (SNR), and promote the transmission rate. Therefore, by leveraging the RS to transmit packet on the relay path, the bandwidth resource can be well utilized to maximize network throughput.RS can be a specific station in a cellular network or a general mobile station in device-to-device communication [3].

Although the specification of the physical layer and the signaling mechanism in the MAC layer of the RS is well defined, the method to decide the packet transmission path is still vendor-dependent and an open issue. In order to determine the most suitable route between BS and SS for packet transmission, many factors should be considered including the number of hop and channel quality of each link on the selected path. Therefore, some multi-hop relay path selection algorithms are proposed. BFH [4] chooses the path with the best modulation coding scheme of its first hop link, and Markov-BFH [5] selects the relay path according to the future state of the channel quality by using Markov chain model. However, they do not consider the channel quality of links after the first hop. To compensate the lacks of BFH and Markov-BFH, DPS [6] uses the signal-to-noise ratio (SNR), link transmission rate (bits/second), and number of hops of the path as the weight for relay path selection.

On the other hand, in order to transmit relay packets, the BS not only allocates the bandwidth of current frame to the first hop of the relayed connection, but also reserves the bandwidth of the following frame to the remaining hops. This leads that the relay packets will congest and starve on the relay link without coordinating the path selection and bandwidth allocation. Therefore, the path selection algorithm should contemplate the available bandwidth of each link on the selected path that is not reserved for relay packets by the BS. Nevertheless, the proposed path selection algorithms [3-5] described above does not consider this special characteristic.

As a consequence of analyzing the lacks of previous works, a path selection algorithm considering the channel quality and the number of hops of the path and coordinating the bandwidth allocation is needed. Therefore, in this paper, we propose a novel path selection algorithm, which is named Bandwidth Allocation Awareness Greedy Path selection algorithm (BAGP). The goal of our solutions is to maximize the throughput under the considerations of the distinguished features about the path selection. In order to achieve high system throughput, BAGP first picks up the connection that can transmission packet on the path with the best average transmission rate in the network topology, and then BAGP chooses this path as the packet transmission path for the connection. The path selection for the remaining connections follows the same manner. In addition, the selected path will be modified according to the available bandwidth of each link to avoid the congestion and starvation of the relay packets on the relay link.

The organization of this paper is as follows. Section 2 describes the background and Section 3 formally describes the detailed algorithm of BAGP. In Section 4, the effectiveness of the proposed algorithms compared with Round Robin (RR) without of support RS, BFH, and DPS is demonstrated by simulation. Finally, some conclusions are given in Section 5.

2. Background

The protocol stack in a cellular network consists of two layers, physical layer and MAC layer. The former provides the means to transfer raw data, and the latter is designed to support the physical layer to manage radio resource efficiently. With regard to the physical layer adopting OFDMA, three modulations, quadrature phase shift keying (QPSK), 16 quadrature amplitude modulation (16QAM), and 64 quadrature amplitude modulation (64QAM) with different coding rate are used according to the channel quality, i.e. carrier-to-interference-plus-noise ratio (CINR). Three duplex modes are used in OFDMA including TDD (Time Division Duplex), FDD (Frequency Division Duplex), and H-FDD (Half-duplex Frequency Division Duplex). TDD is usually the more attractive duplex mode because of its flexibility.

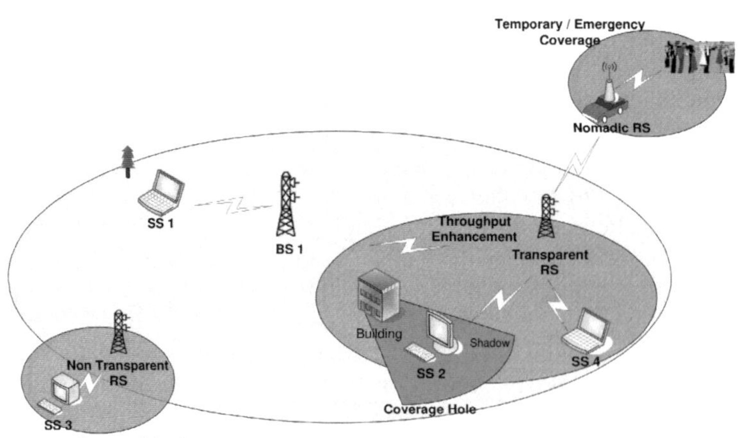

Figure 1. Multi-hop relay model.

To improve the robustness of the network, an amendment of introducing multi-hop technology and a Relay Station (RS), is developed [2]. RS is also regarded as a mobile station in D2D communication. By adopting multi-hop technology, the RS can enhance scalability and coverage of different deployment scenarios according to the characteristics and the requirements of the network environment.

Figure 1. shows the usage models of the RS that can be categorized as the following applications: (1) the RS can reduce the network constructions cost by deploying in stationary areas to improve the SNR and transmission rate of the SS; (2) the RS can fill the coverage hole shadowed by the building, tunnel, or subway; (3) the RS can be rapid deployed on the place for temporary or emergency events where a large group of people are densely packed into a small area.

3. Proposed algorithm

In this section, we first elaborate the notations and formal problem definition on the multi-hop relay path selection problem. Then, a Bandwidth Allocation Awareness Greedy Path selection algorithm (BAGP) that considering the channel quality of overall path and coordinating the bandwidth allocation is proposed.

Table 1. The notations for the Path Selection Problem and the proposed algorithm.

Notation Name	Explanation
F	The frame structure
DLC	The sets of all DL connections where $\lvert DLC \rvert = n$
$R_{m,n}$	The matrix of each connection's modulation coding schemes where $R_{m,n}$ specifies the transmission data size (bytes/slot) from node m to node n
B_i	The data size of i-th connection requested for downlink transmission
BWR	The set containing all B_i
P_i^j	The diverse path between BS and SS that selects one path to send the data
$W_{P_i^j}^h$	The number of pre-allocation slots of each hop that connection i transmits packets on path j by the certain bandwidth allocation algorithm Round Robin (RR)
BWA	The set containing all $W_{P_i^j}^h$
FPS	Stands for the number of frames the BS sends per one second
$Th(F)$	Throughput of F which denotes the set $A = \{(P_i^j, W_{P_i^j}^h) \mid i \in DLC\}$ allocates the minimum number of slots to gain the maximum throughput

3.1. The problem definition

As mentioned above, the multi-hop relay path selection algorithm has to decide the route between BS and SS and the amount of bandwidth to allocate for each connection in order to maximize the overall throughput, where the former is called the path selection and the latter is bandwidth allocation. As the notations shown in table 1, the formal definition of the path selection problem is that given a Frame F, a DLC set, a modulation coding scheme matrix R, a BWR set, and a BWA set allocated by the certain bandwidth allocation algorithm, find a set of path and allocate bandwidth $A = \{(P_i^j, W_{P_i^j}^h) \mid i \in DLC\}$ that the total throughput $Th(F)$ is maximized.

3.2. Bandwidth Allocation Awareness Greedy Path selection algorithm (BAGP)

The detailed algorithm of BAGP is presented in Figure 2 and explained as follows:

Intuitionally, each connection should transmit packet on the path with best transmission rate to promote the throughput. Thus, in the first step, we list all possible transmission paths of each connection and calculate the average transmission rate respectively.

$R_{m,n}$: the modulation coding scheme of each hop from node m to node n
B_i : the bandwidth requirement of i-th connection
BWR : the set of connections
for each connection i arrives
{
//Step1: Decide the suitable path form the path set
for each diverse path
{
calculate $A_{P_i^j} = \dfrac{1}{\sum_{R_{m,n} \in P_i^j} \dfrac{1}{R_{m,n}}}$
}
Sort($A_{P_i^j}$) in descending order
//Step2: Pre-Allocation the minimun number of slots
for each $A_{P_i^j}$ to allocation bandwidth
{
if (P_i^j = Relay)
{
$\sum_{h=1}^{H} W_{P_i^j}^{h} = require\_slot(B_i, R_{m,n}, FPS)$
$saving = \sum_{h=1}^{H} W_{P_i^j}^{h}$
$N_{relay\_zone}^{free} = N_{relay\_zone} - saving$
//Step 3: Modify the selected path to avoid the congestion
if (the remaining bandwidth of relay is exhausted or less than the require slot of connection)
{
Mark the P_i^j and do not assign it to the connection
Find next P_i^j
if (P_i^j fulfills require slot)
{
Assign P_i^j to this connection and allocate the suitable bandwidth to the connection
$W_{P_i^j}^{1} = require\_slot(B_i, R_{m,n}, FPS)$
}
}
}
}
while(all connections of BWR have been satisfied)
Terminate the execution of BAGP
}

Figure 2. The BAGP algorithm.

The average transmission rate of the path j for the connection i is calculated by:

$$A_{P_i^j} = \frac{1}{\sum_{R_{m,n} \in P_i^j} \frac{1}{R_{m,n}}} \quad (1)$$

In (1), $1/R_{m,n}$ is the number of slot to transmit one bit per hop, $\sum_{R_{m,n}\in P_i^j} 1/R_{m,n}$ means the number of slot to transmit one bit on the path P_i^j, therefore $A_{P_i^j}$ is the average transmission rate that connection i transmits packets on path j. Then, BAGP selects the connection that owns the path with best average transmission rate among the topology, and this path is set as the packet transmission path for the selected connection.

In the second step, BAGP uses Round Robin (RR) as the bandwidth allocation algorithm to allocate suitable number of slots to this chose connection according to its requested bandwidth and the modulation coding scheme of the first hop of the selected path.

$$\sum_{h=1}^{H} W_{P_i^j}^h = require\_slot(B_i, R_{m,n}, FPS) \tag{2}$$

In (2), $require\_slot(B_i, R_{m,n}, FPS)$ is the overall number of slots necessary to ensure the bandwidth requirements then divide it by the number of frames, therefore, $\sum_{h=1}^{H} W_{P_i^j}^h$ aggregates the number of pre-allocation slots of each hop that connection i transmits packets on path j we have to allocate in current and following frame.

In addition, BAGP will reserve the bandwidth saving for the remaining hops of the selected path to avoid that packets are congested and starved in the relay path. Furthermore, if the selected path does not have enough bandwidth of the each hop for the requested bandwidth of this connection, BAGP will not assign this path to the connection, and the bandwidth scheduling for this connection will be detained until being picked up by BAGP in the same fashion described above.

For the other connections, the path selection follows the same manner depicted above.

4. Simulations

In this section, we compare the proposed algorithm BAGP with the Round Robin (RR) without supports of RS, BFH, and DPS in terms of downlink throughput, where the downlink throughput is the summation of throughput of each downlink connection. Because BFH and DPS only depict the path selection algorithm, we use RR as the bandwidth allocation algorithm to combine with BFH and DPS.

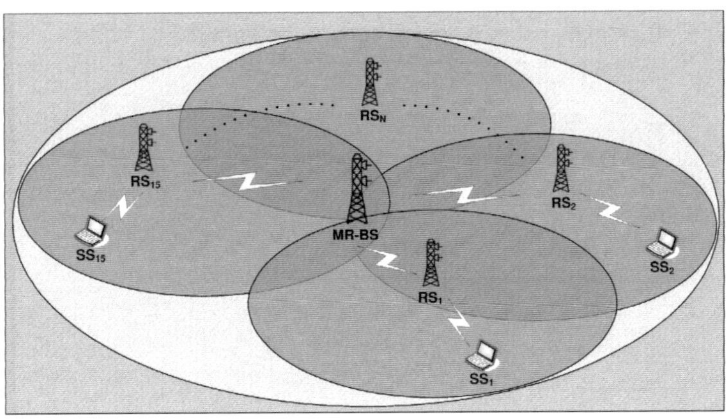

Figure 3. Network topology.

4.1. Simulation model

For the simulation environment, an OFDMA system with 20MHz bandwidth is used. The number of sub channels and symbols for a downlink sub frame are set to 60 and 36, which is 18 downlink slot width where one downlink slot occupies two symbols. Figure 3 shows the network topology that includes one BS and fifteen SSs. In order to prevent bandwidth allocation in the SSs from influencing the simulation results, each SS owns only one connection. The packet interarrival time of the connections follows the Exponential distribution. Different modulation coding schemes of each connection are adopted according to the received Carrier to Interference-plus-Noise Ratio (CINR) as shown in 2 [1]. In addition, the received CINR follows the Normal distribution with different means and standard deviations according to each simulation scenario.

Table 2. The CINR / Modulation coding scheme mapping table.

Modulation	Coding Rate	Receiver CINR (dB)	Bytes Per Slot
QPSK	1/2	6.0	6
	3/4	8.5	9
16QAM	1/2	11.5	12
	3/4	15	16
64-QAM	2/3	19	24
	3/4	21	27

4.2. Traffic load

To observe the influence of different traffic load on downlink throughput, 15 downlink connections with different data transmission rate from 100kbps to 800kbps are measured. The mean CINR between each SS and RS and between RS and BS are 15dB, and the standard deviation is 5dB. The mean CINR between each SS and BS is 8dB and the standard deviation is 1dB.

Observing from Figure 4, the downlink throughput improved by BAGP increases as the requested bandwidth increases. The reason is that BFH selects the path by considering the channel quality of the first hop, leading that the relay packet will congest on the relay link if the channel quality of relay link is worse. On the other hand, although DPS selects the path according to the SNR value, transmission rate of the link, and the number of hops, but it does not consider the available bandwidth of each link in the following frame, causing that the packets congest and starve on the relay link. With regard to RR without supports of RS, it suffers worst performance because the connection cannot relay the packets by RS if the channel quality of the access link is bad.

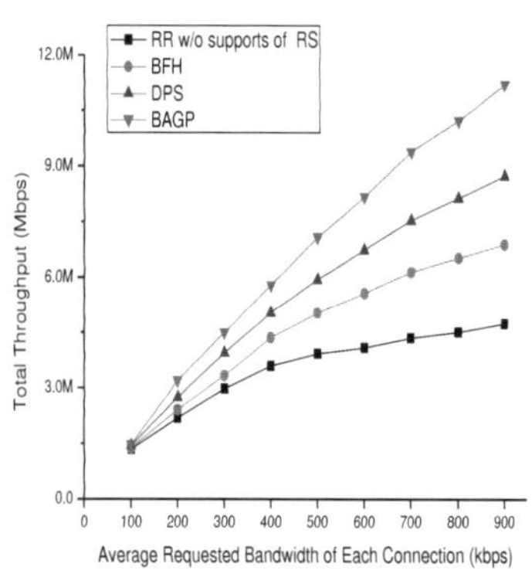

Figure 4. Total throughput vs. requested bandwidth.

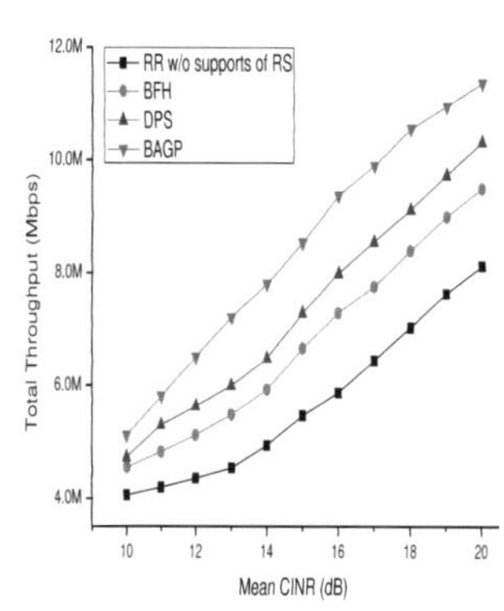

Figure 5. Total throughput vs. CINR.

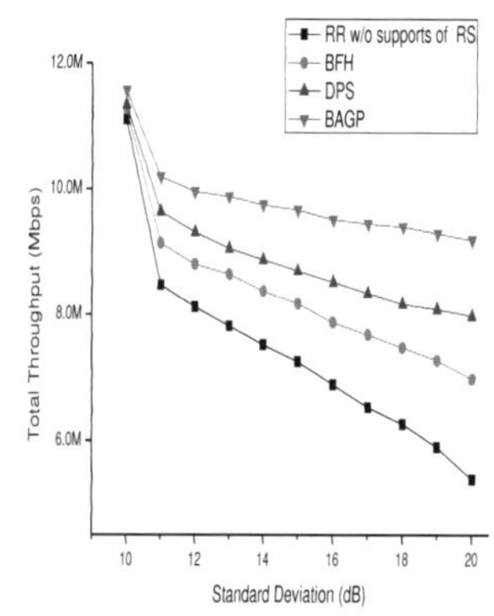

Figure 6. Total throughput vs. standard deviation.

4.3. Channel quality

We investigate the effects of the channel quality for BAGP on the downlink throughput. In this simulation, 15 downlink connections with 800kbps transmission rate are measured. The mean CINR of the subcarriers received by each SS is varied from 10dB to 20dB, and the standard deviation is 5dB.

From Figure 5, the downlink throughput BAGP still outperforms other algorithms, as the channel quality is getter better. However, the improve of BAGP on downlink throughput is small while the mean CINR is between 10 db to 12 db, this is because packets have extra bandwidth to relay RS and the influence to relay packet through RS will begin to shrink if the channel quality of the relay link is close to the access link.

4.4. The variation of channel quality
We observe the effects of different variation of the channel quality on the total throughput in this section. In this simulation, the mean CINR of the subcarriers received by the SS is 15dB, and the standard deviation is varied from 0dB to 10dB. Figure 6 shows that all algorithms have the same performance when standard deviation is 0, this is because each algorithm decides to transmit each packet directly to each SS from BS without any help of RS, since the channel quality of access link and relay link are the same and transmitting packet through the relay path costs extra bandwidth for the relay link. Furthermore, the improvement of BAGP increases as the standard variation increase, the reason is as the same as above. On the other hand, the downlink throughput of all algorithms decreases as the standard deviation of the CINR increases, this is because the more the channel condition of overall path is varied, the more chance that the connection will adopt worse modulation coding scheme, leading that the downlink throughput suffers worse performance.

5. Conclusion
In this paper, the design issues and current works with regard to the multi-hop relay path selection in the wireless network are investigated and analyzed. We also propose a path selection algorithm, Bandwidth Allocation Awareness Greedy Path selection algorithm (BAGP), which considers the channel quality, the number of hop, and available bandwidth of the path for the lacks of the current studies. The simulation results show that BAGP provides higher throughput comparing with the previous works, BFH and DPS.

Acknowledgements
The authors would like to thank Ministry of Science and Technology of Republic of China, Taiwan, for financially supporting this research under Contract No. MOST 106-3114-E-011-003 and MOST 106-2221-E-011-013.

References
[1] IEEE 802.16 Working Group 2007 Oct *Draft standard for local and metropolitan area networks—Part 16: Air interface for broadband wireless access systems* IEEE P802.16Rev2/D1
[2] IEEE 802.16 Working Group 2008 Jan *Part 16: Air interface for fixed and mobile broadband wireless access systems—multihop relay specification* IEEE P802.16j/D5
[3] Kim J and Lee J W 2014 Oct The mode selection scheme for group Device-to-Device communications underlay cellular networks *Proc. IEEE ICTC* pp 259–260
[4] Wang S S, Yin H C, Tsai Y H and Sheu S T 2007 July 1–4 An effective path selection metric for IEEE 802.16-based multi-hop relay networks *Proc. IEEE Symp. on Computers and Communications* (Aveiro) pp 1051–56
[5] Wang S S, Yin H C and Sheu S T 2008 May 26 Symmetric path selection in IEEE 802.16 multi-hop relay networks with error prone links *Proc. IEEE Int. Conf. on Circuits & Systems for Communications* (Shanghai) pp 807–11
[6] Ann S, Lee K G and Kim H S 2008 Aug 25–31 A path selection method in IEEE 802.16 j mobile multi-hop relay networks *2nd Int. Conf. on Sensor Technologies and Applications* (Cap Esterel) pp 808–12

Leaf App: Leaf recognition with deep convolutional neural networks

T L I Sugata and C K Yang

Department of Information Management, National Taiwan University of Science and Technology, Taipei 106, Taiwan

E-mail: triluhur04@gmail.com

Abstract. In this paper, a very deep convolutional neural network is used to do leaf recognition. In order to predict location of leaves, some pre-processing technique is adopted to extract regions in the image before doing classification. To improve the accuracy, we enlarge the dataset by data augmentation, i.e., doing several transformations such as horizontal reflection, contrast enhancement and rotations. Experimental results show that by using deep convolutional neural network with data augmentation, our system can achieve accuracy close to the state-of-the-art systems.

1. Introduction

Plants are very important in our lives because they provide us oxygen which is needed by all living things. In order to identify variety of plants much work has been done, including in the field of computer vision. Extraction of leaf shapes, geometric features (diameter, physiological length, physiological width, etc.) and morphological features (aspect ratio, perimeter ratio, rectangularity, etc.) [1], [2] are usually what involved to automatically identify a leaf.

Convolutional neural network (convnet or CNN) has recently become popular because it can achieve state-of-the-art result in image classification by feeding the network with a huge number of images. Alex Krizhevsky et al. [3] used five convolutional, three fully-connected layers and utilized drop out in ImageNet [5] dataset to achieve an error rate of 16.4%. Zeiler et al [4] proposed a technique to visualize convnet by using a deconvolutional network (deconvnet). They visualize AlexNet [3] to examine the convnet and do several improvements to achieve better results.

In this paper, we make the following contributions. First, we proposed a mobile leaf recognition system using a well-known deep convolutional neural network called VGG-16 architecture to achieve a better result without extracting leaf features such as shape, diameter, aspect ratio, etc. from images. Second, we extract the region of images to let the system recognize multiple and overlapped leaves together with their bounding box. Figure 1 shows the overall process of the system.

The rest of the paper is organized as follows. In section 2, we review some of the previous work. In section 3, we describe the pre-processing technique we have done before doing the classification. Section 4 describes the network architecture. Section 5 presents the experimental result. Section 6 will be the discussion and conclusion of the paper.

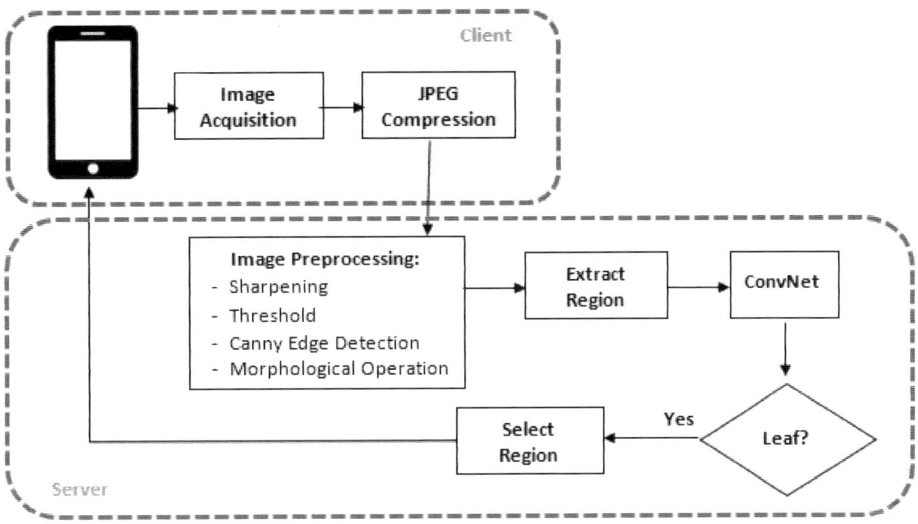

Figure 1. The overall process flow.

2. Related Works

Leaf Recognition. In recent years, many researches have been conducted on leaf recognition. Kumar et.al [11] proposed mobile app for recognize 184 kind of trees by extracting the curvature features. Wang et al. [12] use Pulse-coupled neural network (PCNN) to extract leaf features. They achieve accuracy above 90% in three different datasets. While Hu et al. [13] applies Multiscale Distance Matrix to get geometric structure of the leaf shape.

Deep Learning for Leaf Recognition. Recently, several papers start to utilize CNN to do the leaf recognition [9][10]. Data augmentation was added [9] in order to improve the accuracy. Lee et al. [10] used the AlexNet [3] model to identify leaf and visualize its feature to analyze which features are important for the leaf identification.

3. Image Pre-Processing

This process is aiming at extracting regions in the image. Because the input image from a mobile phone may not contain a single leaf. So applying some simple image preprocessing techniques on the image is needed, especially in the condition where a leaf is overlapped with each other.

3.1. Image Sharpening

For edge detection is it necessary to sharpen the image, so that edges can be fully retrieved. Image sharpening is achieved by convolving the following kernel K with the input image:

$$\mathrm{K} = \frac{1}{8}\begin{bmatrix} -1 & -1 & -1 & -1 & -1 \\ -1 & 2 & 2 & 2 & -1 \\ -1 & 2 & 8 & 2 & -1 \\ -1 & 2 & 2 & 2 & -1 \\ -1 & -1 & -1 & -1 & -1 \end{bmatrix} \quad (1)$$

3.2. Thresholding

To generate a binary image, a black pixel is used to indicate the foreground and white pixel the background, and the Otsu method [6] is applied to an image that is made grayscale beforehand. Otsu method will automatically select threshold for image segmentation by utilizing the 0-th and the first-order cumulative moment of the gray-level histogram.

3.3. Edge Detection
The well-known edge detector Canny edge detection is used [7]. This algorithm is widely adopted because it can have good detection results. The canny algorithm itself runs by the following steps:

a. Smoothen an image using a Gaussian filter to remove the noise
b. Calculate the image gradient
c. Apply non-maximal suppression to filter out response
d. Apply double thresholds to remove edge pixels with a weak gradient
e. Use hysteresis to track the edge which is not extracted from noise

3.4. Morphological Operation
Morphological operators are applied to images to remove the noise inside the image. Mathematical morphology operators consist of four basic operations which are opening, closing, dilation and erosion. The following are opening (2) and closing (3) operators:

$$□□□ \circ\circ I_{\circ} \; H = (□I \ominus H□) \oplus H□ \quad (2)$$

$$□I \bullet H□ = (□I \oplus H□) \ominus H□ \quad (3)$$

And the following are the dilation (4) and erosion (5) operators:

$$I \oplus H = \max_{(i,j) \in H} \{I(u+i, v+j) + H(i,j)\} \quad (4)$$

$$I \ominus H = \min_{(i,j) \in H} \{I(u+i, v+j) + H(i,j)\} \quad (5)$$

3.5. Image Segmentation
For extracting individual leaves in images that contain overlapping leaves or multiple leaves, using a thresholding image is not enough because deep convnet will detect it as a “single-leaf” or “non-leaf” image. So the following steps shown in Figure 2 are used.

a. Get the threshold map using Otsu method
b. Get the edge map using Canny edge
c. Sum up the threshold map and edge map
d. Invert the map
e. Remove noise/small holes by using contour hierarchy

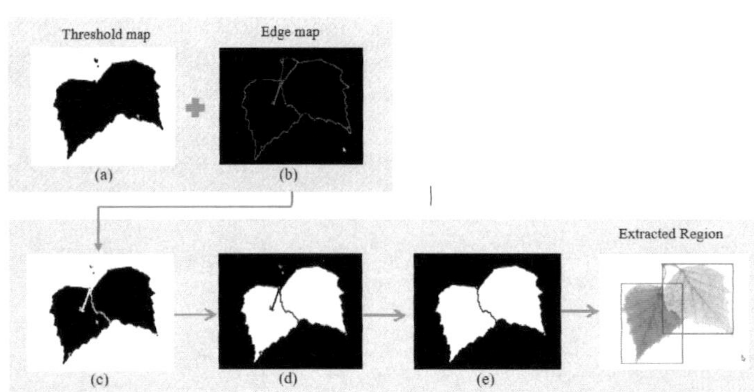

Figure 2. Overview of the segmentation approach.

4. Network Architecture
The deep convolutional neural network model used in this paper is VGG16 [8]. VGG16 as shown in **Fig. 3** consists of 16 weight layers 13 convolutional layers and 3 fully connected layers with 138million total number of parameters and takes 224 x 224 pixels RGB image as the input.

Fine-tuning. In deep convnet, we simply perform a fine tuning in the pre-trained model from ImageNet-1K dataset into Flavia dataset to classify 32 leaf classes. The fine tuning is chosen because to train the dataset from scratch will be time-consuming and the weights learned from ImageNet-1K pre-train model can be transferred to VGG16 with Flavia dataset.

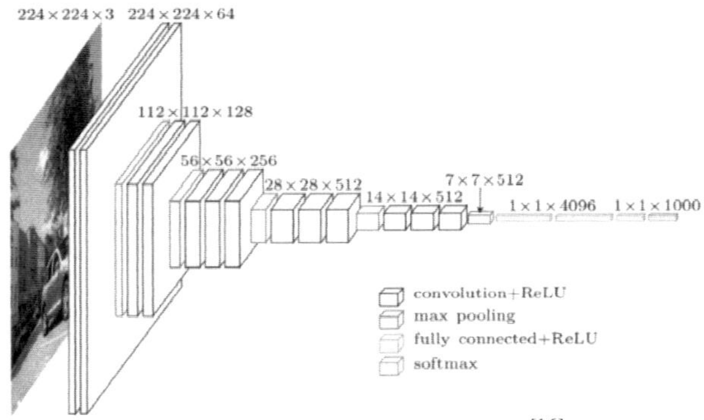

Figure 3. VGG16 architecture [16]

5. Experimental Result

Experiments were conducted on NVIDIA GeForce GTX 1080 8GB GPU using an open source framework developed by Berkeley Vision called Caffe [9] run as a server. The model training is set with an initial learning rate of 0.001, batch size of 25 examples, weight decay of 0.0005, momentum 0.9 and we use a stochastic gradient descent (SGD) as the solver type.

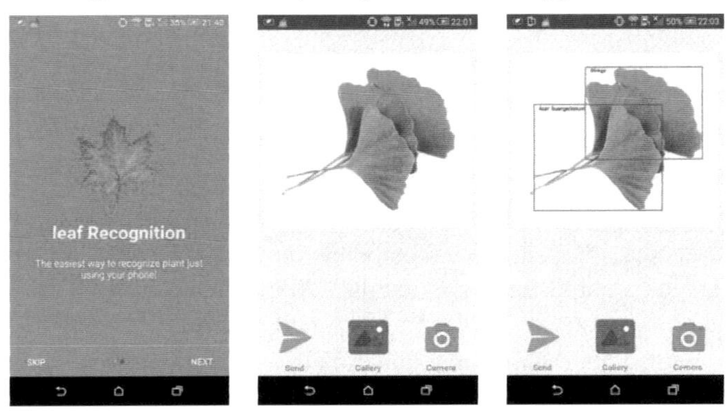

Figure 4. User interface of leaf app.

5.1. Dataset

We train the network using Flavia dataset [2], which is a popular dataset for leaf recognition. It is publicly available and consists of 32 plants classes. Each image contains a single leaf and the number of images in the dataset is 1907. To make it work with previous model, we rescaled all image dimension to 224x224.

5.2. Data Augmentation

Since we are using a deep network, we need to have a huge number of training images to get the better results and also to reduce the overfitting effect. Because the Flavia dataset only contains 1907 images, we enlarge the dataset by doing several transformations: horizontal reflection, contrast enhancement and rotations (by rotating the image with the interval 10 degrees).

5.3. Comparative Results

Table 1 shows the accuracy comparison when different datasets are used. Images are selected randomly for testing. Our system shows that by using the initial images in the Flavia dataset which contains 1907 images, it can already achieve the accuracy of 99.7%. While using a bigger dataset (initial image is expanded twenty times through data augmentation) [9], the accuracy improves to 99.9%, as shown in Figure 5.

Table 1. Accuracy in dataset.

Test set	Accuracy (%)
Initial dataset	99.7
Initial + data augmentation	99.9

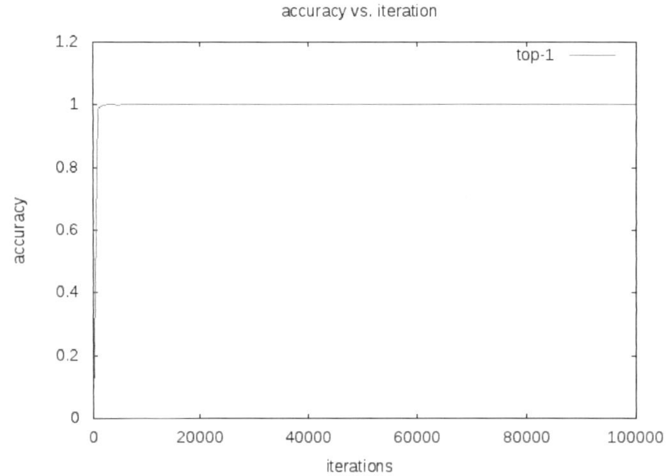

Figure 5. Accuracy according to iterations.

In Table 2, we can see that by using deep convnet, our system can achieve accuracy 99.9%. It means that deep convnet can outperform the results from other schemes. By using convnet, leaf recognition can be done automatically. Extracting the leaf shape, diameter, aspect ratio, etc. is no longer needed.

Table 2. Comparison with other schemes

Scheme	Accuracy (%)
PNN in [2]	90
PCNN in [12]	96.67
MLNN in [15]	94
CNN in [9]	94.6
Deep CNN in [10]	99.6
VGG+data augmentation	99.9

6. Discussion and Conclusion

This paper presents deep convolutional neural network for leaf recognition. We develop a mobile application, where a user can capture an image from the camera or load an image from the phone's gallery, send it to the server and the server will return the class which has the highest score along with its locations. From the experimental results, it can be shown that our system can achieve a better result when compared with the other methods. However, to be able to extract the region and locate the leaf image it is highly dependent on the image conditions (e.g., shadow, illumination, etc.). For future works, we will develop a system that can recognize leaf image under more general conditions.

Acknowledgment
This work was supported in part by the Ministry of Science and Technology of Taiwan under the grants: MOST 104-2221-E-011-083-MY2, MOST 105-2218-E-011-005, MOST 105-2218-E-001-001 and MOST 106-3114-E-011-003.

References
[1] Zhao C, Chan S S F, Cham W-K and Chu L M 2015 Plant identification using leaf shapes—A pattern counting approach *Pattern Recogn.* pp 3203–15
[2] Wu S G, Bao F S, Xu E Y, Wang Y-X, Chang Y-F and Xiang Q-L 2007 A leaf recognition algorithm for plant classification using probabilistic neural network *IEEE Int. Symp. on Signal Processing and Information Technology* pp 11–16
[3] Krizhevsky A, Sutskever I and Hinton G E 2012 Imagenet classification with deep convolutional neural networks *NIPS* **25** pp 1097–1105
[4] Zeiler M D and Fergus R 2014 Visualizing and understanding convolutional networks *Computer Vision – ECCV 2014*, *Lect. Notes Comp. Sc.* **8689** eds Fleet D, Pajdla T, Schiele B and Tuytelaars T (Cham: Springer) pp 818–833
[5] Russakovsky O, Deng J and Su H *et al* 2015 ImageNet Large Scale Visual Recognition Challenge *Int. J. Comput. Vision* **115** pp 211–252
[6] Otsu N 1979 A threshold selection method from gray level histogram *IEEE T. Syst. Man Cyb.* **9** pp 62–66
[7] Canny J 1986 A computational approach to edge detection *IEEE T. Pattern Anal.* **PAMI-8** pp 679–698
[8] Simonyan K and Zisserman A 2015 Very deep convolutional networks for large-scale image recognition *Int. Conf. on Learning Representations*
[9] Zhang C, Zhou P, Li C and Liu L 2015 A convolutional neural network for leaves recognition using data augmentation *IEEE Int. Conf. on Computer and Information Technology*
[10] Lee S H, Chan C S, Wilkin P and Remagnino P 2015 Deep-plant: Plant identification with convolutional neural networks *IEEE Int. Conf. on Image Processing (ICIP)* pp 452–456
[11] Kumar N, Belhumeur P, Biswas A, Jacobs D W, Kress W J, Lopez I C and Soares J V B 2012 Leafsnap: A computer vision system for automatic plant species identification *ECCV* pp 502–516
[12] Wang Z, Sun X, Zhang Y, Ying Z and Ma Y 2016 Leaf recognition based on PCNN *Neural Comput. & Applic.* **27** pp 899–908
[13] Hu R, Jia W, Ling H and Huang D 2012 Multiscale distance matrix for fast plant leaf recognition *IEEE T. Image Process.* **21** pp 4667–72
[14] Jia Y, Shelhamer E, Donahue J, Karayev S, Long J, Girshick R, Guadarrama S and Darrell T 2014 Caffe: Convolutional architecture for fast feature embedding *arXiv preprint arXiv: 1408.5093*
[15] Du J, Huang D, Wang X and Gu X 2005 Shape recognition based on radial basis probabilistic neural network and application to plant species identification *Advances in Neural Networks – ISNN 2005*, *Lect. Notes Comp. Sc.* **3497** eds Wang J, Liao X F and Yi Z (Berlin, Heidelberg: Springer) pp 281–285
[16] https://blog.heuritech.com/2016/02/29/a-brief-report-of-the-heuritech-deep-learning-meetup-5/

Recycled asphalt pavement – fly ash geopolymer as a sustainable stabilized pavement material

S Horpibulsuk[1], M Hoy[2], P Witchayaphong[3], R Rachan[4] and A Arulrajah[5]

[1] Professor and Chair, School of Civil Engineering, and Director, Center of Excellence in Innovation for sustainable Infrastructure Development, Suranaree University of Technology, Thailand
[2] PhD Scholar, School of Civil Engineering, Suranaree University of Technology, Thailand
[3] Researcher, Center of Excellence in Innovation for Sustainable Infrastructure Development, Suranaree University of Technology, Thailand
[4] Department of Civil Engineering, Mahanakorn University of Technology, Thailand
[5] Professor, Department of Civil and Construction Engineering, Swinburne University of Technology, Australia

E-mail: suksun@g.sut.ac.th

Abstract. Strength, durability, microstructure and leachate characteristics of Recycled Asphalt Pavement and Fly Ash (RAP-FA) geopolymers and RAP-FA blends as a sustainable pavement material are evaluated in this paper. The strength development of the stabilized materials with and without effect wetting-drying (w-d) cycles was determined by Unconfined Compression Strength (UCS) test. The mineralogical and microstructural changes of the stabilized material were analyzed by X-Ray Diffraction (XRD) and Scanning Electron Microscopy (SEM). The leachability of the heavy metals were measured by Toxicity Characteristic Leaching Procedure (TCLP) and compared with international standard. The results show that both RAP-FA blend and RAP-FA geopolymer increase with increasing the number of w-d cycles (*C*), reaching its peak at 6 w-d cycles. The XRD and SEM analyses indicate that the strength development of RAP-FA blend occurs due to stimulation of the chemical reaction between the high amount to Calcium in RAP and the high amount of Silica and Alumina in FA leaching to production of Calcium Aluminium (Silicate) Hydrate, while the geopolymerization reaction is observed in RAP-FA geopolymer. For $C > 6$, the significant macro- and micro-cracks developed during w-d cycles cause strength reduction for both RAP-FA blend and geopolymer. The TCLP results demonstrate that there is no environmental risk for these stabilized materials. Furthermore, FA-geopolymer can reduce the leachability of heavy metal in RAP-FA blend. The outcome from this research confirms the viability of using RAP-FA blend and RAP-FA geopolymer as alternative sustainable pavement materials.

1. Introduction

Recycled Asphalt Pavement (RAP), a combination of aggregate and asphalt binder, is obtained from spent asphalt extracted from roads that have reached the end of their design life [1]. Roads are a central component of many nations' infrastructure and present a wide array of opportunities for the usage of vast quantities of recycled materials such as RAP. Several researchers [2, 3] reported that

RAP could be used as pavement material when it was stabilized with cement or chemical stabilization. However, the use of cement leads to negative environmental impacts, which immediately requires alternative green material. Fly Ash (FA) based geopolymer is an environmentally friendly additive, which potentially produce better mechanical properties than Portland cement[4].

The usage of waste by-products in civil infrastructure enables a more durable alternative to quarried materials resulting in conservation of natural resources and decreased energy use. Hence, the durability of the stabilized material under severe climatic condition is a crucial parameter when used in road construction application. Cyclic wetting-drying (w-d) test simulates weather changes over a geological age and is considered to be one of the most appropriate simulation that can induce damage to pavement material [5]. Though the utilization of recycled waste materials in highway construction can be considered as having significant impacts on resource management, the hazardous compounds that can leach out and pollute the water resource should also be considered [6].

Therefore, this research attempts to study the possibility of using a green stabilizer (FA-geopolymer) to improve engineering properties of RAP to be a sustainable pavement material. The engineering properties of RAP-FA blend as a control material are also investigated to examine the effect of geochemical agent. First, the strength development of RAP-FA geopolymer and RAP-FA blend with and without effect of w-d cycles are determined by unconfined compression strength (UCS) test. The mineralogical and microstructural changes before and after the w-d cycles were examined by the application of X-Ray Diffraction (XRD) and Scanning Electron Microscopy (SEM) analyses. Finally, an environmental assessment was carried out by the leachate test to estimate the contaminate concentration from the RAP-FA geopolymer and RAP-FA blend and compared with international standard. The outcomes of this research will have significant impact on construction guidelines and specification for using RAP-FA blend and RAP-FA geopolymer in road construction applications.

2. Materials and methods

2.1. Materials

In this research, RAP samples were collected from a mill asphalt pavement stockpile in Nakhon Ratchasima province, Thailand. The gradation and the engineering properties of air-dried RAP are shown in Figure 1 and Table 1, respectively. The chemical and mineral composition of RAP, obtained by X-Ray Fluorescence (XRF) and XRD analyses, are presented in Table 2 and Figure 2, respectively.

FA, obtained from the largest lignite power plant in the northern region of Thailand, was used in this study. The grain size distribution curve of FA, obtained by a laser particle analyzer, is shown in Figure 1. Table 2 summarizes the chemical composition of FA using XRF analysis. The XRD pattern of FA is also shown in Figure 2. Liquid alkaline activator (L) is a mixture of sodium silicate (Na_2SO_3) and sodium hydroxide (NaOH) solution with a 10 M concentration.

2.2. Sample preparation

The RAP-FA geopolymer is a combination of RAP, FA, and L ($NaOH+Na_2SiO_3$). The $NaOH/Na_2SiO_3$ ratios studied were 100:0, 90:10, 60:40, and 50:50. The RAP-FA blend, which is a mixture of RAP, FA, and water, was prepared as a control material to investigate the effect of L on strength development. FA replacement ratios are 10%, 20%, and 30% by weight of RAP.

The mixtures were compacted in a cylindrical mold (101.6 mm in diameter and 116.3 mm in height) under the modified Proctor energy [7] to determine the optimum water content (OWC) and optimum liquid alkaline activator content (OLC). The sample at the OWC and OLC were prepared for UCS test following the ASTM D1633 [8]. The samples after 7 and 28 days of curing at room temperature (RT), were soaked in water for 2 hours and then were air-dried for 1 hour prior to UCS test according to the specification of the Department of Highway, Thailand [9].

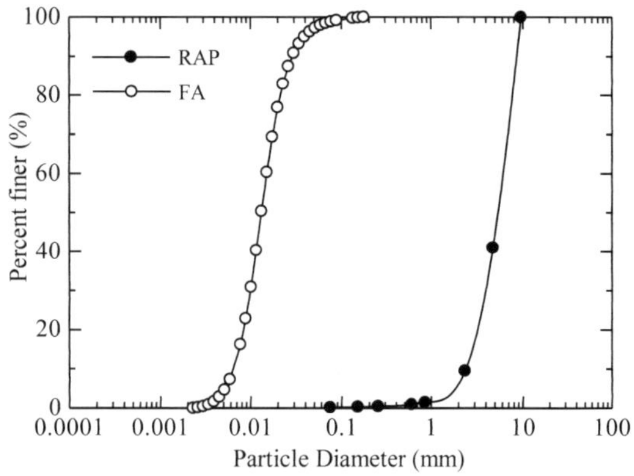

Figure 1. Grain size distribution of RAP and FA.

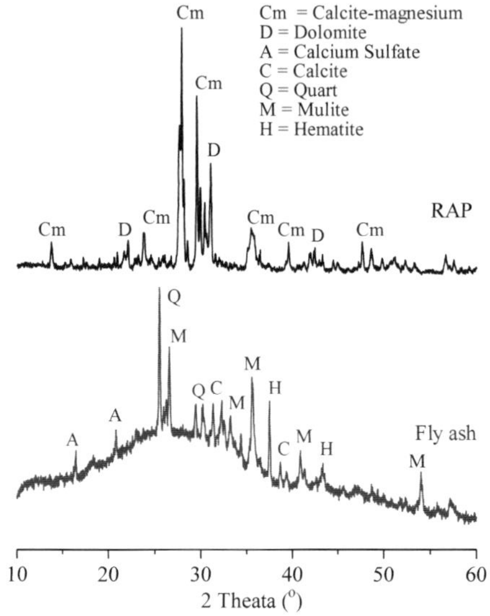

Figure 2. XRD pattern of RAP and FA.

2.3. Wetting and drying (w-d) test

Standard w-d test method for compacted soil-cement mixture [10] was adopted for this study. 28-day samples were submerged in water for 5 hours then dried in an oven at 70°C for 42 hours and air-dried for 1 hour. This is counted for 1 cycle. The UCS of the samples were measured at 1 to 20 w-d cycles and compared with that of the samples without w-d cycle to investigate the effect of w-d cycles on the UCS.

Table 1. Geotechnical properties of RAP.

Parameter	Values	ASTM
USCS Classification	SP	D2487-11
Specific gravity	2.70	D1883-07
CBR (%)	10-15	D577-12
Water absorption (%)	6.80	-
Swelling ratio (%)	0.20	-
Max. dry unit weight (kN/m^3)	17.50	D1557-12
Optimum water content (%)	4.10	D1557-12

Table 2. Chemical composition of RAP and FA.

Chemical formula	RAP	FA
SiO2	3.15	40.13
Al2O3	4.78	20.51
FeO3	0.10	5.83
CaO	41.93	12.45
MgO	36.18	3.11
SO3	0.89	0.42
LOI	-	0.40

2.4. Mineralogical and microstructural analyses
UCS development of the samples before and after w-d cycles test were examined by using XRD and SEM analyses, which performed on the small fragments taken from the broken portion of the UCS samples to investigate the mineralogical and microstructure changes, respectively.

2.5. Toxicity characteristic leaching procedure (TCLP) test
The Toxicity Characteristic Leaching Procedure (TCLP) test is the method prescribed by the U.S. EPA guidelines (Method 1311) to determine if the solid waste is hazardous. The TCLP tests were carried out on 100% RAP, RAP-FA blend and geopolymers for different types of heavy metal.

3. Results and discussion

3.1. Unconfined compression strength (UCS)
Figure 3 summarizes the UCS results of the RAP-FA blend and RAP-FA geopolymers for various $NaOH/Na_2SiO_3$ ratios, curing condition (7 and 28 days at RT). It is noted that the UCS values of both RAP-FA blends and geopolymers increase with increasing curing time. This result is similar to that reported in previous research on the strength development of cement-stabilized RAP and fly-treated RAP [3].

The 7-day UCS value of both RAP+20%FA blend and RAP+30%FA blend are greater than the strength requirement specified by the Thailand national road authorities in which UCS > 1,723 kPa and UCS > 2,413 kPa for both low and high volume roads, respectively [9, 11]. The UCS of RAP-FA geopolymer at $NaOH/Na_2SiO_3$<10:90 is higher than that of RAP-FA blends and the UCS values of RAP-FA geopolymers increase when $NaOH/Na_2SiO_3$ ratio is decreased. However, the UCS of RAP-FA blend and geopolymer improves insignificantly when the FA replacement ratio exceeds 20%, indicating this to be the optimal ratio. Hence, the RAP+20%FA blend and geopolymer samples curing age of 28 days were selected for durability test.

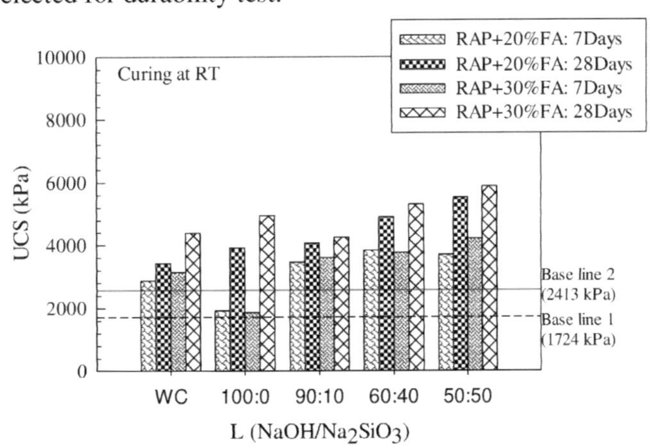

Figure 3. UCS of RAP-FA blends and RAP-FA geopolymers cured for 7 days and 28 days.

3.2. Wetting-drying cycled strength
The UCS of RAP+20%FA blend and RAP+20%FA geopolymer at various number of w-d cycles, *C* is presented in Figure 4. The UCS of RAP+20%FA blend and RAP+20%FA geopolymer increases with increasing *C*, up to $C = 6$ and then decreased when $C > 6$. Previous research on effect of w-d cycles on strength development of FA stabilized with lime and gypsum indicated the strength increase due to the development of cementitious compounds during w-d process [12]. The UCS of RAP+20%FA geopolymer at $NaOH/Na_2SiO_3$ = 100 and 50:50 increases sharply after the first w-d cycle and is much higher than that of RAP+20%FA blend for all *C* tested. Although the RAP+20%FA geopolymer sample with $NaOH/Na_2SO_3$ = 50:50 possesses higher UCS than the sample with $NaOH/Na_2SiO_3$ = 100:0 within the first 3 w-d cycles, it has lower UCS when $C > 6$.

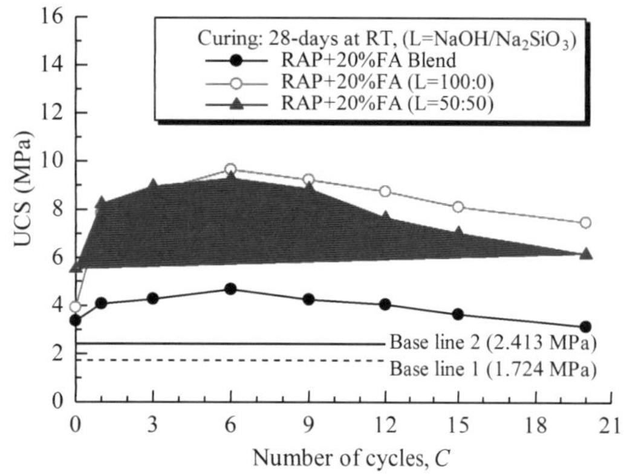

Figure 4. Relationship between strength and number of w-d cycles of RAP+20%FA blend and RAP+20%FA geopolymers.

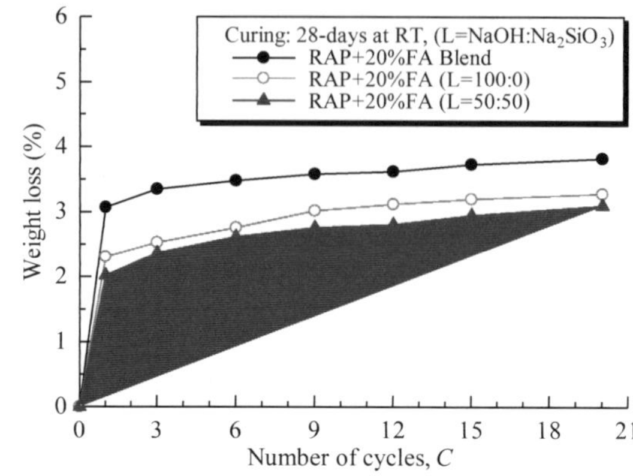

Figure 5. Relationship between weight loss and number of w-d cycles of RAP+20%FA blend and RAP+20%FA geopolymers.

The relationship between the weight loss of RAP+20%FA geopolymer and the RAP+20%FA blend versus number of w-d cycles, *C*is illustrated in Figure 5. The weight loss of both RAP+20%FA blend and geopolymer remarkably increases within the first w-d cycle and thereafter gradually increase with an increase in *C*. It is noted that the weight loss of RAP+20%FA geopolymer for NaOH/Na_2SiO_3 ratios of 100:0 and 50:50 is lower than that of RAP+20%FA blend due to the stronger RAP-FA geopolymer bonding structure. The effect of cyclic w-d cycles on the external surface of the RAP+20%FA blend and RAP+20%FA geopolymer at NaOH/Na_2SiO_3 ratios of 100:0 and 50:50 is evident in Figure 6a, b and c, respectively at a particular $C = 20$.

Large macro-cracks and surface deterioration on the RAP+20%FA blend are clearly observed, which leads to strength loss. On the other hand, Figure 6b obviously shows the minimum cracks on the surface of the RAP+20%FA geopolymer at NaOH/Na_2SiO_3 = 100:0, while more cracks are observed for NaOH/Na_2SiO_3 = 50:50 (Figure 6c). This imply that the samples at NaOH/Na_2SiO_3 = 100:0 has stronger bonding structure than the sample at NaOH/Na_2SiO_3 = 50:50. From the cyclic w-d results and the photos, it is evident that RAP+20%FA blend provides a fairly good durability when subjected to w-d cycles. FA geopolymer can enhance the durability of RAP-FA material, especially for the sample at NaOH/Na_2SiO_3 = 100:0.

Figure 6. Photos of (a) RAP+20%FA blend, (b) RAP+20%FA geopolymer (NaOH/Na_2SiO_3 = 100:0), and (c) RAP+20%FA geopolymer (NaOH/Na_2SiO_3 = 50:50) at $C = 20$.

3.3. Mineralogical and microstructural changes

The XRD patterns of RAP+20%FA blend at various *C* are shown in Figure 7. Without w-d cycle (C = 0), the RAP+20%FA blend (Figure 7a) contains the amorphous phases of Calcium Magnesium as the predominant minerals (Silica- and Alumina- products), such as Anorthite, Diopsite, Ladradorite, and Ettringite. These new minerals are formed when RAP is mixed with FA (RAP-FA blend). In other words, the chemical reaction between the high amount of Calcium of RAP results in the formation of

Calcium Silicate Hydrate (C-S-H) and Calcium Aluminate Hydrate (C-A-H), similar to the hydration of Portland cement [13], that can enhance the strength development.

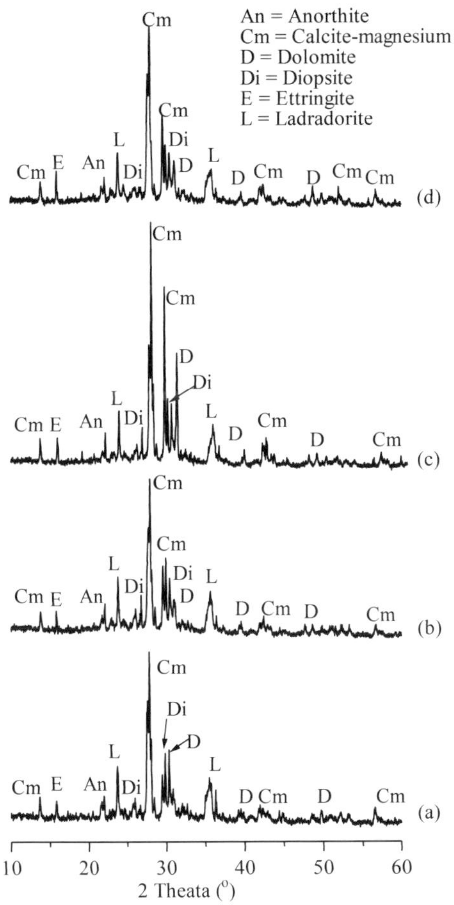

Figure 7. XRD patterns of RAP+20%FA blend samples at: (a) C = 0, (b) C = 1, (c) C = 6, and (d) C = 12.

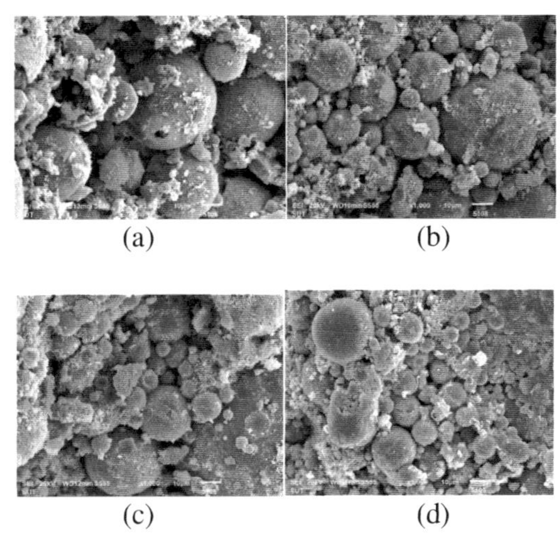

Figure 8. SEM images of RAP+20%FA blend samples at: (a) C = 0, (b) C = 1, (c) C = 6, and (d) C = 12.

The increase in peaks corresponded to Anorthite, Diopsite, and Ladradorite with increasing C to 6 is observed by comparing Figure 7b (C = 1) with Figure 7c (C = 6), that indicates the increase of C-A-S-H. Drying at 70°C for w-d test evidently enhances the cementitious products (C-A-S-H) [14]; i.e., an increased temperature results in a faster moisture diffusivity of the cementitious materials and hence cement hardening [15]. The same is however not true for C> 6. The temperature affects the water physical properties (density and surface tension) [16] and causes the coarsening of the pre structure in relation to Ettringite dissolution and C-S-H alteration [14]. Figure 7d indicates the presence of Ettringite and the decreased intensity of Anorthite and Diopsite minerals when the samples are subjected to 12 w-d cycles. Ettringite is a hydrous mineral that exhibits expansive behavior upon wetting [17] and makes the RAP-FA blends potentially volumetrically unstable [18].

Beside the XRD results, SEM images of RAP+20%FA blend at various $C$are illustrated in Figure 8. The growth of C-A-S-H gel inner and on the spherical surface of FA with increasing C (C = 0 to 6, see Figure 8a-c) is observed while reduction in cementitious gel at the C = 12 (Figure 8d) is detected, which confirms the XRD results.

Figure 9 and 10 present the XRD patterns of RAP+20%FA geopolymer at $NaOH/Na_2SiO_3$ ratios of 100:0 and 50:50 and at various w-d cycles, respectively. As a result of the alkaline activation, new geopolymerization products inclusive of Albite, Nepheline, and Analcime are observed at the broad hump between 22-33°2θ in the XRD patter of the sample at $NaOH/Na_2SiO_3$ = 100:0 (Figure 9a) and at $NaOH/Na_2SiO_3$ = 50:50 (Figure 10a) for C = 0. A good performance of geopolymer was observed when the precursor was in contact with the alkali activator at curing temperature of 40–75°C [19]. Therefore, the C-A-S-H from RAP-FA blend co-exist with the geopolymer products and contribute to the additional strength development of RAP+20%FA geopolymer in the first 6 w-d cycles. The silica present in sodium silicate is highly soluble and consequently incorporated immediately into the N-A-S-H gel, hence more geopolymer products are detected in the XRD patterns of the geopolymer sample at $NaOH/Na_2SiO_3$ = 50:50 (Figure 10b and c) compared with those at $NaOH/Na_2SiO_3$ = 100:0 (Figure 9b and c).Consequently, the UCS values of RAP+20%FA geopolymer at $NaOH/Na_2SiO_3$ = 50:50 are higher than those at $NaOH/Na_2SiO_3$ = 100:0 within the first 6 w-d cycles (see Figure 4).

The SEM images of RAP+20%FA geopolymer at $NaOH/Na_2SiO_3$ ratios of 100:0 and 50:50 (Figure 11 and 12) were examined to support the XRD results. The geopolymerization products in the sample at $NaOH/Na_2SiO_3$ = 50:50 are more than those in the samples at $NaOH/Na_2SiO_3$ = 100:0 for C< 6. For C> 6, at a particular C = 12, the micro-cracks are clearly observed in RAP+20%FA geopolymer samples (Fig. 11d and 12d) due to the loss of moisture content result in external surface cracks leading to a strength loss.

It is noted that the sample at $NaOH/Na_2SiO_3$ = 50:50 (Figure 12d) has more micro-cracks than the sample at $NaOH/Na_2SiO_3$ (Figure 11d). Consequently, the rate of strength reduction of the sample $NaOH/Na_2SiO_3$ = 50:50 is higher than that of the sample at $NaOH/Na_2SiO_3$ = 100:0 when C> 6. This indicates that the RAP+20%FA geopolymer with $NaOH/Na_2SiO_3$ = 100:0 has higher durability against w-d cycles than that with $NaOH/Na_2SiO_3$ = 50:50. This high durability is attributed to the formation of stable cross-linked alumino-silicate polymer structure.

3.4. Toxicity characteristic leaching procedure results

Table 3 shows the leachate analysis of 100%RAP and RAP+20%FA blends using acetic leachate extraction. According to benchmark mandated by the U.S. Environmental Protection Agency (EPA) for storm-water sampling, pH values should be in the range of 6 to 9 [20].

Leachate results show that pH level in 100%RAP is 512 and 5.59 for RAP+20%FA blend, which are within allowable limits. Table 4 presents the prescribed limits for drinking water and the threshold for hazardous waste defined by the U.S. EPA. Wartman et al. [21] reported that a material is designated as a hazardous waste in according to U.S. EPA if any detected metal is present in concentrations > 100 times the drinking water standards. Based on this criterion, the comparison of TCLP results between Tables 3 and 4 indicated that all metal contaminates are within acceptable limits.

From a geotechnical engineering perspective, the research results indicate that RAP is mechanically and economically viable for use in pavement base application when it stabilized with 20% of FA. Besides having a high UCS, the RAP-FA blend exhibits good durability against w-d cycles, which can be attributed to the growth of C-A-S-H. FA geopolymer can significantly improve the strength and durability of RAP-FA blend. Furthermore, these materials provide a positive environmental impact as environmental test results show no significant risk to the ground water or stream water line.

IOP Conf. Series: Materials Science and Engineering **273** (2017) 012005 doi:10.1088/1757-899X/273/1/012005

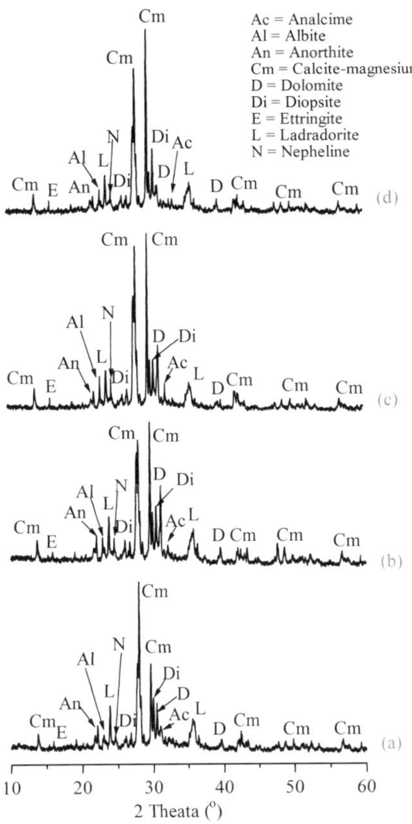

Figure 9. XRD patterns of RAP+20%FA geopolymer ($NaOH/Na_2SiO_3$ = 100:0) samples at: (a) C = 0, (b) C = 1, (c) C = 6, and (d) C = 12.

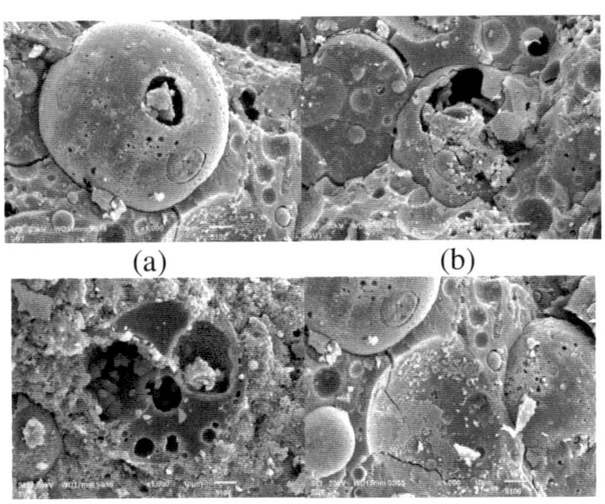

Figure 11. SEM images of RAP+20%FA geopolymer ($NaOH/Na_2SiO_3$ = 100:0) samples at: (a) C = 0, (b) C = 1, (c) C = 6, and (d) C = 12.

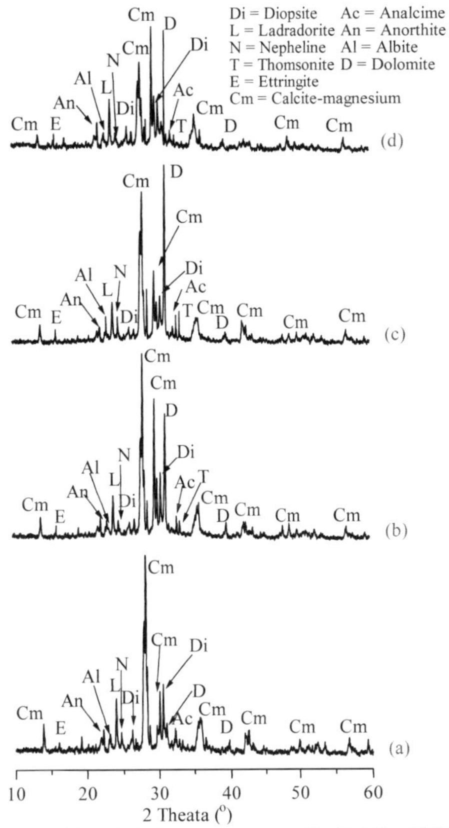

Figure 10. XRD patterns of RAP+20%FA geopolymer ($NaOH/Na_2SiO_3$ = 50:50) samples at: (a) C = 0, (b) C = 1, (d) C = 6, and (d) C = 12.

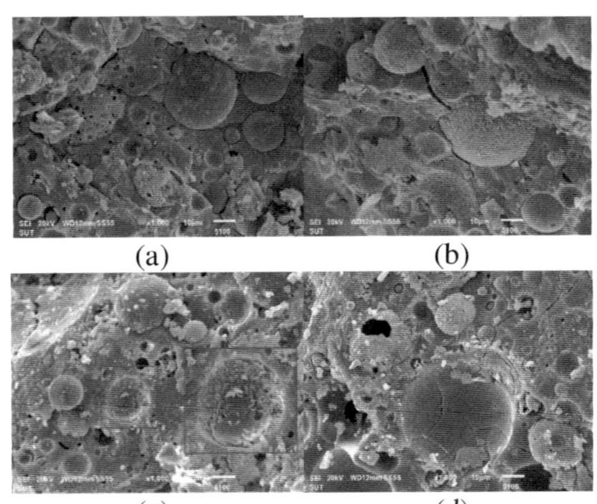

Figure 12. SEM images of RAP+20%FA geopolymer ($NaOH/Na_2SiO_3$ = 50:50) samples at: (a) C = 0, (b) C = 1, (c) C = 6, and (d) C = 12.

4. Conclusions

The present study investigated the possibility of using the RAP-FA blend as a sustainable pavement material. The 7-day UCS of the compacted RAP-FA blend at OWC meets the strength requirement for base course specified by national road authorities for both 20% and 30%FA replacement. The UCS improves insignificantly when the FA replacement ratio exceeds 20%, indicating this to be the optimal blend. FA-geopolymer can significantly improve the UCS and durability of RAP-FA blend. The XRD and SEM analyses of RAP-FA blends indicate the growth of C-A-S-H, hence its UCS increase over time. The co-exist of C-A-S-H from RAP-FA blend with the N-A-S-H from RAP-FA geopolymer can enhance the durability of RAP-FA blend remarkably.

The TCLP results indicate that RAP-FA blends and RAP-FA geopolymer scan be safely used as a sustainable pavement base application as these materials pose no significant environmental and leaching hazard into soil and ground water sources. In addition, FA geopolymer used can reduce the leachability of metal concentration from RAP-FA blend. This study can confirm the utilization of these recycled materials results in significant energy saving and reduction in greenhouse gas emission.

Table 3. Leachate analysis of 100%RAP, RAP+20%FA blend and RAP+20%FA geopolymer.

	Sample of acid leachate extraction (mg/L)			
Parameter	100%RAP	RAP-FA blend	RAP-FA geopolymer ($NaOH/Na_2SiO_3$ = 100:0)	RAP-FA geopolymer ($NaOH/Na_2SiO_3$ = 50:50)
pH	5.12	5.59	7.59	7.44
Arsenic	<0.01	<0.01	BDL	BDL
Cadmium	BDL	BDL	BDL	BDL
Chromium	<0.05	<0.05	<0.05	<0.05
Copper	BDL	BDL	BDL	BDL
Lead	BDL	BDL	BDL	BDL
Mercury	BDL	BDL	BDL	BDL
Nickel	<005	0.051	<0.05	<0.05
Zinc	1.348	0.657	BDL	BDL

Note: BDL = Below Detection Limit

Table 4. Comparison of TCLP data analysis with U.S. EPA Requirements.

Contaminant	Drinking water standard (EPA, 1999) (mg/L)	Threshold for solid waste (EPA, 2009) (mg/L)
Arsenic	0.05	0.35
Barium	2.0	35.0
Cadmium	0.005	0.1
Chromium	0.1	2.5
Lead	0.015	0.5
Mercury	0.002	0.05
Selenium	0.05	0.5
Silver	0.05	5.0

Acknowledgments

This work was financially supported by the Thailand Research Fund under the TRF Senior Research Scholar program Grant No. RTA5980005, Suranaree University of Technology and the Higher

Education Research Promotion and National Research University Project of Thailand, Office of Higher Education Commission.

References

[1] Arulrajah A, Piratheepan J and Disfani M 2013 Reclaimed asphalt pavement and recycled concrete aggregate blends in pavement subbases: Laboratory and field evaluation *J. Mater. Civil Eng.* **26** pp 349–357

[2] Puppala A J, Hoyos L R and Potturi A K 2011 Resilient moduli response of moderately cement-treated reclaimed asphalt pavement aggregates *J. Mater. Civil Eng.* **23** pp 990–998

[3] Taha R, Al-Harthy A, Al-Shamsi K and Al-Zubeidi M 2002 Cement stabilization of reclaimed asphalt pavement aggregate for road bases and subbases *J. Mater. Civil Eng.* **14** pp 239–245

[4] Salih M A, Ali A A A and Farzadnia N 2014 Characterization of mechanical and microstructural properties of palm oil fuel ash geopolymer cement paste *Constr. Build. Mater.* **65** pp 592–603

[5] Allam M M and Sridharan A 1981 Effect of wetting and drying on shear strength *J. Geotech. Eng.-ASCE* **107** pp 421–438

[6] Dawson A 2009 *Water in Road Structures: Movement, Drainage & Effects* (Nottingham UK: Springer)

[7] ASTM-D1557 2012 *Standard test methods for laboratory compaction characteristics of soil using modified effort (56,000 ft-1bf/ft3 (2,700 kN/m3)* in West Conshocken, PA

[8] ASTM-D1633 2007*Standard Test Methods for Compressive Strength of Molded Soil-Cement Cylinders* (ASTM International)

[9] DOH *DH-S204/2000 Standard of soil cement base: Department of Highways* Thailand

[10] ASTM-D559/D559M-15 2015 *Standard Test Methods for Wetting and Drying Compacted Soil-Cement Mixtures*

[11] DRR *DRR244-2013 Standard of soil cement base: Department of Rural Roads*

[12] Sivapullaiah P V and Moghal A A B 2010 Role of gypsum in the strength development of fly ashes with lime *J. Mater. Civil Eng.* **23** pp 197–206

[13] Cristelo N, Glendinning S, Fernandes L and Pinto A T 2012 Effect of calcium content on soil stabilisation with alkaline activation *Constr. Build. Mater.* **29** pp 167–174

[14] Brue F, Davy C A, Skoczylas F, Burlion N and Bourbon X 2012 Effect of temperature on the water retention properties of two high performance concretes *Cement Concrete Res.* **42** pp 384–396

[15] Wang Y, Fall M and Wu A 2016 Initial temperature-dependence of strength development and self-desiccation in cemented paste backfill that contains sodium silicate *Cement Concrete Comp.* **67** pp 101–110

[16] Wu M, Johannesson B and Geiker M 2014 A study of the water vapor sorption isotherms of hardened cement pastes: Possible pore structure changes at low relative humidity and the impact of temperature on isotherms *Cement Concrete Res.* **56** pp 97–105

[17] Celik E and Nalbantoglu Z 2013 Effects of ground granulated blast furnace slag (GGBS) on the swelling properties of lime-stabilized sulfate-bearing soils *Eng. Geol.* **163** pp 20–25

[18] Aldaood A, Bouasker M and Al-Mukhtar M 2014 Impact of wetting–drying cycles on the microstructure and mechanical properties of lime-stabilized gypseous soils *Eng. Geol.* **174** pp 11–21

[19] Sukmak P, Horpibulsuk S, Shen S-L, Chindaprasirt P and Suksiripattanapong C 2013 Factors influencing strength development in clay–fly ash geopolymer *Constr. Build. Mater.* **47** pp 1125–36

[20] EPA 2005 *Benchmarks for storm-water sampling* (Sacramento, CA: Business Environmental Resource Center)

[21] Wartman J, Grubb D G and Nasim A 2004 Select engineering characteristics of crushed glass *J. Mater. Civil Eng.* **16** pp 526–539

IOP Conf. Series: Materials Science and Engineering **273** (2017) 012006 doi:10.1088/1757-899X/273/1/012006

Effects of glass scraps powder and glass fiber on mechanical properties of polyester composites

K Sonsakul and W Boongsood

School of Manufacturing Engineering, Suranaree University of Technology, Thailand

E-mail: wanwanut@sut.ac.th

Abstract. One concern in bus manufacturing is the high cost of glass fiber reinforced in polyester composites parts. The composites of glass fiber and polyester are low elongation and high strength, and glass scraps powder displays high hardness and good chemical compatibility with the polymer matrix and glass fiber. This research aimed to study the effects of glass scraps powder and glass fiber on mechanical performance of polyester composites. Glass fiber was randomly oriented fiber and used as new. Glass scraps were obtained from a bus factory and crushed to powder sizes of 120 and 240 μm by a ball mill. Polyester composites were prepared using Vacuum Infusion Process (VIP).Polyester reinforced with 3 layers of glass fiber was an initial condition. Then, one layer of glass fiber was replaced with glass scraps powder. Flexural strength, tensile strength, impact strength and hardness of the polyester composites were determined. Hardness was increased with a combination of smaller size and higher volume of glass scraps powder. Pictures of specimens obtained by using scanning electron microscope (SEM) confirmed that the powder of glass scraps packed in the layers of glass fiber in polyester composites.

1. Introduction

A composite, one of the most significant inventions of the materials sciences (matrix and reinforcement), is made from distinct materials combined to make a better material [1]. Composite materials are used in assembly boards, furniture, fencing, paneling to civil constructions, automobile and even space or aircraft manufacturing. Composite materials can also be used in many fields with high quality and low cost applications [2]. Fiber restressing techniques were developed to improve the properties of continuous fiber reinforce composites [3, 4]. Fiber reinforced composites were successfully used for many decades in engineering applications. In this study, the matrix was polyester resin and the reinforcements were glass fiber and glass scraps powder. The glass fiber was chopped mat No. 450. The glass scraps powder were laminated glass recycling which obtained from terminated buses or crack during operation in bus fleet and manufacturing. Then, the scraps were grinded and passed through sieve mesh No. 60 and 140. The glass fiber had the density of 450 g/m^2.

Vacuum Infusion Process (VIP) is an efficient manufacturing process for complex laminates with many plies of fibers and core materials. The method is used for producing composites and has many advantages over hand lay-up method such as higher quality, higher specific strength and stiffness, better consistency, good interior finish, faster cycle time and lower cost [5].

Mechanical property of a combination of glass fiber and polyester is characterized by high toughness leading to resistance to crack propagation [6].

The main objective of this research was to study the mechanical properties of the polyester composites reinforced with glass fiber and the powder of glass scraps which the composites were fabricated using VIP. Initial composite material composition was polyester resin matrix reinforced with 3 layers of glass fiber. Next, one layer of glass fiber was replaced by glass scrap powder. The weights of one layer of glass fiber and glass scrap powder were identical. Tests were performed to determine flexural, tensile, impact and hardness properties of the polyester composites.

2. Methods

2.1. Materials

Materials used for this project were;

- Matrix: Polyester resin
- Catalyst: Methyl Ethyl Ketone Peroxide Norox KP-100
- Reinforcements: Glass fiber chopped stand mat No. 450 and glass powder sizes of 60 and 140 Mesh.
- Others: Gel coat, Styrene monomer and Cobalt in solution as accelerator.

Glass scraps were milled into powder by using a ball mill and screened with sieves No.60 and 140 Mesh where it was decided that the particles could be filled in between the fibers. The sizes were measured by a laser particle counter before used. The results of size measurement are shown in Table 1 and Figure 1. Table 2showssample preparation that was PGF, polyester resin reinforced with 3 layers of pure glass fiber, and D12/W45, D12/W55, D24/W45, D24/W55, the resin reinforced with 2 layers of glass fiber and the amount of glass powder. Initially, one layer of glass fiber had weight of 450 g/m^2. Therefore, the weight of the glass scraps powder in the experiment were 450 and 550 g/m^2 which the former was identical with the weight of one layer of glass fiber and the latter was expected to have more glass powder compacted in the composite.

Table 1. Sizes of glass scraps powder.

	Particle sizes of glass scraps powder (μm)	
Mean size	240.66	120.45
Std. Dev.	100.60	47.02
Median size	221.35	117.03

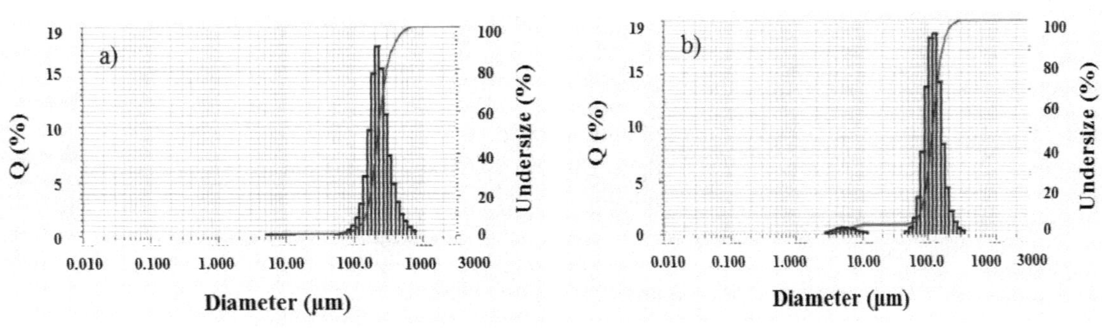

Figure 1. Particle size distribution of glass scraps particle through sieve (a) No. 60 (240 μm) and (b) No. 140 (120μm).

Table 2. Samples preparation.

Sample	Polyester resin	Catalyst	Gel coat (0.5 mm thick)	Glass fiber	Glass scraps powder	
					Size	Weight
	(g/m^2)	(g/m^2)	(g/m^2)	(No. of layer)	(μm)	(g/m^2)
PGF	2,430	24.3	815.1	3	-	-
D12/W45	2,430	24.3	815.1	2	120	450
D12/W55	2,430	24.3	815.1	2	120	550
D24/W45	2,430	24.3	815.1	2	240	450
D24/W55	2,430	24.3	815.1	2	240	550

2.2. Fabrication using VIP

The VIP was used as a manufacturing technique in this study. The technique applies vacuum to pull resin into a laminate (Figure 2). The fabric fibers and core materials are loaded into the mold. Then, a vacuum bag is covered on the top to close the mold and creates a vacuum seal. After that, a vacuum pump is used to remove all of the air in the cavity and consolidate the fiber and core materials. Finally, resin is infused under vacuum into the mold cavity to wet out the fiber.

In the present study, the process was followed the practical steps. The glass powder was sprayed after laying the fiber layer by layer. The glass powder was sprayed with a controlled feed flow rate. The graphic of rectangular mold used in the experiment is presented in Figure 3. The locations of the vacuum ports and the resin insertion points need to be carefully planned to ensure full resin infusion. Thus, the resin infusion inlet and vacuum outlet were placed at the mid edge of the mold.

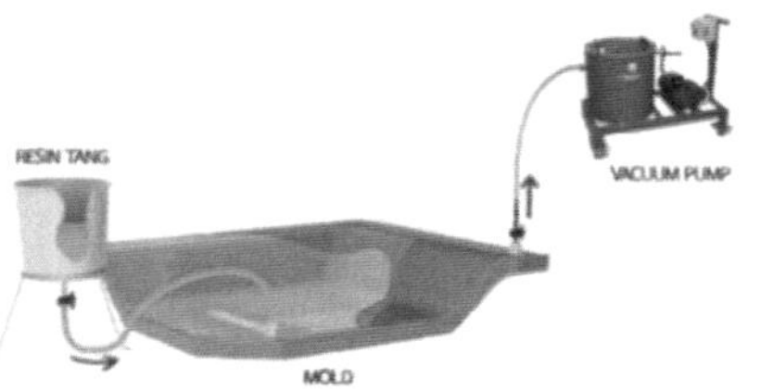

Figure 2. Vacuum infusion process (VIP).

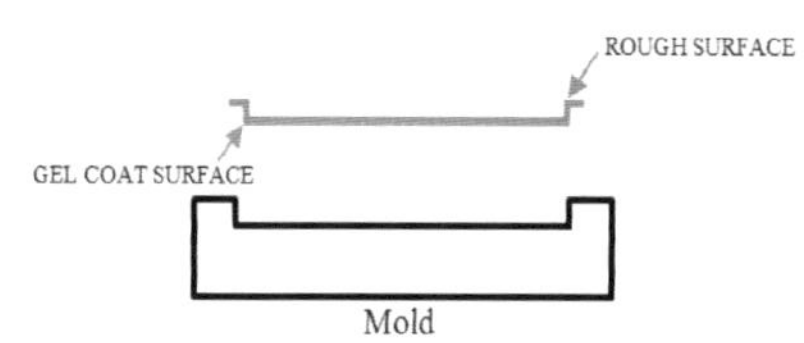

Figure 3. Composite plate forming of VIP.

2.3. Mechanical properties testing

Each composition in Table 2 was fabricated with a size of 50x50 cm^2 then cut into small specimens as shown in Figure 4 for determining mechanical properties effected by longitudinal (L) and transverse (T) infusion flow, and locations at the edge (E), middle (M) and corner (C) on the workpiece.

Flexural and tensile tests were performed using a universal testing machine, with a speed of 5 mm/min at room temperature. The test was according to ASTM D790 [2, 3] and ASTM D3039, respectively.

Izod impact test (unnotched) was conducted using INSTRON (CEAST)/CEAST 9050 impact according to ASTM 4812 apparatus with a hammer of 11 J, at room temperature. Then, the fracture surfaces of the specimens subjected to Izod impact test were taken to scanning electron microscope (SEM) to analyze their microstructure which the specimens were coated with gold [1].

Hardness test was carried out using the shore D hardness according to ASTM D2240 during 10s [1].

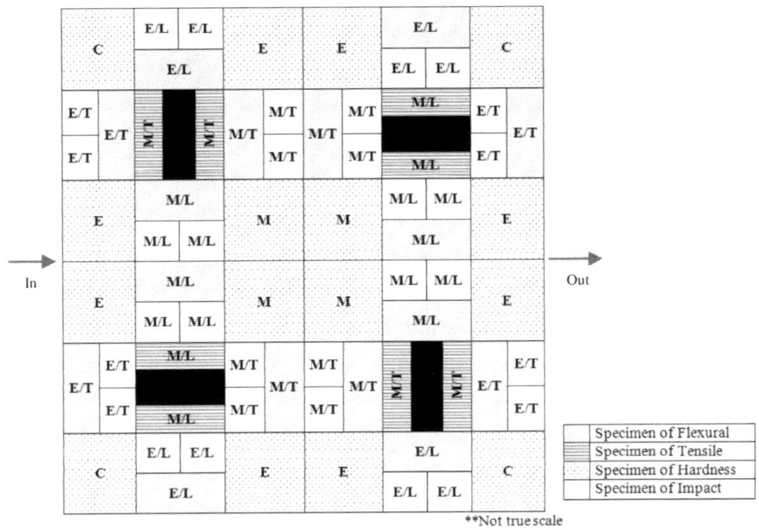

Figure 4. Specimens.

3. Result and discussion
The test results were graphically and statistically analyzed and discussed as following.

3.1. Flexural test
Figure 5a shows that either the directions of infusion flow or locations of the specimens had no effect with the flexural strength. The diverse compositions affected the strength. The strengths obtained from D24/W45 and D24/W55 were only slightly lower than PGF, see Figure 5b, that might be resulted from the larger particle size in the matrix.

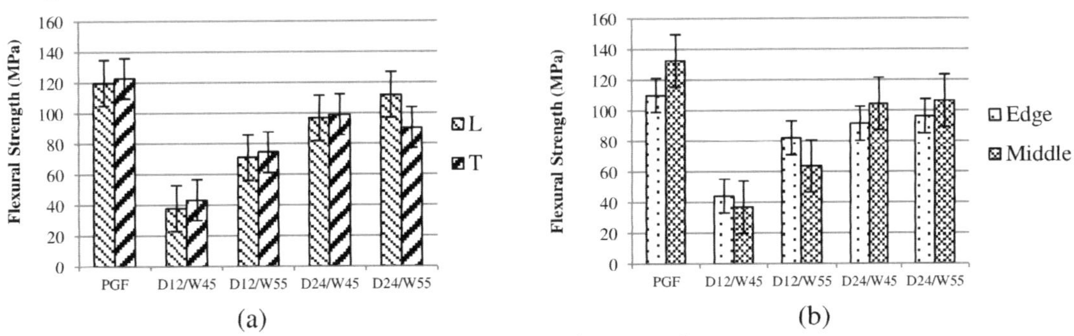

Figure 5. Flexural test results.

3.2. Tensile test
The results indicate that the directions of infusion had no effect on the tensile resistance (Figure 6). The tensile strength diminished with adding glass scraps powder weight of 450 g/m^2 into the glass fiber and polyester.

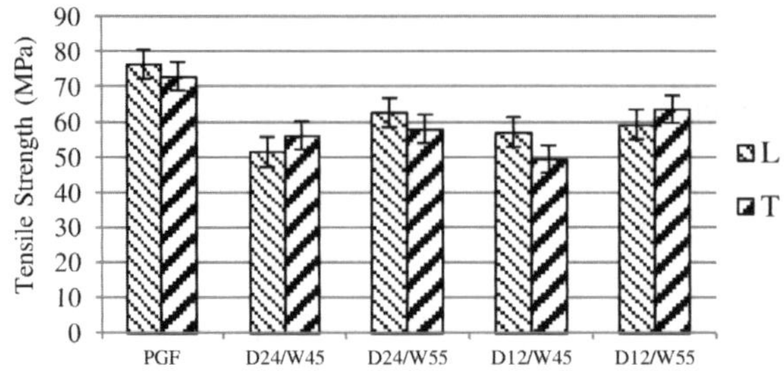

Figure 6. Tensile test results

3.3. Impact test

As well as in the flexural test, the infusion directions had no effect on the impact resistance (Figure 8a). All proposed composite compositions showed significantly lower impacting energy absorption (Figure 8b). Averagely, impacting energy was absorbed by PGF 39%, composition, D12/W45 22%, D12/W55 19%, D24/W45 28% and D24/W55 26%. That might be the consequence of reinforcement with short and tiny particles instead of long and strong fibers.

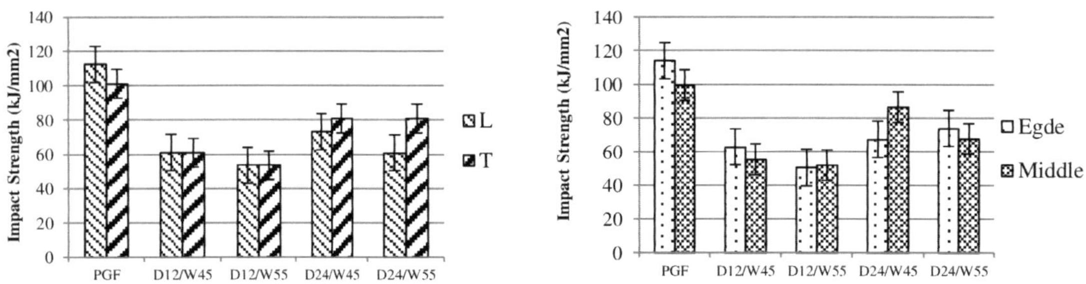

(a) (b)

Figure 7. Impact test results

3.4. Hardness test

From statistical analysis, it was found that the hardness of the specimens at the corners was significantly lower than at the edge or in the mid plate (Figure 7) but not so affecting with PGF. This may result from the flow at the corner not so compromise as other positions. Though, the D12/W55 (the mix between smaller particle size and higher weight) gained a higher hardness than, and D12/W45 (the mix between smaller particle size and an identical weight of glass fiber), PGF resulted by the rigid and a numbers of particles spreading thoroughly over the work to absorbing the compression.

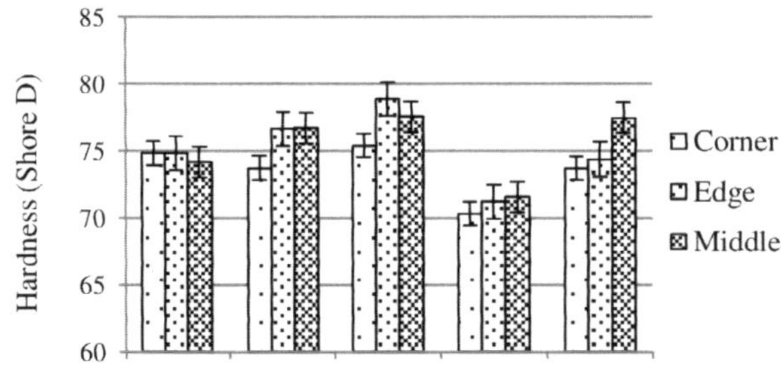

Figure 8. Hardness test results

3.5. SEM analysis

The fracture surfaces of the specimen subjected to impact test were analyzed using SEM. Figure 8 presents the proof that particles of glass scraps trapped as expected in the space between the glass fabrics to reinforce the composite.

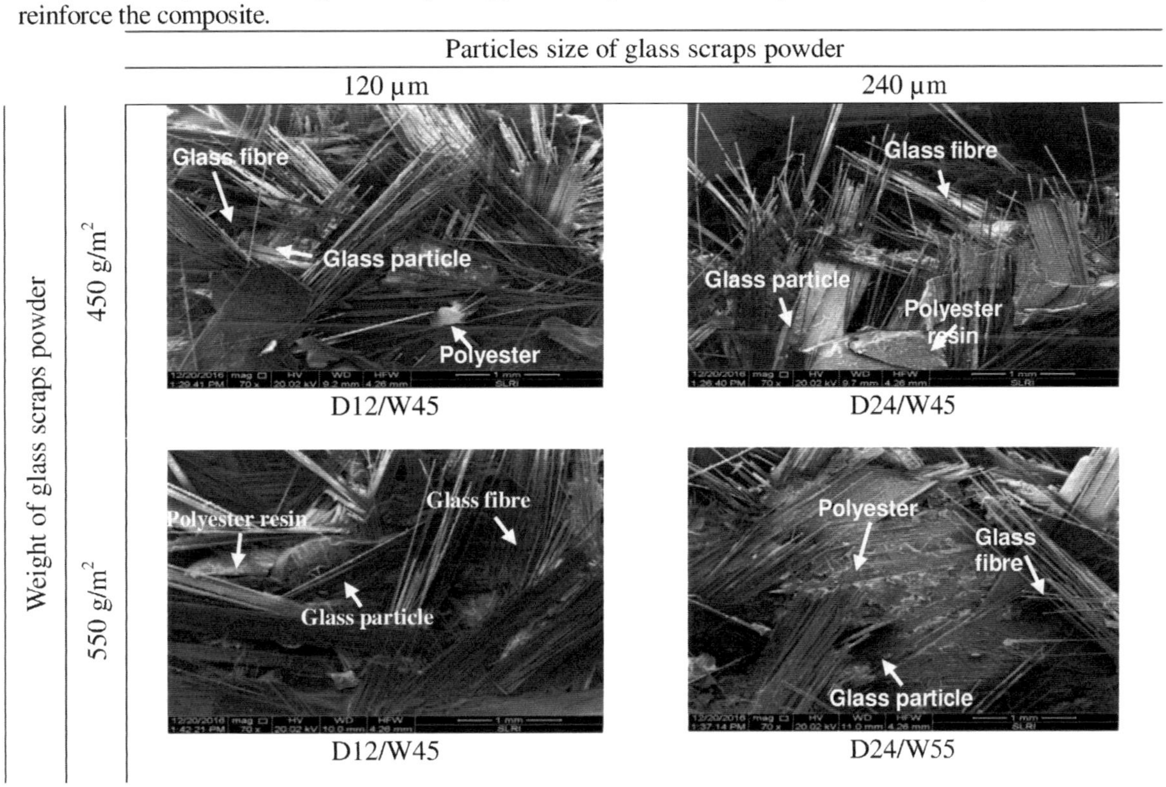

Figure 9. SEM photomicrographs of impact test fractured surfaces (70X).

Table 3 shows the summary of test results obtained by flexural, tensile, impact and hardness tests on polyester composites reinforced with pure glass fiber and glass scraps powder. The values shown in the table was averaged from the total numbers of all specimens in each type of tests. The point estimation in the table agreed well with the graphical and statistical analysis.

Table 3. Summary of mechanical properties.

Sample	Particles size of glass scraps powder (μm)	Weight of glass scraps powder (g/m^2)	Flexural strength (MPa)	Tensile strength (MPa)	Impact Strength (kJ/mm^2)	Hardness (Shore D)
PGF	-	-	121.52	74.56	106.85	74.6
D12/W45	120	450	40.71	53.79	61.05	75.7
D12/W55	120	550	72.90	60.32	53.56	77.3
D24/W45	240	450	98.12	53.28	77.06	71.0
D24/W55	240	550	101.38	61.33	70.92	75.2

IOP Conf. Series: Materials Science and Engineering **273** (2017) 012006 doi:10.1088/1757-899X/273/1/012006

4. Conclusion
From the results and discussion, the conclusion can be made here.

- The resistance to bending, flexural strength, was maintained comparing to composite reinforced with pure glass fiber by composing the glass scraps powder size of 240 μm.
- Adding glass scraps powder weight of450 g/m^{2}decreased the tensile strength.
- Absorbed energy, impact strength, of the composite samples was reduced according with powder reinforcement as oppose to fiber strengthening.
- Hardness was increased through spreading of glass particles over the composite by increasing proportion of glass powder together with a refinement of particle size.
- Pictures from SEM shows that the powders of glass scraps were inserted in the gap of glass fibers.

References
[1] Araújo E M, Araújo K D, Pereira O D, Ribeiro P C and Melo T J A 2006 Fiberglass wastes/polyester resin composite: Mechanical properties and water sorption *Polímeros: Ciência e Tecnol* **16** pp 332–335

[2] Elahi A H M F, Hossain M M, Afrin S and Khan M A 2014 Study on the mechanical properties of glass fiber reinforced polyester composites *Int. Conf. on Mechanical Industrial and Energy Engineering*

[3] Cao Y and Cameron J 2006 Flexural and shear properties of silica particle modified glass fiber reinforced epoxy composite *J. Reinf. Plast. Comp.* **25** pp 347–359

[4] Cao Y and Cameron J 2006 Impact properties of silica particle modified glass fiber reinforced epoxy composite *J. Reinf. Plast. Comp.* **25** pp 761–769

[5] Performance Composites Inc Vacuum Infusion Process Guide Technical Information California USA http://www.performancecomposites.com

[6] Aramide F O, Atanda P O and Olorunniwo O O 2012 Mechanical properties of a polyester fiber glass composite *International Journal of Composite Materials* **2** pp 147–151

Phenol hydroxylation on Al-Fe modified-bentonite: Effect of Fe loading, temperature and reaction time

R K Widi, A Budhyantoro and A Christianto

Department of Chemical Engineering, University of Surabaya, Raya Kalirungkut, Surabaya 60293, Indonesia

E-mail: restu@staff.ubaya.ac.id

Abstract. The present work reflects the study of the phenol hydroxylation reactions to synthesize hydroquinone and catechol on Al-Fe modified-bentonite. This study started with synthesizes the catalyst material based on the modified bentonite. Natural bentonite from Pacitan, Indonesia was intercalated with Cetyl-TetramethylammoniumBromida (CTMA-Br) followed by pillarization using Alumina. The pillared bentonite was then impregnated with Fe solution (0.01 M, 0.05 M, and 0.1 M). The solid material obtained was calcined at 723 K for 4 hours. All the materials were characterized using BET N_2 adsorption. Their catalytic activity and selectivity were studied for phenol hydroxylation using H_2O_2 (30%). The reaction conditions of this reaction were as follows: ratio of phenol/ H_2O_2 = 1:1 (molar ratio), concentration of phenol = 1 M and ratio of catalyst/phenol was 1:10. Reaction temperatures were varied at 333, 343 and 353 K. The reaction time was also varied at 3, 4 and 5 hours. The result shows that the materials have potential catalyst activity.

Keywords: betonies; phenol; Fe; hydroxylation

1. Introduction

Hydroquinone is used extensively as a photography chemicals, antioxidants, polymerization inhibitors, flavoring agents and drug intermediates [1]. Several utilizations reflect that hydroquinone is one of the important chemicals that have gained interest from many researchers. There are a lot of efforts on catalyst synthesis development for improving the phenol conversion and hydroquinone selectivity. The reaction might be oxidizing the reaction products, catechol and hydroquinone, synthesizing undesirable heavy molecules, like benzoquinones. Phenol hydroxylation has become an important industrial process to synthesize hydroquinone and catechol. This process involves the reaction of phenol with H_2O_2 which is simple and environmental friendly. The reaction has been widely investigated using various homogenous and heterogeneous catalysts. The heterogeneous catalysts are preferred over the homogeneous ones because they give a high phenol conversion with high product selectivity and can be separated easily from the reaction mixture. Some researchers use titanium silicate [2–4], TiO_2 in supports such as silica [5], alumina [6], zeolite [7, 8], and activated carbon [9], zeolite MCM [10], copper-iron complexes [11, 12] as a heterogeneous catalyst for phenol hydroxylation. The other researchers also examined the activity of phenol hydroxylation on zeolite MCM [10] and copper-iron complexes [11, 12].

Among several heterogeneous catalysts, supported Fe is widely used. Fe has been a catalyst of interest for the reaction over other metals because it gives high phenol conversion at mild reaction conditions and high product selectivity. There are several ways to improve the catalytic performance of Fe catalysts, such as immobilization of Fe on MCM [13, 14] and on wood-based activated carbon [15].

In our previous work we described that use of modified betonies as a Fe supporting material to facilitate Fe dispersion depicted promising catalysts for the phenol hydroxylation with hydrogen peroxide [16, 17]. We used betonies as it is abundant material in Indonesia and it has special characteristic in its porous structure to facilitate Fe dispersion. In this paper we report more focus on the effect of Fe loading into the modified betonies, and the effect of temperature and reaction time toward catalyst activity on phenol hydroxylation.

2. Experimental

2.1. Materials

Clay (natural Pacitan bentonite), $FeCl_2$ (Merck), Cetyl-Tetramethylammonium Bromida, CTMA-Br (Merck), Noah (Merck), $Al(OH)_3$ (Merck).

2.2. Instrumentation

The chemical analysis of the solids was carried out by Fourier Transform Infra Red Spectrophotometer (FTIR, Broker Tensor 27, Germany). The specific surface areas were obtained from the BET method. Nitrogen adsorption was determined with an ASAP 2010 Adsorption Analyzer from Micrometrics. The concentration of phenol, hydroquinone and cathecol was measured by High Performance Liquid Chromatography (HPLC) from Knauer.

2.3. Procedure

The starting clay was a natural Pacitan bentonite, extracted from Pakistan, Indonesia. Noah and Al (molar ratio of OH to Al = 0.8) was mixed to generate the pillaring agent solution. Modification of betonies was carried out by mixing betonies, pillaring agent solution and intercalating agent (Cetyl-TetramethylammoniumBromida, CTMA-Br) with ratio [gram bentonite/ volume of solution] = 1 gram/50 mL, and it was heated at 343 K for 5 hours. The obtained solid was dried and calcined at 773 K for 4 hours with nitrogen and oxygen stream. This modified bentonite was then impregnated with Fe solution. The Fe concentrations were 0.01 M, 0.05 M, and 0.1 M. After impregnation process, Al-CTMA bentonite was calcined at 773 K for 4 hours with nitrogen and oxygen stream. The catalytic oxidation reaction of phenol in a diluted aqueous medium was carried out in a batch glass reactor equipped with thermometer and reflux condenser. It was open to the atmosphere and thoroughly stirring with magnetic stirrer. The temperature was varied at 333 K, 343 K and 343 K for 3, 4 and 5 hours. The hydrogen peroxide solution was added stepwise during a reaction time. The molar ratio of phenol to hydrogen peroxide was 1:1. The weight ratio of catalyst to phenol was 1:10. The course of the phenol conversion and selectivity was followed by high performance liquid chromatography (HPLC). For all chromatographic separations, a Knavery HPLC System was used, including Euros her II 100-3 C18, 100 x 3 mm ID column. The mobile phase was methanol:water (40:60 v/v) with the flow rate was 0.5ml/min at room temperature. UV 254 nm (10 mm cell, 5 Hz, 0.2 sec.) was used as detector. Conversion is defined as the ratio of consumed phenol over the fed phenol for the reaction. Selectivities of products are the ratio of produced product over the consumed phenol. The desired product was hydroquinone.

3. Results and discussion

3.1. Catalyst characterization

Figure 1 shows the FTIR spectra of fresh (unmodified) bentonite and CTMABr-Al-Fe modified bentonite before and after calcinations. The absorption peak in the region 3623.05 cm^{-1} indicates –OH asymmetric stretching from Al–OH, Si–OH and H–O–H in bentonite. The presence of alkyl groups were shown by absorption peak in the region 2920.25 cm^{-1} ($-CH_3$ *streching asymmetric*), 2851.20 cm^{-1} ($-CH_2$ *stretching symmetric*), and supported by absorption peak of 1471.28 cm^{-1} (-C-H bending vibration). These peaks proved that CTMABr molecules might lay on the bentonite surface to form pillared bentonite. The absorption peak in the region 1030 – 1040 cm^{-1} indicated asymmetric stretching O-Si-O, O-Fe-O and O-Al-O group in material. These bands exhibited a slight shift to the higher wave numbers due to the formation of new system polymerization of shifted to the O-Al-O and O-Si-O in the pillared bentonite. The difference of width absorption region in the range of 950 – 1200 cm^{-1} proved that the Fe pillars are formed permanently in the bentonite. The FTIR spectra and X-Ray difractogram characterizations and its analysis have been described in more detail before [16, 17].

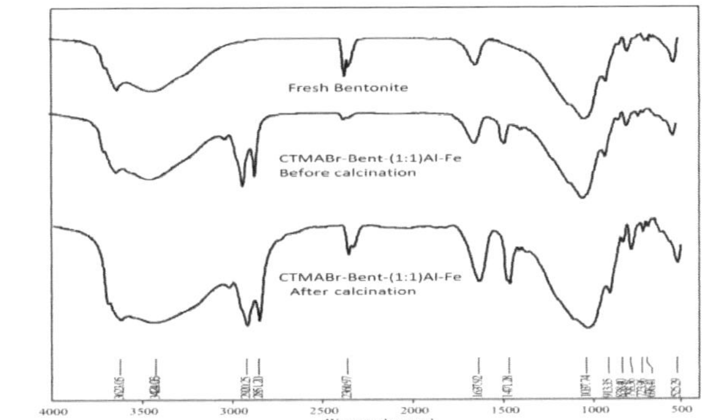

Figure 1. FTIR spectra fresh (unmodified) bentonite and CTMABr-Al-Fe modified bentonite before and after calcinations.

The determination of distribution of pore size and surface area of pillared bentonite can be done by using adsorption N_2 gas analysis. The observation focused on the influence of existing Fe metal to pillar height and porous size of bentonite. The success of intercalating, pillarization and impregnation process can be indicated by increasing of porous size and its distribution. The created material will be better, if the number of mesopore (> 20 Å) is higher. The high surface area and mesopore size distribution allowing the dispersion of a large number of catalytically active species [18]. The effect of Fe concentration to created pore is shown by figure 2 and table 1.

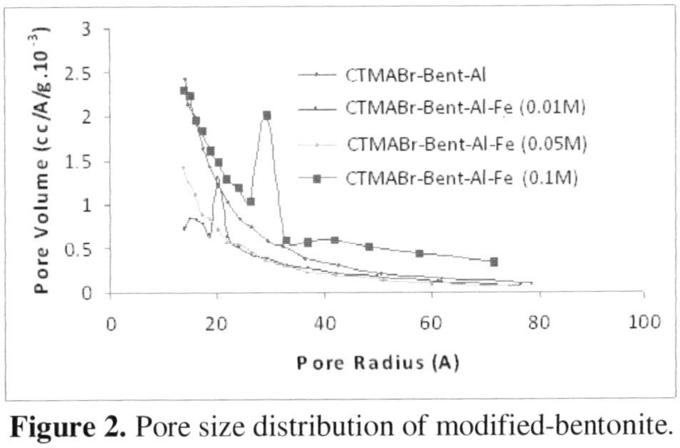

Figure 2. Pore size distribution of modified-bentonite.

Table 1. The effect of Fe concentration on average pore radius of modified-bentonite.

Sample	Average Pore Radius (Å)	Specific Surface Area (m^2/g)
Fresh bentonite	14,380127	59,746942
CTMABr-Bent-Al	24,203416	38,039192
CTMABr-Bent-Al-Fe (0,01 M)	26,112431	19,024643
CTMABr-Bent-Al-Fe (0,05 M)	21,235114	22,589425
CTMABr-Bent-Al-Fe (0,1 M)	35,919653	44,333773

BET analysis which is listed on table 6 shows that fresh bentonite pores are distributed in the micro size (≤ 20 Å) and has high specific surface area (59.75 m^2/g). The disadvantage of fresh bentonite is that it has impermanent porosity. Fresh bentonite will swell when it is hydrated and collapse when it is dehydrated. Increasing Fe concentration along impregnation process will affect the increase of the average pore radius. Figure 2 shows that pore size distribution tend to increase in the range of micropore to mesopore. Fe impregnated bentonite that has been calcinated, the bentonite structure change into amorphous. It has lower crystallinity but the pore volume is getting higher as shown by XRD [16, 17] and BET analysis. Table 1 show that the biggest average pore radius is that of CTMABr-Bent-Al-Fe (0.1 M). This phenomenon was supported by data in figure 3 which showed that CTMABr-Bent-Al-Fe (0.1 M) had pore size distribution in the range of mesopore.

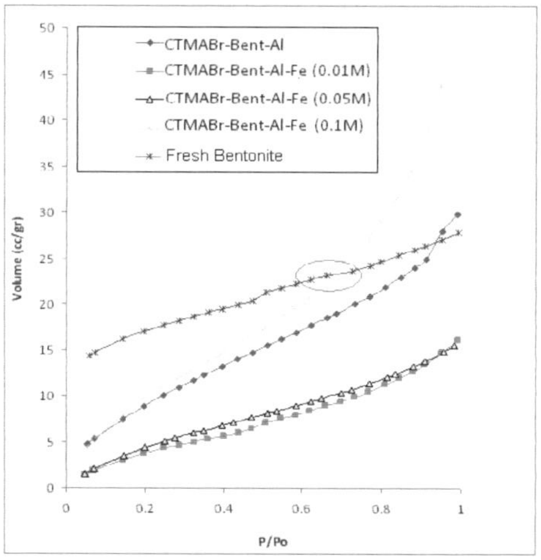

Figure 3. N_2 isotherm adsorption on fresh and modified bentonite.

3.2. Catalytic activity test

The phenol hydroxylation reaction follows the proposed reaction mechanism below (scheme 1).

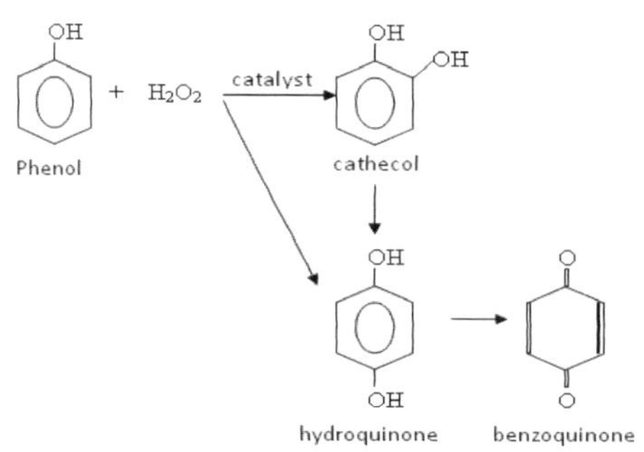

Figure 4. Proposed reaction pathway of phenol hydroxylation

Figure 5 depicts the result of phenol hydroxylation reaction at 333 K. This result describes that the most active catalyst was Comair-Bent-Al-Fe (0.1M), since this catalyst gave the highest propane conversion and hydroquinone selectivity. This result is consistent with the BET data showing that the catalyst CTMABr-Bent-Al-Fe (0.1M) has the biggest pore radius and specific surface area. This characteristic of the catalyst leads the phenol reactant to be adsorbed and could be contacted in longer time with peroxide. In addition, this data are also consistent with the previous work which described that the more Fe is loaded, the more active is the catalyst since the catalytic activity needs the presence of ion Fe to oxidize phenol molecules. The mechanism of oxidation of phenol is as follows [19]:

$Fe^{2+} + H_2O_2 \rightarrow Fe^{3+} + OH^{\cdot} + OH^-$

Phenol radical + $Fe^{3+} \rightarrow$ Product + Fe^{2+}

Catechu has been postulated to be produced in the catalyst external surface while hydroquinone is formed inside the pores [20]. The barrier of reactants to diffuse into the porous network of the pillared bentonite can be the reason for reduced hydroquinone selectivity, although thermodynamically *o*- and *p*-products have equal possibility.

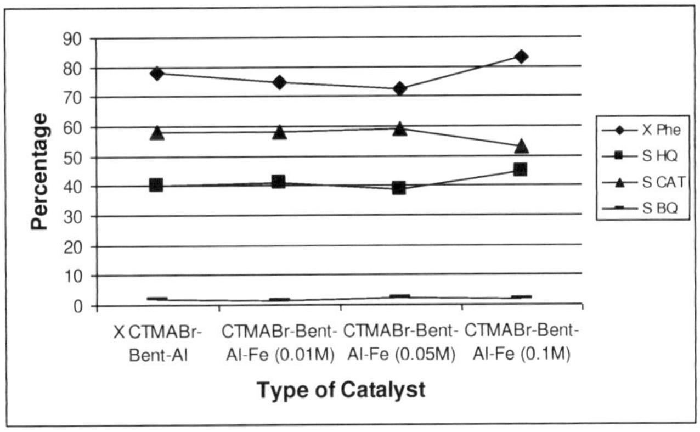

Figure 5. Phenol hydroxylation on different type of catalyst at 333K for 3 hours (X Phe = Conversion of Phenol; S HQ = Selectivity of Hydroquinone; S CAT = Selectivity of Cathecol; S BQ = Selectivity of Benzoquinone).

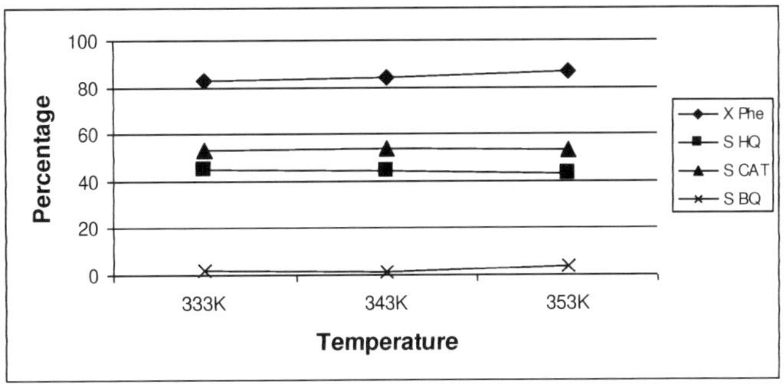

Figure 6. Effect of reaction temperature on Phenol hydroxylation on CTMABr-Bent-Al-Fe (0.1M) for 3 hours (X Phe = Conversion of Phenol; S HQ = Selectivity of Hydroquinone; S CAT = Selectivity of Cathecol; S BQ = Selectivity of Benzoquinone).

Figure 6 shows that phenol conversion increased, in line with increasing reaction temperature. Increasing reaction temperature affected the phenol conversion. This is due to the fact that energy given to the system is high enough resulting in the increase in the reaction rate.

The effect of reaction time is shown in figure 7. It can be observed the phenol conversion reached a maximum value of 78% with the reaction time of 4 hours. However, the phenol conversion slightly decreased when reaction time further increased to 5 hours. It can also be described that selectivity of hydroquinone decreased when reaction time increased, which may be attributed to the fact that hydroquinone was further oxidized to generate more by-product, i.e. 1,4-benzoquinone, in a longer time. Over-oxidation of hydroquinone occurred because more energy was needed to form 1,4-benzoquinone due to its low structure stability. The results show that higher phenol conversion did not tend to produce high amount of hydroquinone. It is probably because of the energy given to the system was excessive and enhanced the over-oxidation to produce 1,4-benzoquinone.

4. Conclusion

In general, the modification of bentonite shows prospective support material for Fe as catalyst. The catalytic activity tests for phenol hydroxylation show that Fe loading plays an important role to the catalytic performance. Reaction conditions are very important to get the best result in term of the highest phenol conversion and also the highest selectivity to hydroquinone. Besides the composition of the catalyst precursor, synthesis method, drying method, activation method, all reaction parameters (such as, reactant ratio, reactant-catalyst ratio, reaction time and reaction temperature) are important variables that require further research in obtaining optimized combination for the best phenol hydroxylation to hydroquinone.

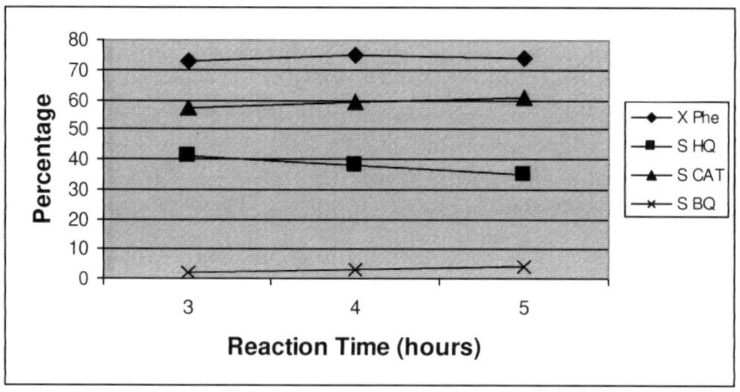

Figure 7. Effect of reaction time on Phenol hydroxylation on CTMABr-Bent-Al-Fe (0.1M) at 333K (X Phe = Conversion of Phenol; S HQ = Selectivity of Hydroquinone; S CAT = Selectivity of Cathecol; S BQ = Selectivity of Benzoquinone).

Acknowledgment

The authors acknowledge: Ministry of Research, Technology and Higher Education of the Republic of Indonesia, Research Grant (Hibah Berbasis Kompetensi 2017) for funding.

References

[1] Siegel H, Voges H-W, Hamamoto T, Umemura S, Iwata T, Miki H, Fujita Y, Buysch H-J, Garbe D and Paulus W 2000 *Phenol Derivatives (Ullmann's Encyclopedia of Industrial Chemistry* vol 26*)* ed B Elvers *et al* (New York: Wiley-VCH) p 521

[2] Liu H, G. Lu G, Guo Y, Guo Y and Wang J 2006 Chemical kinetics of hydroxylation of phenol catalyzed by TS-1/diatomite in fixed-bed reactor *Chem. Eng. J.* **116** pp 179–186

[3] Kumar A and Srinivas D 2013 Hydroxylation of phenol with hydrogen peroxide catalyzed by Ti-SBA-12 and Ti-SBA-16 *J. Mol. Catal. A-Chem.* **368-369** pp 112–118

[4] Makuch E and Wroblewska A 2013 Preparation of titanium silicate Ti-SBA-15 catalyst *Chemik* **67** pp 811–816

[5] Dietrich L A S, Sahu M, Biswas P and Fein J B 2012 Experimental study of TiO_2 nanoparticle adhesion to silica and Fe(III) oxide-coated silica surfaces *Chem. Geol.* **332-333** pp 148–156

[6] Hankare P P, Patil R P, Jadhav A V, Garadkar K M and Sasikala R 2011 Enhanced photocatalytic degradation of methyl red and thymol blue using titania–alumina–zinc ferrite nanocomposite *Appl. Catal. B-Environ.* **107** pp 333–339

[7] Durgakumari V, Subrahmanyam M, Rao K V S, Ratnamala A, Noorjahan M and Tanaka K 2002 An easy and efficient use of TiO_2 supported HZSM-5 and TiO_2+HZSM-5 zeolite combinate in the photodegradation of aqueous phenol and *p*-chlorophenol *Appl. Catal. A-Gen.* **234** pp 155–165

[8] Yube K, Furuta M, Aoki N and Mae K 2007 Control of selectivity in phenol hydroxylation using microstructured catalytic wall reactors *Appl. Catal. A-Gen.* **327** pp 278–286

[9] Shaban Y A, El Sayed M A, El Maradny A A, Al Farawati R K and Al Zobidi M I 2013 Photocatalytic degradation of phenol in natural seawater using visible light active carbon modified (CM)-n-TiO_2 nanoparticles under UV light and natural sunlight illuminations *Chemosphere* **91** pp 307–313

[10] Yang P, Yu J, Xu N, Wang J and Wu T 2007 Influence of synthesis method on the properties and catalytic performance of MCM-49 zeolites *Catal. Commun.* **8** pp 997–1002

[11] Karakhanov E A, Maximov A L, Kardasheva Y S, Skorkin V A, Kardashev S V, Ivanova E A, Luke E L, Seeley J A and Cron S L 2010 Hydroxylation of Phenol by Hydrogen Peroxide Catalyzed by Copper(II) and Iron(III) Complexes: The Structure of the Ligand and the Selectivity of ortho-Hydroxylation *Ind. Eng. Chem. Res.* **49** pp 4607–13

[12] Zhao Y, He G, Dai W and Chen H 2014 High catalytic activity in the phenol hydroxylation of magnetically separable $CuFe_2O_4$–reduced graphene oxide *Ind. Eng. Chem. Res.* **53** pp 12566–74

[13] Choi J, Yoon S, Jang S and Ahn W S 2006 Phenol hydroxylation using Fe-MCM-41 catalysts *Catal. Today* **111** pp 280–287

[14] Liang X, Yang R and Li G 2013 Phenol hydroxylation over Fe-incorporated mesoporous materials prepared by coprecipitation *Micropor. Mesopor. Mat.* **182** pp 62–72

[15] Yang R, Li G and Hu C 2015 The preparation of Fe/wood-based activated carbon catalyst for phenol hydroxylation from Fe^{2+} and Fe^{3+} precursors *Catal. Sc. Technol.* **5** pp 2486–95

[16] Widi R K, Budhyantoro and Savitri E 2009 Hydroxylation of phenol with hydrogen peroxide catalyzed by Fe- and AlFe-Bentonite *J. Chemistry and Chemical Engineering* **3** pp 48–52

[17] Widi R K, Budhyantoro A and Savitri E 2014 Catalytic performance of Al-HDTMA bentonite impregnated Fe on phenol hydroxylation *Int. J. Appl. Eng. Res.* **9** pp 7521–29

[18] Xiong H, Zhang Y, Liew K and Li J 2008 Fischer–Tropsch synthesis: The role of pore size for Co/SBA-15 catalysts *J. Mol. Catal. A-Chem.* **295** pp 68–76

[19] Tanarungsun G, Kiatkittipong W, Praserthdam P, Yamada H, Tagawa T and Assabumrungrat S 2008 Hydroxylation of benzene to phenol on Fe/TiO_2 catalysts loaded with different types of second metal *Catal. Commun.* **9** pp 1886–90

[20] Kurian M and Sugunan S 2006 Wet peroxide oxidation of phenol over mixed pillared montmorillonites *Chem. Eng. J.* **115** pp 139–146

Equilibrium study for ternary mixtures of biodiesel

S Doungsri, T Sookkumnerd, A Wongkoblap and A Nuchitprasittichai

School of Chemical Engineering, Institute of Engineering, Suranaree University of Technology, Nakhon Ratchasima 30000, Thailand

E-mail: terasut@sut.ac.th, atichat@sut.ac.th, aroonsri@sut.ac.th

Abstract. The liquid-liquid equilibrium (LLE) data for the ternary mixtures of methanol + fatty acid methyl ester (FAME) + palm oil and FAME + palm oil + glycerol at various temperatures from 35 to 55°C, the tie lines and binodial curves were also investigated and plotted in the equilibrium curve. The experimental results showed that the binodial curves of methanol + FAME + palm oil depended significantly with temperature while the binodial curves of FAME + palm oil + glycerol illustrated insignificant change with temperatures. The interaction parameters between liquid pair obtained for NRTL (Nonrandom Two-Liquid) and UNIQUAC (Universal Quasi-Chemical Theory) models from the experimental data were also investigated. It was found that the correlated parameters of UNIQUAC model for system of FAME + palm oil + glycerol, denoted as a13 and a31, were 580.42K and -123.69K, respectively, while those for system of methanol + FAME + palm oil, denoted as a42 and a24 , were 71.48 K and 965.57K, respectively. The ternary LLE data reported here would be beneficial for engineers and scientists to use for prediction of yield and purity of biodiesel for the production. The UNIQUAC model agreed well with the experimental data of ternary mixtures of biodiesel.

1. Introduction

The development of alternative fuels from renewable resources, like biomass, has received considerable attention in all countries. It is considered that biodiesel contributes much less to global warming than fossil fuels, being produced from renewable natural sources such as vegetable oils and animal fats consisting of the alkyl esters of fatty acids [1, 2]. Tranesterification of triglyceride (TG) with methanol (MeOH) to produce biodiesel or fatty acid methyl ester (FAME) and glycerol is as following equation:

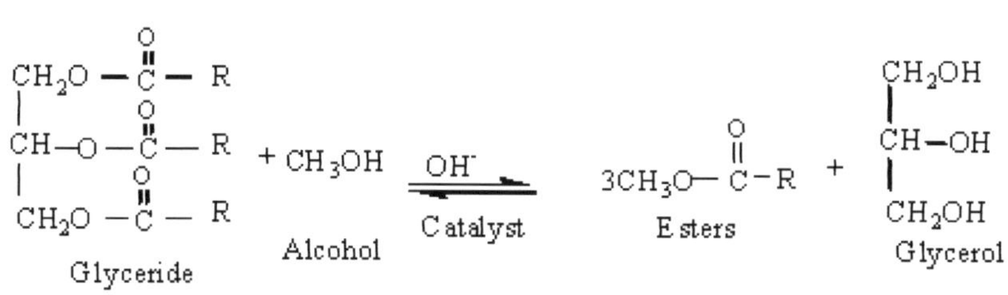

(1)

The most widely used vegetable oils for the production of biodiesel are taken from rape seeds, sunflowers, and soybeans [3]. In Thailand, the most available feedstock is palm oil.

Knowledge about the phase equilibrium in these systems is essential for better understanding of the process and improvement of the reaction rate, the selectivity of the desired product, and the separation process for the product mixture [4]. Moreover, simulations for developing new processes for such systems require an activity coefficient model that can adequately describe these multiphase systems. Some authors have used UNIFAC for describing similar systems [4].

In this study, we present a phase behavior of liquid-liquid equilibrium by using an activity coefficient model to predict the new processes developing, however it is not investigated in kinetic study. In this paper, the experimental results of phase behavior for the methanol + FAME + palm oil and that for FAME + palm oil + glycerol systems are presented. The UNIQUAC interaction parameters w, a_{ij}, (with the combinatorial parameters r_i and q_i, calculated by UNIFAC-Dortmund) were estimated from the mutual solubility data.

2. Materials and methodology

2.1. Chemical

Chemical used in this experiment were palm oil (> 99.5%), methanol (99.5%), glycerol (> 99.0%) and FAME obtained from Bioenergy Plus Ltd, in Thailand, the composition of FAME is shown in Table 1 and its density is 0.88 g.cm-3.

Table 1. The composition of FAME.

Components	Composition (mass %)
Palmitic Acid ME	52.88
Stearic Acid ME	4.670
Oleic Acid ME	33.98
Linoleic Acid ME	8.050
Linolenic Acid ME	0.420

2.2. The binodial curve study

The binodial curves for both systems were investigated and then the thermodynamic model was used to analyst the biodiesel system.

2.2.1. The binodial curves of methanol + FAME + palm oil

The phase boundary or the binodial curve was determined by turbidimetric analyzing using the titration method under isothermal conditions at 35, 45 and 55°C. The experiment was conducted in a water bath equipped with a temperature controller within a fluctuation of ± 0.1°C, and the measurement of each phase is as followings:

Palm oil-rich phase: titrate methanol with a microburet into mixture of palm oil and FAME while the mixture was stirred by a mechanical agitator. The end point where the mixture changed from transparent to turbid was considered to be the saturation point of methanol in palm oil + FAME solution. The amount of methanol dissolved in the palm oil + FAME solution was calculated from the volume of methanol used. The solubility of methanol was measured in different compositions of the palm oil + FAME solution, and the equilibrium data were plotted in the triangular graph.

Methanol-rich phase: palm oil was titrated into a mixture of methanol + FAME until the turbidity was observed as described in the procedure of *Palm oil-rich phase*. The solubility of palm oil in the methanol + FAME solutions was obtained for various compositions of methanol and FAME.

2.2.2. The binodial curves of FAME + palm oil + glycerol

The same procedures as those for the binodial curves of methanol + FAME + palm oil were used for FAME + palm oil + glycerol.

2.3. Tie lines developing for methanol + FAME + palm oil system
Mixtures of methanol, FAME, and palm oil with various concentrations were intensively agitated for 2 hours after that they were left until they reached the two-phase equilibrium. The mixture was kept in an isothermal water bath for 12 hours until the ternary mixture was separated into two phases. The upper phase was enriched with methanol, while the lower layer was enriched with palm oil. The upper phase, methanol and FAME contents were measured by gas chromatography and the lower phase was determined from material balance. After obtaining the composition of the palm oil-rich phase and the methanol-rich phase, the tie lines were obtained by plotting the data in the phase diagram and connecting the two points with beeline.

2.4. Thermodynamic correlations
The LLE calculation is basically defined by equation 2.

$$\gamma_i^I x_i^I = \gamma_i^{II} x_i^{II} \quad (2)$$

$$Fw_i = F^I w_i^I + F^{II} w_i^{II} \quad (3)$$

Where γ_i and x_i represent the activity coefficient and mole fraction of component i presented in both phases I and II, respectively.

The NRTL and the UNIQUAC models [5] were then applied in this study to correlate the LLE data with the material balance shown in equation 3, w_i and F represent the total feed mass fraction of component i and the total mass feed, respectively. The binary interaction parameters of the NRTL and UNIQUAC models were calculated by using nonlinear regression analysis. The objective function used for correlating of experimental data was in the following form of equation 4.

$$\sum_{k=1}^{N}\sum_{i=1}^{3}\left(x_{ik,cal} - x_{ik,\exp}\right)^2 + \sum_{k=1}^{N}\sum_{i=1}^{3}\left(y_{ik,cal} - y_{ik,\exp}\right)^2 \quad (4)$$

where x_{ik} is the mole fraction of component i in methanol-rich phase for the number of tie line k, y_{ik} is the mole fraction of component i in palm oil-rich phase for the number of tie line k, and the subscript *cal* and *exp* refer to the calculated and experimental values of the liquid phase concentrations, respectively.

3. Results and discussion

3.1. The solubility curves
Table 2 shows experimental data of methanol + FAME + palm oil ternary system at temperature of 35, 45, and 55°C. The data were plotted on a triangular phase diagram (Figure 1) to represent behavior of partially miscible ternary system of methanol + FAME + palm oil at each temperature. Any mixture whose overall composition falls in area below the equilibrium curve separates into two liquid phases coexist, while a single liquid phase occurs above the solubility curve. Methanol-FAME pair and FAME-palm oil pair are completely mutually soluble to form binary solutions as shown at the edges of the triangular phase diagram, but methanol-palm oil pair is partially soluble in each other.

Table 3 shows experimental data of FAME + palm oil + glycerol ternary system at temperature of 35, 45, and 55°C. The triangular phase diagram of this system at each temperature is illustrated in Figure 2. Most mixture is formed two phases. The results reveal that the binodial curves of methanol + FAME + palm oil depend significantly with temperature while the binodial curves of FAME + palm oil + glycerol change insignificantly with temperature as shown in Figures 1 and 2, respectively.

Generally, polar molecules, such as methanol and glycerol, tend to have high solubility parameters for example dielectric constant of fluids, while a nonpolar molecule, such as palm oil and FAME, has low solubility parameters. As a rough approximation, substances with solubility parameters difference by 3 $(cal./cm^3)^{1/2}$ or more will generate two liquid phases [6]. In the system of FAME + palm oil + glycerol, glycerol has very high dielectric constant while FAME or biodiesel and palm oil have low dielectric constant. As a result, the different polarity between FAME and palm oil pair is not much and then they can completely dissolve while the difference between FAME-glycerol and palm oil-glycerol pairs are high, the coexist can be observed.

Table 2. Experimental compositions of Methanol+FAME+Palm oil system along the Binodal Curve based on mass fraction

35 °C			45 °C			55 °C		
Palm oil	FAME	Methanol	Palm oil	FAME	Methanol	Palm oil	FAME	Methanol
0.9418	0.0000	0.0582	0.9172	0.0000	0.0828	0.9080	0.0000	0.0920
0.8836	0.0465	0.0699	0.8645	0.0454	0.0901	0.8589	0.0452	0.0959
0.8338	0.0925	0.0737	0.8141	0.0904	0.0955	0.7984	0.0889	0.1127
0.7335	0.1834	0.0831	0.7172	0.1774	0.1054	0.7000	0.1750	0.1250
0.6352	0.2715	0.0933	0.6191	0.2637	0.1172	0.6022	0.2582	0.1396
0.5337	0.3564	0.1099	0.5204	0.3471	0.1325	0.5054	0.3369	0.1577
0.4354	0.4319	0.1327	0.4250	0.4252	0.1498	0.4081	0.4076	0.1843
0.3363	0.5046	0.1591	0.3236	0.4853	0.1911	0.3078	0.4612	0.2310
0.2402	0.5605	0.1994	0.2735	0.5078	0.2188	0.2744	0.4743	0.2513
0.1056	0.5778	0.3166	0.2247	0.5241	0.2513	0.2056	0.4798	0.3147
0.0892	0.5487	0.3621	0.1328	0.5225	0.3447	0.2104	0.4757	0.3139
0.0742	0.5169	0.4089	0.1208	0.4907	0.3885	0.1762	0.4597	0.3641
0.0593	0.4825	0.4582	0.0929	0.4652	0.4419	0.1324	0.4449	0.4227
0.0523	0.4247	0.5230	0.0733	0.4153	0.5113	0.1069	0.4002	0.4929
0.0333	0.3891	0.5777	0.0590	0.3796	0.5614	0.0812	0.3698	0.5490
0.0212	0.3484	0.6304	0.0436	0.3405	0.6159	0.0609	0.3342	0.6049
0.0179	0.2910	0.6911	0.0223	0.2898	0.6879	0.0344	0.2861	0.6796
0.0154	0.2457	0.7389	0.0092	0.2000	0.7908	0.0120	0.1993	0.7887
0.0092	0.1998	0.7909	0.0055	0.0985	0.8961	0.0091	0.0983	0.8927
0.0046	0.0985	0.8969	0.0005	0.0000	0.9995	0.0005	0.0000	0.9995
0.0005	0.0000	0.9995						

Table 3. Experimental compositions of FAME+Palm oil+Glycerol system along the Binodal Curve based on mass fraction

35 °C			45 °C			55 °C		
Glycerol	Palm oil	FAME	Glycerol	Palm oil	FAME	Glycerol	Palm oil	FAME
0.9968	0.0032	0.0000	0.9975	0.0025	0.0000	0.9977	0.0023	0.0000
0.9987	0.0000	0.0013	0.9987	0.0000	0.0013	0.9991	0.0000	0.0009
0.0016	0.0000	0.9984	0.0016	0.0000	0.9984	0.0013	0.0000	0.9987
0.0019	0.1005	0.8976	0.0019	0.1004	0.8977	0.0025	0.0998	0.8977
0.0038	0.2987	0.6975	0.0025	0.2995	0.6980	0.0019	0.2995	0.6986
0.0025	0.3990	0.5985	0.0031	0.3990	0.5979	0.0025	0.3989	0.5986
0.0025	0.4986	0.4989	0.0025	0.4986	0.4989	0.0025	0.4990	0.4985
0.0031	0.5981	0.3988	0.0025	0.5985	0.3990	0.0025	0.5985	0.3990
0.0050	0.6967	0.2983	0.0044	0.6959	0.2997	0.0019	0.6986	0.2996
0.0044	0.7914	0.2042	0.0025	0.7973	0.2002	0.0019	0.7985	0.1996
0.0038	0.8967	0.0995	0.0025	0.8978	0.0997	0.0025	0.8977	0.0998
0.0041	0.9959	0.0000	0.0044	0.9956	0.0000	0.0044	0.9956	0.0000

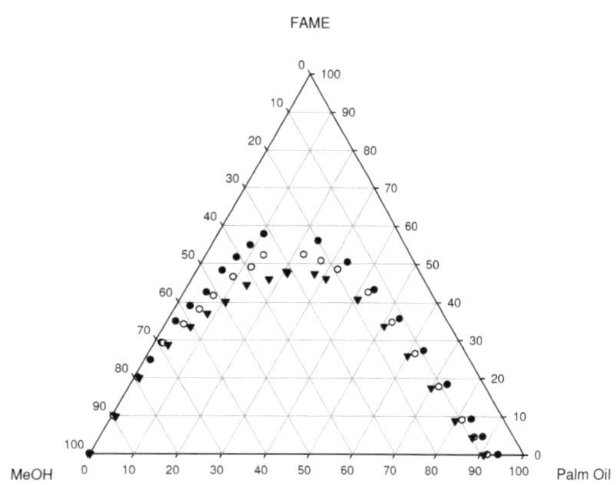

Figure 1. The binodial curves of methanol + FAME + palm oil system based on mass fraction: •; 35°C, ○; 45°C and ▼; 55°C.

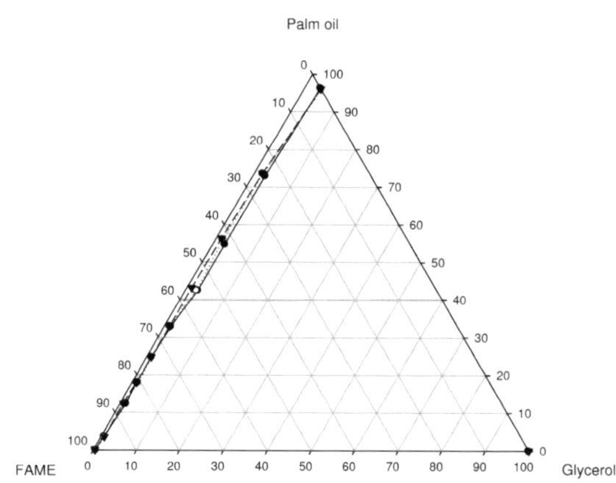

Figure 2. The binodial curves of FAME + palm oil + glycerol system based on mole fraction: •; 35°C, ○; 45°C and ▼; 55°C.

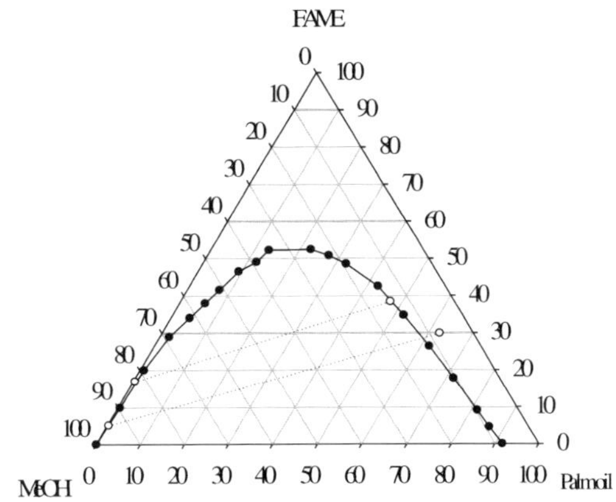

Figure 3. The binodial curve with the tie lines of methanol + FAME + palm oil system based on mass fraction at 45°C.

3.2. Tie lines

The lines connect phases in equilibrium with each other shown in phase diagram are called tie lines, the binodial curve with tie lines for methanol + FAME + palm oil system 45°C is shown in Figure 3. However, the system of FAME + palm oil + glycerol, only the binodial was constructed without experimental tie lines due to its immiscible property. Two tie lines connect between two equilibrium compositions as open circle shown in Figure 3, as one can see, one tie line agrees well on the equilibrium curve while the other one shows a bit difference. The tie lines calculated by UNIQUAC Dortmund were used to verify experimental tie lines, and it was found that they agreed quite well. Therefore, the equilibrium curves and tie lines obtained from experimental data at various temperatures were used to evaluate the thermodynamic parameters for LLE of these two systems in section 3.3.

3.3. Activity coefficient model

The volume (r) and surface area (q) obtained from the UNIFAC-Dortmund [7, 8] are used to optimize values of model parameters by minimization method from equation 4 [9]. The r and q parameters of the NRTL and UNIQUAC models for each component are listed in Table 4. The interaction parameters of mutual solubility of the UNIQUAC model at 35, 45 and 55°C for these two systems are listed in Table 5 and Table 6.

Table 4. Parameters r_i and q_i for the mutual solubility.

Components	r_i	q_i
FAME	11.3629	13.3311
Palm oil	34.3817	32.3022
Glycerol	5.5881	4.4497
Methanol	0.8585	0.9938

Table 5. The UNIQUAC interaction parameters of mutual solubility of methanol + FAME + palm oil.

Temp.	a_{42}	a_{24}
°C	K	K
35	61.40	1028
45	71.92	938.7
55	75.15	929.3
Average	71.48	965.6

Table 6. The UNIQUAC interaction parameters of mutual solubility of FAME + palm oil + glycerol.

Temp.	a_{13}	a_{31}
°C	K	K
35	556.54	-121.73
45	592.67	-129.63
55	592.04	-123.69
Average	580.42	-123.69

The interaction parameters listed in Table 5 and Table 6 are also used to correlate the tie line data as discussed in section 3.2 in order to obtain the interaction parameters or the completely miscible chemical pairs. The parameters obtained from this study will be used to evaluate the equilibrium data for biodiesel production.

4. Conclusions

For the ternary mixture of methanol + FAME + palm oil system, the solubility curves change significantly with temperature in the range from 35°C to 55°C. On the other hand, the equilibrium curves for ternary mixture of FAME + palm oil + glycerol system change insignificantly with temperature and this is due to the polar properties of liquids. The activity coefficient models obtained from the UNIQUAC and NRTL model are used to correlate with the tie lines obtained from experimental data. They agree quite well, although some fluctuation can be observed.

Acknowledgements

The authors acknowledge Suranaree University of Technology for the financial support to Sujitra Doungsri.

References

[1] Demirbas A 2007 Progress and recent trends in biofuels *Prog. Energ. Combust.* **33** pp 1–18
[2] Asakuma Y, Maeda K, Kuramochi H and Fukui K 2009 Theoretical study of the transesterification of triglycerides to biodiesel fuel *Fuel* **88** pp 786–791
[3] Čerče T, Peter S and Weidner E 2005 Biodiesel-transesterification of biological oils with liquid catalysts: Thermodynamic properties of oil–methanol–amine mixtures *Ind. Eng. Chem. Res.* **44** pp 9535–41
[4] Negi D S, Sobotka F, Kimmel T, Wozny G and Schomacker R 2006 Liquid–liquid phase equilibrium in glycerol-methanol-methyl oleate and glycerol-monoolein-methyl oleate ternary systems *Ind. Eng. Chem. Res.* **45** pp 3693–96
[5] Prausnitz J M, Lichtenthaler R N and de Azevedo E G 1986 *Molecular Thermodynamics of Fluid-Phase Equilibria* (New Jersey: Prentice-Hall)
[6] Csernica S N and Hsu J T 2011 Reverse-phase ternary phase diagram, tie lines, and plait point for commercial biodiesel-glycerol-methanol *Ind. Eng. Chem. Res.* **50** pp 1012–16
[7] Gmehling J, Li J and Schiller M 1993 A modified UNIFAC model. 2. Present parameter matrix and results for different thermodynamic properties *Ind. Eng. Chem. Res.* **32** pp 178–193
[8] Mohsen-Nia M and Dargahi M 2007 Liquid– liquid equilibrium for systems of (corn oil+ oleic acid+ methanol or ethanol) at (303.15 and 313.15) K *J. Chem. Eng. Data* **52** pp 910–914
[9] Lee M-J, Lo Y-C and Lin H-M 2010 Liquid–liquid equilibria for mixtures containing water, methanol, fatty acid methyl esters, and glycerol *Fluid Phase Equilibr.* **299** pp 180–190

Identification, measurement, and assessment of water cycle of unhusked rice agricultural phases: Case study at Tangerang paddy field, Indonesia

N Hartono, Laurence and H P Johannes

Department of Industrial Engineering, University of Pelita Harapan, Tangerang, Indonesia

E-mail: natalia.hartono@uph.edu

Abstract. According to one of UN reports, water scarcity has happened all around the world, including Indonesia. Irrigation sector takes up 70% of world water consumption and potentially increases 20% due to the population explosion. Rice is accounted for 69% of agricultural products contributions in Indonesia's water footprint. Therefore, evaluation of water cycle was essential to raise awareness among practitioners. Data collections were conducted in the functional unit of one-hectare rice field located in Tangerang. This study used CropWat 8.0 and SimaPro software. Identification involved data such as climate, crop, and soil. Nursery became the highest water consumed phase, requiring 419 mm in height. Measurement through water footprint resulted in consumption of green water footprint for 8,183,618.5 liters (62.9%), followed by grey for 4,805,733.2 liters (36.9%) and blue for 23,902.36 liters (0.2%). The grey consumption was exceeding the average, which indicated high doses of pesticides. Life Cycle Assessment showed negative impacts of fertilizers that caused damages like fossil depletion, respiratory health, and eutrophication.

Keywords: Agricultural phases; Life Cycle Assessment; paddy field; water; water footprint

1. Introduction

Due to its significant role in every aspect of life, freshwater has been scarce for several years in many regions. Water scarcity is one of the most crucial environmental problems [1]. This is in part due to changing climates and poor water management strategies, and parts of the scarcity problem find an explanation in natural factors [2]. Freshwater scarcity also happens in Indonesia.

High water consumption is contributed by several sectors, but the highest consumption comes from the agricultural sector in relation to irrigation practices. It takes up 70% of world water consumption and potentially increases 20% without any further improvement in management [3]. Another research said that agriculture was responsible for 85% of the overall global freshwater consumption [4] and caused significant environmental damages [5]. This negative potential is also caused by population explosion as predicted by UN. In 2050, world population will be 9.8 billion people, which from 2017 to 2050, it is expected that half of the world's population growth will be concentrated in just nine countries, including Indonesia [6]. Indonesia considered will be the 9th country that contributed to the total growth of the world's population according to UN reports. This explosion will lead to higher number of water consumption. As an agricultural country, Indonesia produces several crop products, such as rice, maize, cassava, soybeans, groundnut, and many more. Among these products, rice contributes the highest water consumption with 69% [7]. Based on previous facts, it is essential to evaluate the current water flow in rice agricultural phases especially in a paddy field.

Previous researches in Indonesia were mainly about how much water needed for irrigation of paddy field such as Fuadi et al [8] and Indonesian Center for Rice Research, Ministry of Agriculture [9]. There is a need to research to address water scarcity. A research about water footprint was introduced by Hoekstra in 2003. The water footprint of a product is defined as the total volume of freshwater that is used to produce the product [10]. Bulsink in 2009 studied water footprint of Indonesia provinces. He reported that Java is the most water-scarce island [7]. This study tried to identify, measure and assessment on a smaller scale to raise awareness among local farmers. Therefore, this study evaluates water footprint and LCA with identification, measurement, and assessment of agriculture phase at paddy field, a case study at Bojong Indah, Kemuning Village, Tangerang, West Java from September until December 2016 using CropWat 8.0 and SimaPro.

2. Methodology

This research focuses only on agricultural phases of unhusked rice as the main object. Unhusked rice is obtained directly from harvesting without any further process,it is still covered with husks. Data about this object was based on latest process happened in rice field located at Bojong Indah, Kemuning Village, Tangerang. The latest productivity of the rice field was 5.6 tons per hectare. An overallpaddy field is about 25 hectares and led by Mr. Diding, a local farmer.

To achieve the goal of this research, several steps were arranged, starting from preliminary research, problem identification, theoretical background, direct observation, data collecting and processing, data analysis and discussion, and conclusion. Steps in data collection and processing are shown in Fig 1. Data analysis measured impact assessment and evaluation of water cycle.

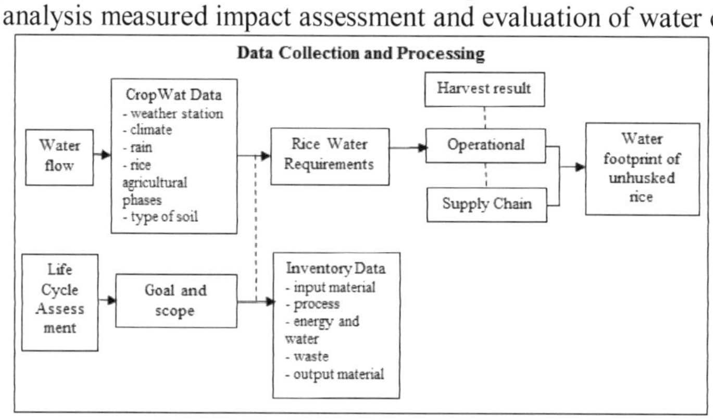

Figure 1. Steps in data collection and processing

This research had three major activities: identification, measurement, and assessment. Identification was used CropWat 8.0 to calculate water requirements in every agricultural phase. This data used a functional unit of one-hectare rice field. Measurement and assessment of the water cycle were examined through LCA using SimaPro. Life Cycle Assessment (LCA) is a tool developed to assist sustainable decision-making processes [2], which in this research, to evaluate the water cycle.

2.1. CropWat data

CropWat is a decision support tool developed by the Land and Water Development Division of FAO, a computer program for the calculation of crop water requirements and irrigation requirements based on soil,climate and crop data [11]. CropWat was used to calculate water requirement. Input data for this software were information about weather station, climate, rain, soil, and crop. Data was taken for 1 year from 1 October 2015 to 30 September 2016. Data about weather station was gathered through

BMKG [12]. The weather station mentioned here is the nearest station that measures some climate statistics in the concerned rice field. Based on search results, the weather station is located in Curug, Tangerang, geographically 6.29 S (latitude) and 106.6 E (longitude) with altitude 42 meters above sea level. Climate statistics needed for CropWat software are temperature, relative humidity, wind speed, the sun, and rain, with different units. Data was converted monthly from October 2015 until September 2016. A complete data about temperature was collected from AccuWeather. Meanwhile, the wind speed was taken from Weather Online. The rests were taken from BMKG (Curug Station) although not all daily data was available [13]. All statistics were then inputted in menu Climate/E To CropWat. For menu Crop, some statistics came from agricultural phases based on an interview with local farmers, especially for the duration of each phase. The rests were taken from CropWat database. Last, menu Soil was fully taken from CropWat database. The type of soil that was chosen was black clay soil according to direct observation and interview.

After all statistics data were inputted, crop water requirements were calculated automatically. The result was total water requirements are 820.1 mm (in height). The first phase (nursery phase) was the highest water required phase.

2.2. *Water footprint*

The component used in water footprint method was indirect consumption, which includes only processes to produce unhusked rice without further processes or consumption. Calculations were both for operational water requirements and supply chain water requirements, classified as green, blue, and grey water footprints. The blue water footprint refers to the volume of surface and groundwater consumed (evaporated) as a result of the production of a good, the green water footprint refers to the rainwater consumed, the greywater footprint of a product refers to the volume of freshwater that was required to assimilate the load of pollutants based on existing ambient water quality standards [14].

In operational, all green consumptions were calculated from CropWat results of water requirements in the previous section. The water quantity was converted into a unit of volume (liter) by multiplying it with the functional unit (one hectare). The blue consumption was calculated from water consumption through the water supply, mostly charged with overhead consumption of local farmers. Meanwhile, the grey consumption is the volume of water to assimilate dangerous substances mostly for pesticides. It was calculated using (1). The value of α is a leaching-runoff fraction which was 0.01 (for pesticides) and 0.03 (for phosphorous) [10]. Appl is the amount of chemical substances released to the environment. Meanwhile, cmax is maximum allowable concentration of chemical substances, which was 0.1 μg/liter (for pesticides) and 200 μg/liter (for phosphorous) [15] [16]. Other values of cmax can also refer to CCME and EPA standards. Value 0 of cnat is given if pesticides are not naturally formed.

$$\text{Grey water footprint} = \frac{\alpha \times \text{Appl}}{c_{max} - c_{nat}} \quad (1)$$

The results of operational water requirements were presented in Table 1. The first six phases were direct consumption through agricultural phases of unhusked rice and the last two aspects were overhead consumption.

Table 1. Operational water requirements.

Phases/ aspects	Operational water footprint (liters)			
	Green	Blue	Grey	Total
Nursery	4,189,000	1,000	0	4,190,000
Land preparation	0	0	0	0
Plantation	382,000	0	0	382,000
Quick growth	1,062,000	0	0	1,062,000
Further growth	1,607,000	0	0	1,607,000
Maturation	941,000	98.38	4,707,222.2	5,648,320.58
Farmers consumption	0	22.500	93.750	116.250
Cleaning	0	0	0	0
Total	8,181,000	23,598.38	4,800,972.2	13,005,570.58

Using the same way, water consumption in the supply chain was also calculated. It includes fabric used in nursery phase, tractors for tilling, fertilizers, pesticides, and fuel to operate tractors (overhead). Several assumptions were stated to adjust calculations so that it can reflect real conditions. The result of supply chain water requirements can be viewed in Table 2. These two water requirements calculations were combined to get the total value of water footprint as presented in Table 3.

Table 2. Supply chain water requirements.

Materials	Supply chain water footprint (liters)			
	Green	Blue	Grey	Total
Fabric	2,618.5	0	357	2,975.5
Tractors	0	302.4	4,392	4,694.4
Fertilizers	0	0	0	0
Pesticides	0	1.58	0	1.58
Fuel	0	0	12	12
Total	2,618.5	303.98	4,761	7,683.48

Table 3. Total water footprint.

Type	Water footprint (liters)			
	Green	Blue	Grey	Total
Operational	8,181,000	23,598.38	4,800,972.2	13,005,570.58
Supply chain	2,618.5	303,98	4,761	7,683.48
Total	8,183,618.5	23,902.36	4,805,733.2	13,013,254.06

From the total water footprint calculation results, it was clear that the highest water requirement was green water footprint (62.9%), followed by grey (36.9%) and blue water footprint (0.2%). The total water footprint is 13,013,254.06 liters.

2.3. Life Cycle Assessment (LCA)

Standard steps in LCA are goal and scope definition, inventory data, impact assessment, and interpretation. This assessment was cradle-to-gate because only includes transformation from raw materials (which are rice seeds) into unhusked rice. This assessment was directed to identify potential damages of unhusked rice agricultural phases. Not only high water consumption that will correlate with water scarcity, damages of the phases have to be evaluated. Functional unit used was one hectare of rice field that produces 5.6 tons of unhusked rice.

To create inventory data, the generic unit process should be drawn based on every agricultural phase, involving energy and water used, also wastes. Several assumptions should be defined following insufficient information.

Impact assessment as the next step was done with two methods, one method from European methods, and another one from water footprint-based methods with the help software SimaPro. Because the focus in this assessment was various damages caused by phases, the European method used was ReCiPe endpoint, which covers damage to human, ecosystem, and resources. Because its correlation with ReCiPe three types of damages, Pfister et al. method [5] was used as water footprint-based impact assessment.

3. Results

Results from CropWat 8.0 showed that nursery phase was the highest water required which happens in parallel time with land preparation. The total height of water needed in the nursery phase is 419 mm (for 2 decades, each decade contains 10 days). It was far different compared to next decades which are around interval 34.6 mm until 44 mm, except the last decade which was only 7.3 mm because contains only 2 days.

This result showed that nursery phase was important to ensure the success of entire processes. The high water requirements for this phase are normal due to the significance of this phase to ensure the good harvest results. Besides its vital role, the effective rain in this phase was also minimum, 0.4 mm (for the first decade) and 0.5 (for the second decade). This value drives to the higher water supply.

The specific contribution of high water requirements can be assessed through water footprint calculation results. Based on total water requirements, operational water requirements take up 13,005,570.58 liters (99.94%), compared to the supply chain which is only 7,683.48 liters (0.06%). It indicates that the agricultural phases mostly consume water in operations, not from material or other inputs. The green consumptions happened in every single operational process. It comes from irrigation process that fully uses rainwater. A relative high operational consumption happened in maturation phase due to pesticides. It consumes 4,707,222.2 liters of water to assimilate dangerous substances. The highest consumption was still the green (62.9%), but the abnormal value of the grey (36.9%) should be anticipated. An indication of high doses of pesticide was highly concerned. To prove this indication, the overall value of water requirements is compared to average. The average value is taken from average water footprint of rice in Banten Province. The difference was converted into meter cubic per ton as presented in Table 4. Both green and blue water footprint were below average, but the grey was exceeding. Although the total water footprint was below average, the value of grey water footprintindicates there werehigh-levelpesticides used.

Table 4. Actual water footprint compared to average.

Value	Water footprint (m^3/ton)			
	Green	Blue	Grey	Total
Average	1,994	1,332	205	3,530
Actual	1,464.57	4.27	858.17	2,327.01

Pesticides can be exposed to human through different mediums. Without protective devices, pesticides can enter the body through direct exposure to skin, as well as a respiratory and digestive tract. Pesticides that touch skin can absorb into the body and cause toxicity. Pesticides can also enter respiratory tract. Gas and spray particles which are less than 10 microns get into lungs, whereas larger particles stuck in mucous membrane or throat. Finally, exposure canalso go through digestive tract while eating foods that contain pesticides or when pesticides carried by the wind into the mouth. Moreover, pesticides used in this agricultural phases are dangerous. It contained carbamate and pyrethroid types. A carbamate type called carbofuran (forbidden pesticide based on US EPA since 2008) was used. Also, the application of lambda-cyhalothrin (pyrethroid type which was classified as carcinogenic group D) should be concerned.

To explore other impacts of agricultural phases, the next step of Life Cycle Assessment is impact assessment. Further-growth seeds play a significant role because this was the major input in the last phase (maturation). Further-growth seeds are an accumulation of previous processes. According to ReCiPe normalization of damages, damage to resources becomes the highest, followed by human health and ecosystem.

Damage to resources was mainly caused by fossil depletion due to natural gas used in processing fertilizers and diesel used to generate energy for doing land preparation. Damage to human health is due to CO2 emission on fertilizers production processes, which is around 304.5 kg per ton of production. Diesel burning process was also producing dangerous substances such as particulates, CO2, CO, NO, methane, and much more. Meanwhile, damage to theecosystem was caused by the use of fertilizers that generates ammonia and nitric acid. The exceeding concentration of nitrogen and phosphor of fertilizers will lead to water eutrophication and soil acidification.

According to Pfister et al. [5] method, every damage was calculated with different units. Damage to resources was measuring the water consumption related to scarcity level. Damages of human health and ecosystem show that the higher water consumed, the higher its contribution to damages. In this case, the highest water consumption happened in ready-to-plant seeds, which is the main output of the nursery phase, followed by further growth seeds, unhusked rice, and quick-growth seeds.

Overall, farmers around the rice field should give attention and did an evaluation of their agricultural practices. An indication of dangerous use of pesticides should be concerned since the harvest results will be consumed. Standard setting, law enforcement, and direct socialization will be needed to educate farmers. In other sides, fertilizers mainly contribute to several damages. The amount of fertilizer used should be considered because of potential damages like fossil depletion, human respiratory health, water eutrophication, and much more.

4. Conclusion

Identification of water cycle showed the significance of nursery phase that consumed up to 419 mm (in height) water in a20-day period. The water requirements were contributed mostly from operational needs. The green consumption was 8,183,618.5 liters (62.9%), followed by the grey for 4,805,733.2 liters (36.9%) and the blue for 23,902.36 liters (0.2%). The grey consumption exceeded Banten Province average, which indicated the high doses of pesticides used, including carbamate and pyrethroid types that are highly toxic. Meanwhile, the assessment summed that fertilizers played important role in causing damages such as resource depletion from natural gas used, human health problem from emission in the production process, also water eutrophication from fertilizers substances like nitrogen and phosphor.

References

[1] UN-Water, United Nations Educational Scientific and Cultural Organization (UNESCO) 2006 Water: A Shared Responsibility. The United Nations World Water Development Report 2 chapter 1 p 3 http://unesdoc.unesco.org/images/0014/001444/144409E.pdf (retrieved 8 October 2016)

[2] Arnøy S 2012 *Water footprint approaches in Life Cycle Assessment: State-of-the-art and a case study of hydroelectric generation in the Høyanger area* (Norway: Norwegian University of Life Sciences, Ås) p 1

[3] UN-Water: Statistics detail http://www.unwater.org/statistics/statistics-detail/ar/c/246663/ (retrieved 8 October 2016)

[4] Rodda, J C and Shiklomanov I A 2004 *World water resources at the beginning of the Twenty-First Century* (New York: Cambridge University Press)

[5] Pfister S, Koehler A and Hellweg S 2009 Assessing the environmental impacts of freshwater consumption in LCA *Environ. Sci. Technol.* **43** 11 pp 4098–4104

[6] UN DESA. https://www.un.org/development/desa/en/news/population/world-population-prospects-2017.html (retrieved 21 September 2017)

[7] Bulsink F, Hoekstra AY, and Booij MJ 2009 The Water Footprint of Indonesian Provinces Related to The Consumption of Crop Products *Value of Water Research Report Series* **37** (The Netherlands: UNESCO IHE)

[8] Fuadi N A, Purwanto M Y J, Tarigan S D 2016 Study on Water Requirement and Water Productivity of Paddy Field with SRI and Conventional Water Supply System by Using Pipe Irrigation *Jurnal Irigasi* **111** pp 23–32

[9] Indonesian Center for Rice Research, Ministry of Agriculture http://bbpadi.litbang.pertanian.go.id/index.php/en/tahukah-anda/357-kebutuhan-air-untuk-menghasilkan-1-kg-gabah (retrieved 8 October 2016)

[10] Hoekstra AY, Chapagain A, Martinez-Aldaya M, and Mekonnen M 2009 Water Footprint Manual: state of the art 2009 (Enschede, the Netherlands: Water Footprint Network 127p)

[11] FAO http://www.fao.org/land-water/databases-and-software/cropwat/en/ (retrieved 8 October 2016)

[12] Badan Meteorologi, Klimatologi, dan Geofisika http://dataonline.bmkg.go.id/data_iklim (retrieved 8 October 2016)

IOP Conf. Series: Materials Science and Engineering **273** (2017) 012009 doi:10.1088/1757-899X/273/1/012009

[13] Badan Meteorologi, Klimatologi, dan Geofisika http://dataonline.bmkg.go.id/mc_station_metadata (retrieved 8 October 2016)

[14] Mekonnen M and Hoekstra A Y 2010 The Green, Blue and Grey Water Footprint of Crops and Derived Crop Products Volume 1: Main Report, *Value of Water Research Report Series* **47** (The Netherlands: UNESCO IHE, Delft)

[15] Hoekstra A Y and Chapagain A K 2008 *Globalization of Water: Sharing the Planet's Freshwater Resources* (UK: Blackwell Publishing)

[16] Ercin A E, Aldaya M M and Hoekstra A Y 2011 Corporate water footprint accounting and impact assessment: The case of the water footprint of a sugar-containing carbonated beverage *Water Resour. Manag.* **25** pp 721–741

IOP Conf. Series: Materials Science and Engineering **273** (2017) 012010 doi:10.1088/1757-899X/273/1/012010

Performance test of a grid-tied PV system to power a split air conditioner system in Surabaya

E Tarigan

Department of Electrical Engineering, University of Surabaya, Raya Kalirungkut, Surabaya 60293, Indonesia

E-mail: elieser@staff.ubaya.ac.id

Abstract. Air conditioner for cooling air is one of the major needs for those who live in hot climate area such as Indonesia. This work presents the performance test of a grid-tied PV system to power air conditioner under a hot tropical climate in Surabaya, Indonesia. A 800 WP grid-tied photovoltaic (PV) system was used, and its performance was tested to power a 0.5 pk of split air conditioner system. It was found that about 3.5 kWh daily energy was consumed by the tested air conditioner system, and about 80% it could be supplied from the PV system. While the other 20% was supplied by the grid during periods of low solar irradiation, 440 Wh of energy was fed into the grid during operation out of office hours. By using the grid-tied PV system, the energy production by PV system did not need to match the consumption of the air conditioner. However, a larger capacity of PV system would mean that a higher percentage of the load would be covered by PV system.

1. Introduction

Solar energy is one of the most promising renewable energy resources that can be expected to reduce fossil fuels consumption, and in turn to reduce the emission of greenhouse gas in the atmosphere. New innovative solutions need to be attempted in both engineering and policy to promote the solar energy use. Air conditioner for cooling air is one of the major needs for those who live in hot climate area such as Indonesia; even air conditioning is commonly used during all year round. About 50%–60% of buildings energy are commonly from air conditioning needs [1].

For its location that is very close to the equator line, solar irradiation in Surabaya, Indonesia, is relatively high, ranging from 4.8–5.8 kWh/m^2.day [2]. The ambient temperature is also relatively high with an average of 32 °C and 30 °C during dry and wet season, respectively [3]. Therefore, cooling demands for a building is very high throughout the year. On the other hand, solar electricity should be a good option to supply energy for cooling. Surabaya is suitable location to perform the possibility of using air conditioners powered by photovoltaic. The result can be a reference or adopted to other countries or cities with similar weather.

There are several previous works on solar air conditioner have been reported. A study on the design and construction of air conditioning system integrated with a direct current (DC) generated from PV was reported by Daut et al. [4]. To achieve the space cooling, the system design should consider both PV system (such as electrical equivalent, IV characteristic curve and factors affect the output of PV cell) and air conditioner, especially cooling capacity. Tsoutsos et al. [5] reported the performance and economic evaluation of a solar heating and cooling system of a TRNSYS simulation program. It was found that the exploitation of the results of the simulation provided the optimum sizing of the solar system. The studies provided a good guideline in a design process

Aguilar et al. [6] reported a study of employing PV panels to supply electricity to power the compressor of the inverter for an air conditioner unit without batteries or any regulator. Here the inverter of the air conditioner was connected to both PV panels and grid. One advantage of using an inverter air conditioner was that it uses. The advantage of such system is the using of the direct current directly from the photovoltaic without inverter, that means reducing cost and increasing the efficiency energy conversion.

The peak time of using air conditioner for an office is commonly during the day when the hot weather (high solar radiation) and temperature is high, on the other hand at the same time PV panels produces electricity at optimum. In the present work, a grid-tied PV system was tested for its performance to power a split air conditioner under a hot tropical climate in Surabaya, Indonesia. The solar fraction and the advantages and limitation of such system were discussed.

2. Methods

The tested air conditioner unit was a split type with a capacity of 0.5 pk manufactured by Daikin®. The unit pk (stands for paardkracht, Dutch words) was the common commercial unit for air conditioner unit power in Indonesia, where 1 pk = 745.7 Watt. Therefore, the tested air conditioner system in the present work was about 372.85 Watt. The system was used for cooling a small office room with a floor area of 9 m^2.

Grid-tied PV system components consist of: 800 Wp monocrystalline silicon (4 modules of @200 Wp) connected with a 1000 Watt of the grid-tied inverter. The specification of the grid-tied inverter used is as presented in Table 1.

Table 1. Specification of grid-tied inverter.

Specifications	Unit/Explanation
Output Power	1000 W
Inverter Type	DC to AC
Output wave	Pure Sine
Input Voltage	20 – 48 Volt DC
Output Voltage	230Volt AC (with range of 190-260 Volt AC)
Output Frequency	Automatic detection of 45-53Hz / 55Hz-63Hz
Max. Current (DC)	50A
MPPT voltage	26-36VDC
Power Factor	>97%
System type	Grid tie
Max. Input power solar panel	1200 W

The parameters measured for analysis were:
- the DC power from PV modules,
- the AC output power from the inverter,
- solar radiation,
- inside and outside temperatures,
- power consumption air conditioner,
- The start-up power of the air conditioner.

The measurement devices and experiment setup are schematically shown in Figure 1. In the figure, EM1, EM2, and EM3, refers the measurement equipment for energy meter which continuously recording energy in kWh unit. A data logger was used to record the parameters for temperatures, solar radiation, current (A) and voltage (V).

IOP Conf. Series: Materials Science and Engineering **273** (2017) 012010 doi:10.1088/1757-899X/273/1/012010

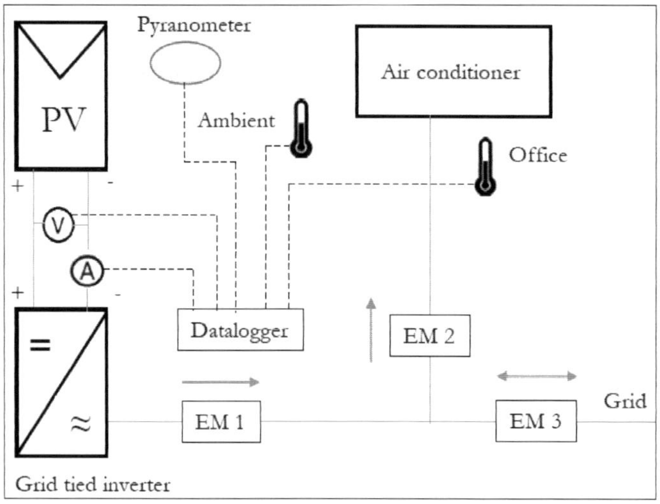

Figure 1. Experiment and measurement devices schematically.

3. Results and discussion

The air conditioner system was set on the temperature of 22 °C and ran during 07:30 to 17:30 fitting with typical office hours period and condition of a typical office in Indonesia. The power consumption of the air during the office hours was approximately 350W and total energy demand is about 3.45 kWh/day. The power rate of air conditioner system over time is shown in Figure 2.

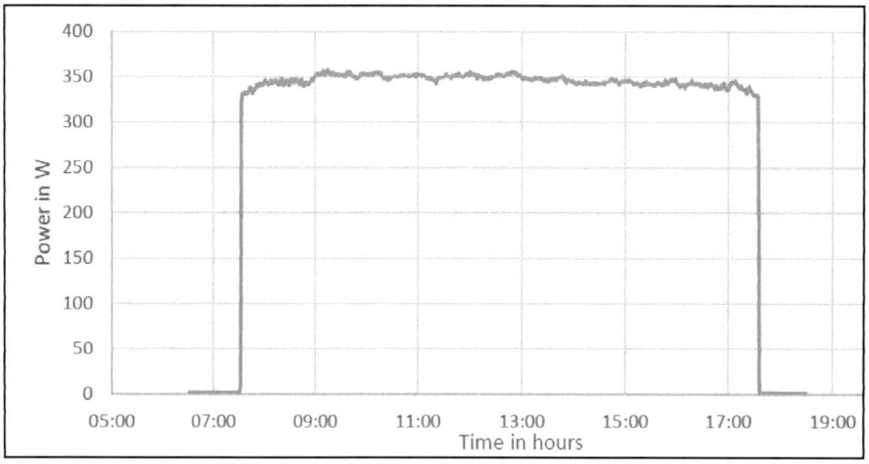

Figure 2. Power consumed by 0.5 pk air conditioner.

The amount of electricity energy produced by the PV system depends on irradiation values. The periods of high energy production are the times of highsolar irradiation. The experiments were conducted during first week of May 2016. The weather was relatively clear sky with average daily radiation of 5.1 kWh/m^2.

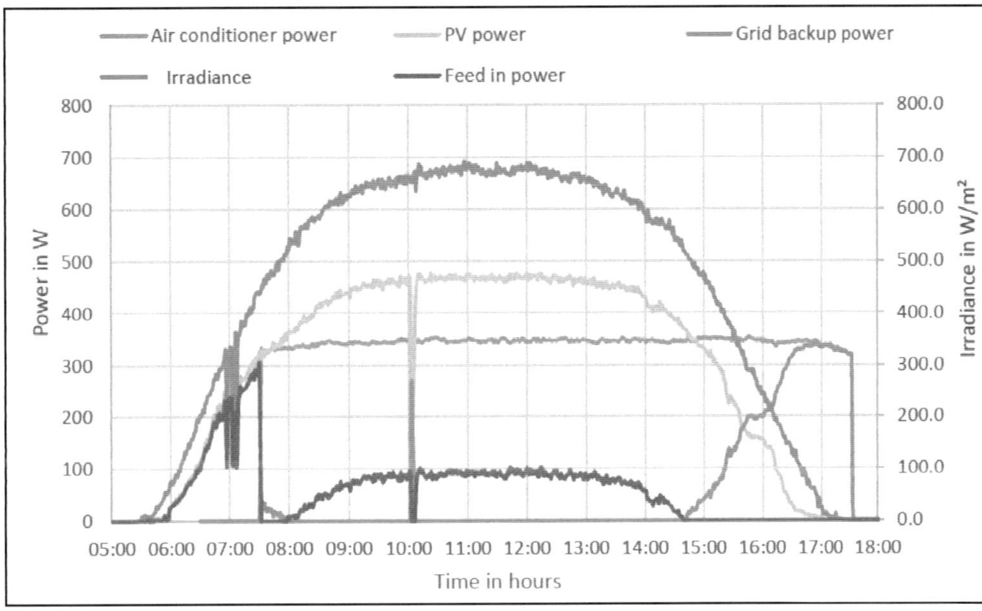

Figure 3. Diurnal variation of AC power, PV power, Backup power, Irradiance and Feed in power.

Figure 3 shows the typical one day measurement results for a sunny day. The figure shows the diurnal variation of the power consumed by the air conditioner, power from PV systems, backup power by the grid, solar irradiance and the feed-in power. It displays how the air conditioning, PV system, and the sun work together. It can be interpreted that during 05:30–07:30 when the air conditioner system was still off, the energy produced by PV was fed into the grid. When theair conditioner was switched on at07:30 the energy production by PV was still too low to fully run the air conditioner. Therefore some portion of power was supplied by the grid. A significant high of power was also needed from the grid for the air conditioner start-up. However, it could not be seen in the graph as it happened in very short of periods (less than a second). From 08:00 irradiance and PV production continued to rise,and the powerproduced reached 450 W. While air conditioner power consumption was at the levelof 350 W means that 100 W excess energy was exported into the grid. This condition was continuously happening until about 14.00. From 14:00–14:40 solar radiation was decreasing and power produced by PV was getting lower, resulting in decreasing of feed in power. From 14:40 the PV system obviously could nolonger fully power the air conditioner, and then the backup power from the grid slowly increased until 17:00 when it was fully powered by the grid, and at 17:30 it switched off.

From results for four consecutive days of experimentas presented in Table 2, it was found that about 3.45 kWh was needed daily by the air conditioner system for running during office hour, and 80% of it (about 2.7 kWh) was supplied by PV panels as called solar fraction. The energy produced by PV system during period of before and after office hour was about 440 Wh. That means that PV panel almost supply all of energy needed by the air conditioner systems.

In such grid-tied PV system, the backup from the grid has two major functions. First, it is to supply power when low PV energy production due to low solar irradiation. In our case, the backup power from the grid was 20% and the majority of use during the last two hours of the office hours. To decrease of power from the grid, it could be attempted by adding more PV panels. However, it would have to be oversized, and it should be considered that from 15:00 onwards the solar irradiation is below 300 W/m^2. To fully power of the air conditioner with PV system during this period was not possible unless using of batteries to store excess power. However, this was an expensive solution. The second function of the backup power from the grid was to power the air conditioner during start-up. This means to eliminate the need to oversize the inverter which is related to the cost and efficiency.

Table 2. Experiment results for four consecutive days for period of 7.30–17.30.

Day	Average power (W)	Energy demand (kWh)	Inverter output useb by AC (kWh)	Solar fraction (%)
1	345	3.47	2.76	79.4
2	342	3.44	2.70	78.3
3	343	3.45	2.82	81.7
4	344	3.46	2.65	76.6

When the grid is available, a grid-tied PV system is considerably the best solution for a PV powered air conditioning. The advantages of the system are its simple system means less complexity of the PV system; low cost as it does not need batteries and charge controller; reliable; no need oversized inverter; the possibility of exporting excess electricity; and adaptable. The government of Indonesia is introducing a feed-in tariff policy for renewable energy resources [7]. If this is applied, a grid-tied system would be more attractive. The limitations of grid-tied system are, it is not 100% powered by PV system, weather dependent, and it is not functioned during blackouts.

4. Conclusion

A 800 WP grid-tied photovoltaic (PV) system has been tested to power a 0.5 pk of split air conditioner system. About 3.5 kWh daily energy was consumed by the air conditioner system, and about 80% of it could be supplied from the PV system, while the other 20% was supplied by the grid during periods of low solar irradiation, and 440 Wh of energy was fed to the grid during out of office hours operation. A grid-tied PV system is considerably the best solution for a PV powered air conditioning where the grid is available. The advantages grid-tied system are the simplicity of the system; low cost in comparison with off grid system; reliable; no need for oversized inverter; the possibility of exporting excess electricity; and adaptable.

References

[1] Fong K F, Lee C K and Chow T T 2012 Comparative study of solar cooling systems with building-integrated solar collectors for use in sub-tropical regions like Hong Kong *Appl. Energy* **90** pp 189–195
[2] Tarigan E, Djuwari and Purba L 2014 Assessment of PV power generation for household in Surabaya using solar GIS-pvplanner simulation *Energy Procedia* **47** pp 85–93
[3] Tarigan E, Djuwari and Kartikasari F D 2015 Techno-economic simulation of a grid-connected PV system design as specifically applied to residential in Surabaya, Indonesia *Energy Procedia* **65** pp 90–99
[4] Daut I, Adzrie M, Irwanto M, Ibrahim P and Fitra M 2013 Solar powered air conditioning system *Energy Procedia* **36** pp 444–453
[5] Tsoutsos T, Aloumpi E, Gkouskos Z and Karagiorgas M 2010 Design of a solar absorption cooling system in a Greek hospital *Energy Build* **42** pp 265–272
[6] Aguilar F J, Quiles P V and Aledo S 2014 Operation and energy efficiency of a hybrid air conditioner simultaneously connected to the grid and to photovoltaic panels *Energy Procedia* 48 pp 768–777
[7] Kementrian ESDM Republik Indonesia 2017 *Peraturan Menteri Energi dan Sumber Daya Energi No 12 Tahun 2017* (in bahasa Indonesia)

IOP Conf. Series: Materials Science and Engineering **273** (2017) 012011 doi:10.1088/1757-899X/273/1/012011

Hydrolysis of alkaline pretreated banana peel

A Fatmawati, K Y Gunawan and F A Hadiwijaya
Department of Chemical Engineering, University of Surabaya, Jl. Raya Kalirungkut, Surabaya 60282, Indonesia

E-mail: akbarningrum@staff.ubaya.ac.id

Abstract. Banana peel is one of food wastes that are rich in carbohydrate. This shows its potential as fermentation substrate including bio-ethanol. This paper presented banana peel alkaline pretreatment and enzymatic hydrolysis. The pretreatment was intended to prepare banana peel in order to increase hydrolysis performance. The alkaline pretreatment used 10, 20, and 30% w/v NaOH solution and was done at 60, 70 and 80°C for 1 hour. The hydrolysis reaction was conducted using two commercial cellulose enzymes. The reaction time was varied for 3, 5, and 7 days. The best condition for pretreatment process was one conducted using 30% NaOH solution and at 80°C. This condition resulted in cellulose content of 90.27% and acid insoluble lignin content of 2.88%. Seven-day hydrolysis time had exhibited the highest reducing sugar concentration, which was7.2869 g/L.

1. Introduction

The environmental damage and fossil fuel reserve decline issues have led to the quest for renewable and environmentally friendly fuel. Bio-ethanol is one of both renewable and environmentally more-friendly fuel compared to fossil fuel. Many carbohydrate-rich agricultural and food waste can be converted biochemically into more economic valuable products through fermentation including bio-ethanol.

Naturally abundant carbohydrate-rich wastes are grouped into lignocellulosic materials as they contain lignin, cellulose and hemicelluloses. These materials are difficult to degrade naturally as they have a strong structure and therefore compose the outer parts of plants and plant products[1]. Nevertheless, with proper processing, lignocellulosic materials are useful to produce fuels, enzymes, organic acids, bio-sorbents, bio-composites, feeds, and medicines [2]. Many researchers have currently focused on the utilization of lignocellulosic waste to produce bio-ethanol as renewable fuel. Banana peel is one of lignocellulosic agricultural food waste and therefore can be used as fermentation substrate including bio-ethanol. This waste is produced by several Banana flour industries in Indonesia as well as other banana processing small business. However, current banana peel utilization in Indonesia seems only limited for livestock feeds.

Bioconversion of lignocellulosic material into bio-ethanol is a three-step process. The steps are pretreatment, hydrolysis and fermentation. The pretreatment process is conducted to disrupt lignin-carbohydrate bond, lower cellulose degree of crystallinity, improve cellulose surface and therefore increase the enzyme access to cellulose [3,4]. Alkaline and acid pretreatment are widely used method to treat lignocellulosic material prior to hydrolysis [5,6]. Alkaline pretreatments results in less hemicelluloses solubilisation and inhibitory compounds formation compared to acid pretreatment [7]. Coconut coir alkaline pretreatment had increase cellulose content from 41.7 to 48.75% [8].

The hydrolysis step breaks β-1,4 glycosidic bonds of cellulose to produce glucose. This process requires the work of cellulose enzymes produced by cellulose-degrading microorganisms, mainly mold. Cellulose is a group of enzymes consisting of exo-1,4-β-d-glucanases or cellobiohydrolases (CBH) (EC 3.2.1.91), which moves along the cellulose chain and cuts cellobiose units from the ends; endo-1,4-β-d-glucanases (EG) (EC 3.2.1.4), which hydrolyses internal β-1,4-glucosidic bonds randomly within the cellulose chain; and 1,4-β-d-glucosidases (EC 3.2.1.21), which hydrolyses cellobiose to glucose [9]. Some cellulose enzymes produced by various microorganisms have been available commercially.

This research investigated the effect pretreatment condition as well as the hydrolysis reaction time on the reducing sugar production from local banana peel in Indonesia. The pretreatment condition varied was alkaline solution concentration and pretreatment temperature.

2. Materials and methods

2.1. Materials

Banana peels used in this research were steam-cooked "*Raja Nangka (Musa paradica)*" banana peels, obtained from local banana flour processing industry. Two commercial enzymes were used in hydrolysis process. The first enzyme was celluclast® 1.5L (EC 3.2.1.4, Sigma Aldrich), an endoglucanase produced by *Trichordermareseei* with activity $\geq$ 700 units/g. The second enzyme was Novozyme 188 (EC 3.2.1.21, Sigma Aldrich), a β-D-glucosidases produced by *Aspergillus niger* with activity $\geq$ 250 units/g.

2.2. Banana peel powder preparation

Banana peels were sundried for six hours prior to grinding. The dry banana peels were then cut into ± 5 cm wide chips. The chips were ground using FFC 23A disc mill machine and then screened to obtain 70 mesh particles.

2.3. Alkaline pretreatment

The alkaline solution used was NaOH solution. The concentration of sodium hydroxide solution was varied into 10, 20, and 30% w/v while banana peel powder slurry concentration was constant at 7.5% w/v. The pretreatment was conducted for one hour and at varied temperature of 60, 70, and 80°C. After the pretreatment process, the powder was filtered and washed until neutral pH of filtrate was obtained.

2.4. Enzymatic hydrolysis

Hydrolysis was conducted in 250 mL flask with total working volume of 150 mL containing 3 g banana powder. The pH was maintained 4.8 by using sterile citrate buffer solution. The reaction was conducted at constant temperature of 50°C with varied reaction time of 3, 5 and 7 days. An Enzyme mixture consisting of 0.495 mL *celluclast®* and 0.495 mL *Novozyme 188* was used. In order to inhibit microbial growth, 40 μg/mL tetracycline was added to the mixture. To ensure homogenous condition, hydrolysis was done on an orbital incubator shaker with rotation speed of 150 rpm. At the end of hydrolysis reaction, the flask was immersed in 95-100°C hot water bath for 10 minutes to inactivate the enzymes.

2.5. Chemical analysis

The cellulose and lignin content of banana peel before and after pretreatment were determined using modified method of NREL/TP-510-42618[10], developed by US National Renewable Energy Laboratory (NREL). The modification was in the omission of HPLC analysis to determine total carbohydrate content. Instead, total carbohydrate content was obtained from the difference weight of initial sample from oven dried powder sample after sulfuric acid treatment. The cellulose content was determined using the same method except that the samples were treated first using an amylase enzyme prior to total carbohydrate and lignin analysis using NREL method. Reducing sugar concentrations measurement after enzymatic hydrolysis was done using Dinitrosalicylic Acid reagent [11].

3. Results and discussion

3.1. Banana peel composition

The dried and milled banana peel powder had been analyzed for its content of lignocellulosic compounds. The percentage of these compounds is presented in Figure 1.

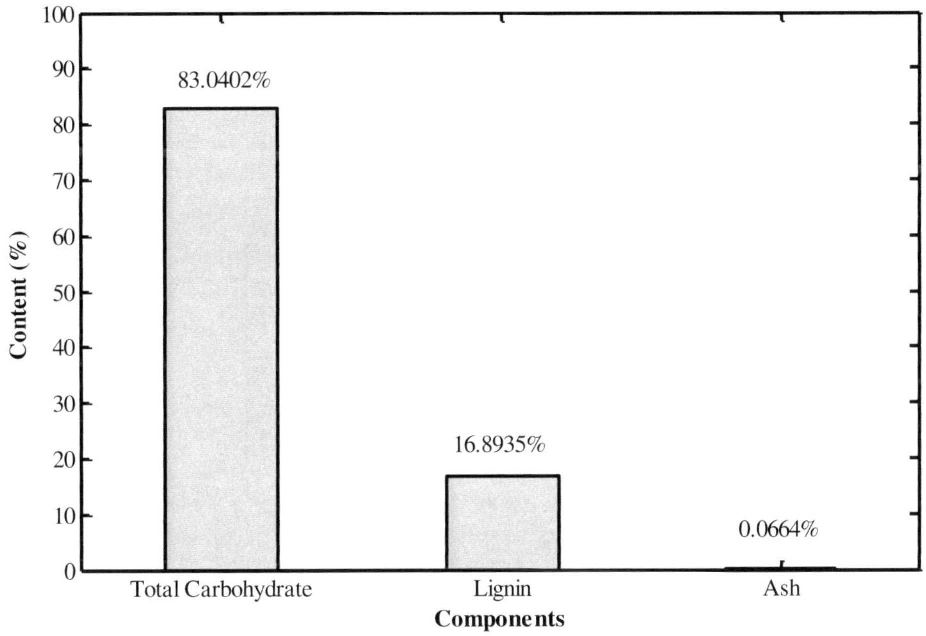

Figure 1. Composition of steam-cooked banana peels.

Figure 1 shows that carbohydrate is the biggest component in banana peels. Chantawongsa (2013) also reported nearly the same content of total carbohydrate and lignin in banana peel [12]. This high carbohydrate content shows banana promising potential to be used for fermentation substrate for valuable chemical production.

3.2. Alkaline pretreatment and hydrolysis

Pretreatment process is known to have effects on the enzymatic hydrolysis of lignocellulosic materials. The effects of alkaline pretreatment on the lignocellulosic components of Raja Nangka banana peel are shown in Figure 2, Figure 3 and Figure 4.According to Figure 2, the highest final cellulose content is obtained at 30%w/v and 80°C pretreatment, which resulted in cellulose content of 90.266%. At higher temperature (70 and 80°C), cellulose concentration tends to increase with alkaline concentration. However, the result at low temperature (60°C) exhibit speaks cellulose content at 20%w/v alkaline concentration. This might be caused by degradation of carbohydrate at higher alkaline concentration.

Figure 3 shows total carbohydrate content after pretreatment using NaOH solution at various temperatures and alkaline solution concentrations. The figure shows that the higher pretreatment temperature the higher total carbohydrate content obtained. In line with the result of cellulose content, total carbohydrate tends to increase with alkaline concentration at higher temperature while at lower one it shows peak at 20%w/v alkaline concentration. The highest total carbohydrate content obtained is 96.87%after pretreatment at 30% w/v and 80°C.

IOP Conf. Series: Materials Science and Engineering **273** (2017) 012011 doi:10.1088/1757-899X/273/1/012011

Figure 4 shows that lignin content has been reduced after alkaline pretreatment. The result also shows that highest reduction is achieved at 30%w/v and 80°C pretreatment condition. This lowest lignin content is 2.884%. Results shown in figure 2-4 shows that alkaline pretreatment had successfully broken lignin carbohydrate linkage in banana peels as other reported results [13,14].

Enzymatic hydrolysis of alkaline pretreated banana peel result is presented in Figure 5. The pretreatment condition was chosen at 80°C and 30%w/v alkaline concentration which resulted in the highest total carbohydrate and cellulose content and hence the lowest lignin content. The figure shows that alkaline pretreatment had increased hydrolysis result compared to unpretreated banana peel. According to the figure, the longer hydrolysis reaction time, the higher reducing sugar concentration produced in the reaction. The highest reducing sugar obtained is 5.9974 g/L after 5 day hydrolysis.

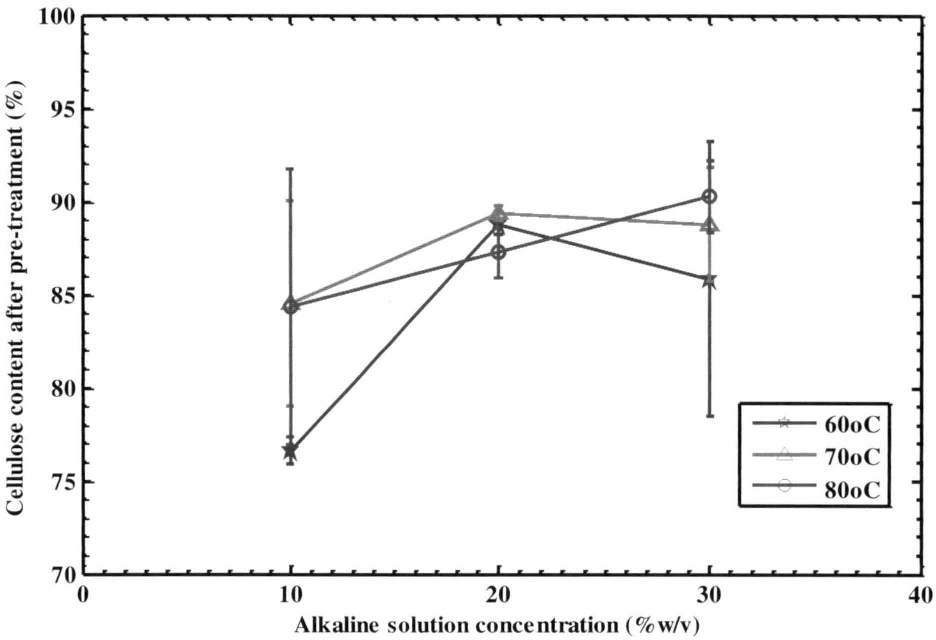

Figure 2. Cellulose content of alkaline-pretreated banana peel.

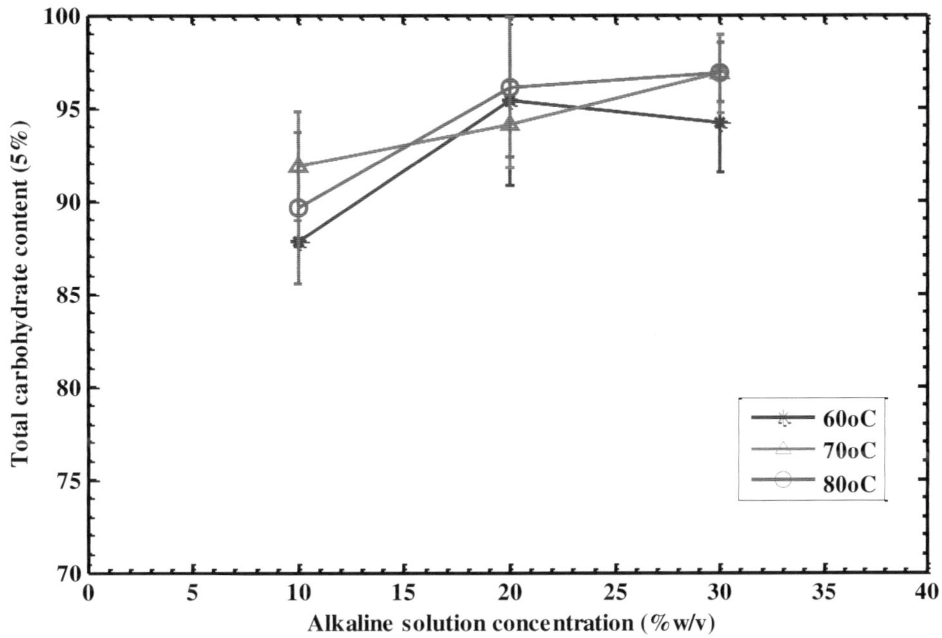

Figure 3. Total carbohydrate content of alkaline-pretreated banana peel.

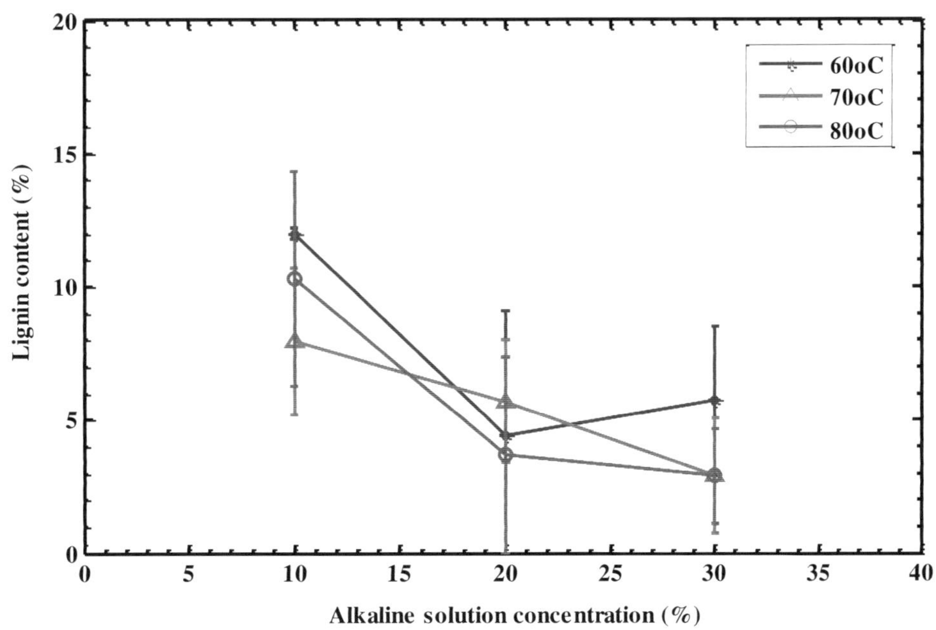

Figure 4. Lignin content of alkaline-pretreated banana peel.

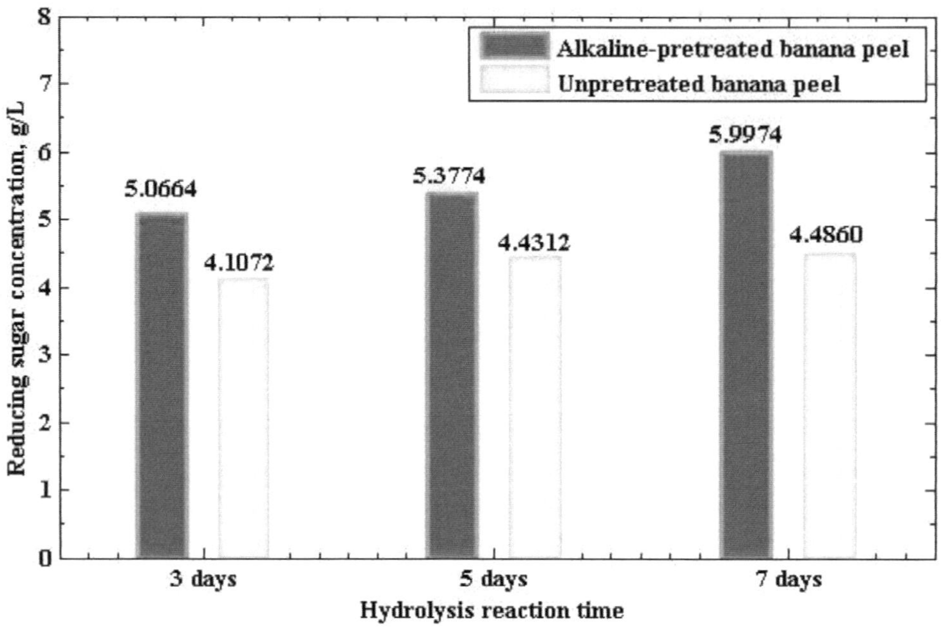

Figure 5. Effect of hydrolysis reaction time on the reducing sugar production.

4. Conclusions
Alkaline pretreatment of banana peel using NaOH solution had been done and resulted in the increase of total carbohydrate and cellulose content of banana peel and the decrease of lignin content. Pretreatment at temperature of 80°C and alkaline concentration of 30%w/v was the best condition in increasing cellulose content up to 90.266%. Pretreated banana peel at the best pretreatment condition was successfully hydrolyzed using two commercial enzymes which yield the highest reducing sugar content 5.9974 g/L after 5 day hydrolysis.

5. References
[1] Velásquez-Arredondo H I, Ruiz-Colorado A A and De Oliveira S 2010 *Energy* **35** pp 3081–87
[2] Ahring B K, Licht D, Schmidt S, Sommer P and Thomsen B 1999 *Bioresour. Technol.* **68** pp 3–9
[3] Cekmecelioglu D and Uncu O N 2013 *Waste Manag.* **33** pp 735–739
[4] Hendriks A T W M and Zeeman G 2009 *Bioresour. Technol.* **100** pp 10–18
[5] Monschein M, Reisinger C and Nidetzky B 2013 *Bioresour. Technol.* **128** pp 679–687
[6] Dawson L and Boopathy R 2007 *Bioresour. Technol.* **98** pp 1695–99
[7] Jönsson L J and Martín C 2016 *Bioresour. Technol.* **199** pp 103–112
[8] Fatmawati A, Agustriyanto R 2015 *Mod. Appl. Sci.* **9** pp 29–35
[9] Jørgensen H, Kristensen J B and Felby C 2007 *Biofuels, Bioprod. Biorefin.* **1** pp 119–134
[10] Sluiter A, Hames B, Ruiz R, Scarlata C, Sluiter J and Templeton D. Determination of Structural Carbohydrates and Lignin in Biomass. National Renewable Energy Laboratory. U.S. Department of Energy; 2011 July [cited 2016 Mar 28]. Report No.: NREL/TP-510-42618. Contract No.: DE-AC36-08-GO28308. Available from: http://www.nrel.gov/biomass/analytical_procedures.html
[11] Miller G L 1959 *Anal. Chem.* **31** pp 426–428
[12] Chantawongsa N and Kongkiattikajorn J 2013 *Int. J. Comput. Internet Manag.* **21** pp 15–19
[13] Rabemanolontsoa H and Saka S 2016 *Bioresour. Technol.* **199** pp 83–91
[14] Sun S, Sun S, Cao X and Sun R 2016 *Bioresour. Technol.* **199** pp 49–58

IOP Conf. Series: Materials Science and Engineering **273** (2017) 012012 doi:10.1088/1757-899X/273/1/012012

Closed-loop simulation of decentralized control using RGA for uncertain binary distillation column

R Agustriyanto[1] and J Zhang[2]

[1] Department of Chemical Engineering, University of Surabaya, Raya Kalirungkut, Surabaya 60293, Indonesia

[2] School of Chemical Engineering and Advanced Materials, Newcastle University, Newcastle upon Tyne, NE1 7RU, UK

E-mail: rudy.agustriyanto@staff.ubaya.ac.id

Abstract. This paper presents the results of closed-loop simulation for the decentralized control of uncertain distillation column. The RGA (Relative Gain Array) RGA analysis was used as the basis for selecting the configuration of the decentralized control system. PI controller obtained was then tuned with optimization methods. The simulation results show that the RGA analysis requires accurate range for uncertain systems. In addition, closed-loop simulation results confirm the RGA analysis.

1. Introduction

Decentralized control is famous in chemical process control even though sophisticated methods for designing centralized multivariable control are now available such as Internal Model Control (IMC), Dynamic Matrix Control (DMC), Linear Quadratic Control (LQC), etc. The advantages of decentralized control for multivariable process are as follows [1]:

- Simple algorithm
- Ease of understanding by plant operating personnel (as a result of the simplicity of control structures)
- The availability of standard control designs for the common unit operations
- Flexibility in operation
- Failure tolerance

Theoretically, a centralized control, which handles the whole plant as a single unit, may give the best performance. However, a complete centralized controller has a number of difficulties such as in controller design, tuning, maintenance and modification. In addition, Skogestad [2] explained the main reason of decentralized controller wide acceptance is that it is related to the high cost associated with obtaining good process models, which serve as prerequisite for applying centralized multivariable control. From plant operator's point of view, a control structure that encompasses a wide range of different units may be beyond understanding. Transparency and intuitiveness of the control structure are important factors for operator acceptance and safe plant operation [3].

RGA (Relative Gain Array) was first proposed by Bristol [4] as a steady state measure of interactions for decentralized control. Properties of RGA matrix were given by Skogestad and Postlethwaite [5].

Chen and Seborg [6] presented an analytical expression for RGA uncertainty bounds. Two types of model uncertainty were considered: worst case bounds, where all elements of the steady state process gain matrix are allowed to change simultaneously within their bounds, and statistical uncertainty bounds. A different method by using the structured singular value (μ) analysis framework was introduced for the calculation of the magnitude of the worst-case relative gain [7].

Agustriyanto and Zhang [8] used an optimization method to calculate RGA range under model uncertainties. The model uncertainty type considered is worst case bounds. The lower and upper bounds of an RGA element are calculated as two constrained optimization problems. The method seeks the minimum (for the lower bound) or maximum (for the upper bound) of an RGA element subject to the constraints that allowable model parameters are within their uncertainty bounds. RGA ranges are shown to be important for control pairing analysis. In this paper, closed loop simulation was then performed to evaluate the RGA analysis.

2. Simulation

The system being studied was the binary distillation column used to separate ethanol and water [9]. It is a 19 plate distillation column with 12 inch diameter copper column having variable feed and side stream draw-off locations. This example was also used by Chen and Seborg [6] to test their method for calculating the uncertain ranges of RGA. A schematic diagram of the binary distillation column is shown in Figure 1.

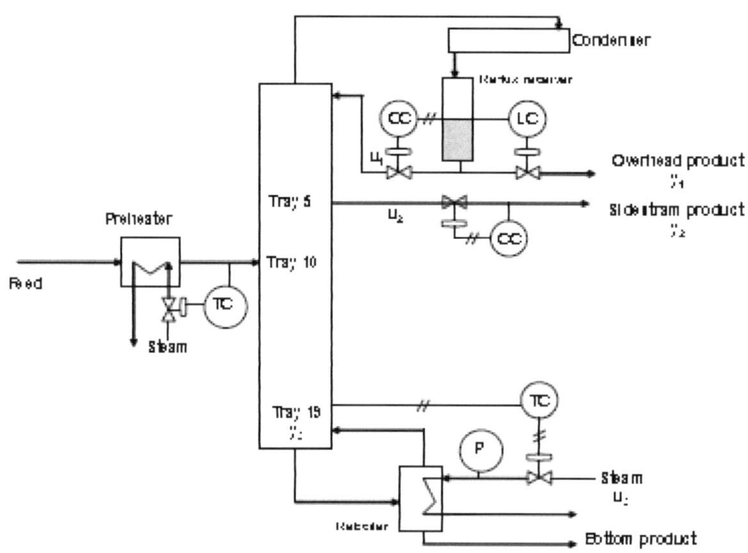

Figure 1. Schematic diagram of the binary distillation column with diagonal control structure

Table 1 shows steady state values of the column and Table 2 shows system constraints. The following Laplace transfer functions have been identified and were used in this study.

$$G = \begin{bmatrix} \dfrac{0.66e^{-2.6s}}{6.7s+1} & \dfrac{-0.61e^{-3.5s}}{8.64s+1} & \dfrac{-0.0049e^{-s}}{9.06s+1} \\ \dfrac{1.11e^{-6.5s}}{3.25s+1} & \dfrac{-2.36e^{-3s}}{5s+1} & \dfrac{-0.012e^{-1.2s}}{7.09s+1} \\ \dfrac{-33.68e^{-9.2s}}{8.15s+1} & \dfrac{46.2e^{-9.4s}}{10.9s+1} & \dfrac{0.87(11.61s+1)e^{-s}}{(3.89s+1)(18.8s+1)} \end{bmatrix} \quad (1)$$

Table 1. Steady state values of binary distillation column

Variable	Description	Value	Units
y_1	Overhead mol fraction of ethanol	0.7	
y_2	The mol fraction of ethanol in the side stream	0.52	
y_3	Temperature on tray 19	92	°C
u_1	Overhead reflux flow rate	0.18	gpm
u_2	Side stream draw-off rate	0.046	gpm
u_3	Reboiler steam pressure	20	psig

Table 2. System constraints

Variable		Lower Constraint	Upper Constraint
u_1	Overhead reflux flow rate	0.068	0.245
u_2	Side stream draw-off rate	0.00694	0.1
u_3	Reboiler steam pressure	15.6	34.0

Generally, the RGA of a non-singular square matrix K is a square matrix and defined as:

$$RGA = K \otimes \left(K^{-1}\right)^T \quad (2)$$

Where $\otimes$ denotes element by element multiplication.
Eq. (2) can be used for 3×3 process and all other systems.

The *ij*th element of the RGA [10] is:

$$\lambda_{ij} = (-1)^{i+j} \frac{K_{ij} \det\left(K^{ij}\right)}{\det(K)} \quad (3)$$

Here, K_{ij} is the element on the *i*th row and *j*th column of *K* and K^{ij} is the sub matrix that remains after the *i*th row and *j*th column of *K* are deleted.

It is obvious that λ_{ij} is a function of *K*, that is

$$\lambda_{ij} = f(K) \quad (4)$$

Now assume that the uncertainty bounds (lower and upper bounds) for steady state gain, K_{ij}, i=1, 2, …, n, j=1,2, …, n, are given, then there will be $2n^2$ constraints for all elements of steady state gains which can be formulated as follows:

$$AX \leq b \quad (5)$$

where

- X is a vector of size $n^2 \times 1$ containing all elements of K as its elements:
$$X = [K_{11} \ldots K_{nn}]^T \quad (6)$$
- b is a vector of size $(2n^2) \times 1$ containing the lower and upper bounds of the corresponding elements of X.
- A is an appropriate matrix of size $(2n^2) \times (n^2)$ satisfying the inequalities in Eq. (5).

Therefore, the lower bound and upper bound of λ_{ij} can be formulated as the following respectively:

$$\text{Lower Bound:} \quad \min_X \lambda_{ij} = f(X) \quad (7)$$

$$\text{Upper Bound:} \quad \max_X \lambda_{ij} = f(X) \quad (8)$$

Subject to constraint of Eq. (5).

Note that λ_{ij} cannot be determined if det(K) = 0. Therefore, in order to use the above method, the range of det(K) should not include 0. The range of det(K) can be calculated by using the same optimization method.

Optimization algorithm that optimized control performance was used to automate trial and error method. Here, *fmincon* function in Matlab Optimization Toolbox was used to obtain lower and upper bounds.

3. Results and discussion

The nominal steady state gain matrix and RGA are given as:

$$K = \begin{bmatrix} 0.66 & -0.61 & -0.0049 \\ 1.11 & -2.36 & -0.012 \\ -33.68 & 46.2 & 0.87 \end{bmatrix} \tag{9}$$

$$RGA = \begin{bmatrix} 1.9454 & -0.6737 & -0.2718 \\ -0.6643 & 1.8991 & -0.2348 \\ -0.2811 & -0.2254 & 1.5065 \end{bmatrix} \tag{10}$$

The nominal RGA values indicate that the diagonal control structure, *(y1-u1, y2 – u2, y3 – u3)*, should be chosen because the sign of the RGA elements for this pairing are all positive and their values are close to 1.

As in Chen and Seborg [6], it is assumed here that the uncertainty for each steady state gain can be expressed as:

$$K_{ij} = K_{ij} \pm \Delta K_{ij} \tag{11}$$

where

$$\left|K_{ij}\right| \le \alpha \left|K_{ij}\right| \tag{12}$$

Case 1: $\alpha = 0.01$

The uncertainty ranges for RGA elements calculated via optimization method described in the above Section are given below:

$$\text{RGA} = \begin{bmatrix} 1.88 \le \lambda_{11} \le 2.02 & -0.74 \le \lambda_{12} \le -0.61 & -0.31 \le \lambda_{13} \le -0.24 \\ -0.73 \le \lambda_{21} \le -0.60 & 1.83 \le \lambda_{22} \le 1.97 & -0.27 \le \lambda_{23} \le -0.20 \\ -0.32 \le \lambda_{31} \le -0.25 & -0.26 \le \lambda_{32} \le -0.20 & 1.48 \le \lambda_{33} \le 1.54 \end{bmatrix} \tag{13}$$

The above results indicate that the recommended controller pairing is obvious and unambiguous: *(y1-u1, y2 – u2, y3 – u3)*.

The results obtained via optimization method had wider range for each element of RGA compared to the results from the analytical method. This was because no approximation had been made in optimization method, while Taylor series expansion was used in analytical method. There exists a combination of K which is still within the constraint Eq. (20) and will give the lower and upper bound of relative gains in Eq. (22). This indicated that the proposed method could find more accurate RGA uncertain bounds than the method proposed previously.

Case 2: $\alpha = 0.1$

The uncertainty ranges for RGA calculated via optimization is given below.

$$\text{RGA} = \begin{bmatrix} 1.48 \le \lambda_{11} \le 3.65 & -1.89 \le \lambda_{12} \le -0.24 & -0.76 \le \lambda_{13} \le -0.02 \\ -1.86 \le \lambda_{21} \le -0.23 & 1.46 \le \lambda_{22} \le 3.42 & -0.66 \le \lambda_{23} \le -0.01 \\ -0.79 \le \lambda_{31} \le -0.03 & -0.66 \le \lambda_{32} \le -0.04 & 1.29 \le \lambda_{33} \le 2.01 \end{bmatrix} \tag{14}$$

Case 3: $\alpha = 0.25$

The uncertainty range for RGA cannot be determined as for α = 0.25, the value of det(K) range will include 0 (K can become singular). While using analytical method[6], the value of α = 0.5 can still be tolerated.

Closed loop simulation was then performed for the recommended controller pairing (*y1 – u1, y2 – u2, y3 –u3*). Table 3 shows the tuning parameters which were obtained via optimization where the sum of absolute errors for setpoint tracking was minimized. The following setpoint changes are used for tuning purpose:

- y1 set-point was changed from 0 to 0.1 at t = 10 min
- y2 set-point was changed from 0 to -0.1 at t = 300 min
- y3 set-point was changed from 0 to -5 at t = 500 min

The output values were recorded for 1000 min simulation time with 1 min sampling time. Therefore, the objective function in the optimization routine is to find a set of controller parameters which will minimize sum of the absolute errors for the specified simulation time.

Table 3. Controller parameters

Controllers	K_c	τ_I
y1-u1	0.0871	0.6053
y2-u2	-0.3829	5.1813
y3-u3	6.1938	3.4343

For the nominal values (i.e. no uncertainties), simulation results for set-point changes are shown in Figure 2. In Figure 2, the solid lines represent the controlled variables and the dashed lines represent the set-points. In all the simulations for this example, *y1* set-point was changed from 0 to 0.1 at $t = 10$ min, *y2* set-point was changed from 0 to -0.1 at $t = 300$ min, and *y3* set-point was changed from 0 to -5 at $t = 500$ min. It can be seen in Figure 2 that the control performance was satisfactory.

Simulations are then performed for the arbitrarily altered process gains (which reflect gain uncertainties) as follows:

$$\overline{G} = \begin{bmatrix} G_{11}(1-\alpha) & G_{12}(1+\alpha) & G_{13}(1-\alpha) \\ G_{21}(1+\alpha) & G_{22}(1-\alpha) & G_{23}(1+\alpha) \\ G_{31}(1-\alpha) & G_{32}(1+\alpha) & G_{33}(1-\alpha) \end{bmatrix} \quad (15)$$

Figures 3 to 5 show the results for three different values of α. Once again, the solid lines represent the controlled variables and the dashed lines represent the set-points. As shown in Figure 5, the controller settings based on the nominal model cannot guarantee the closed loop stability for α = 0.25.

4. Conclusion

A method for calculating RGA ranges for uncertain process models is presented in this paper. Constrained optimization was used to find the uncertain RGA ranges. The proposed method was then applied to the binary distillation column. Closed loop simulation results confirm the analysis based on the proposed method.

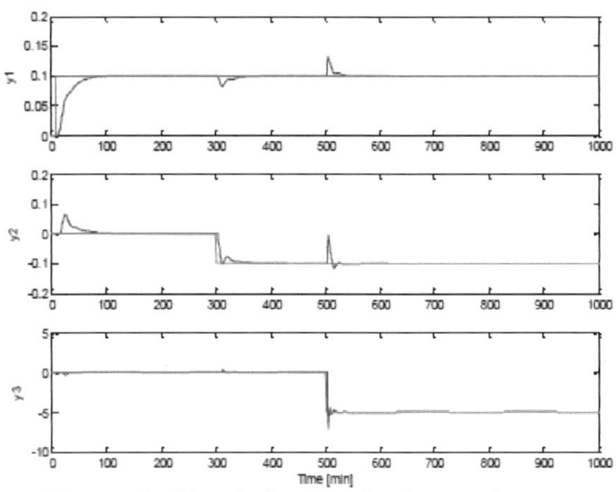

Figure 2. Simulation results for α = 0

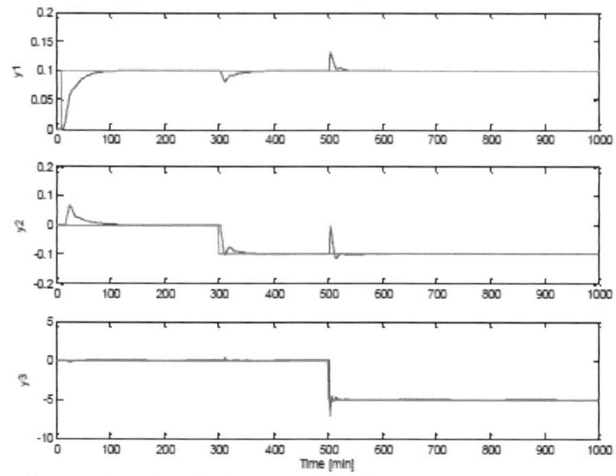

Figure 3. Simulation results for α = 0.01

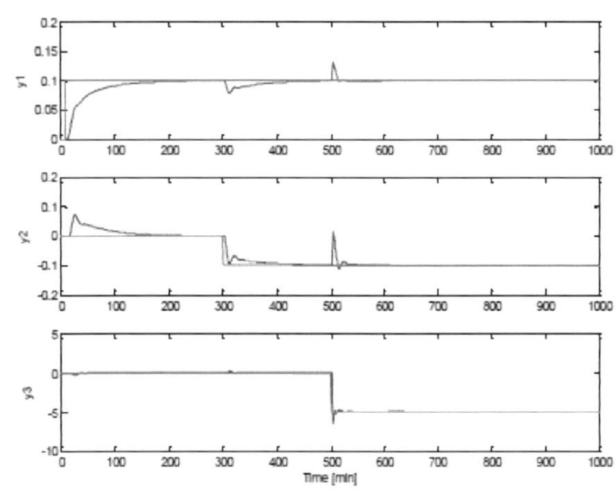

Figure 4. Simulation results for α = 0.1

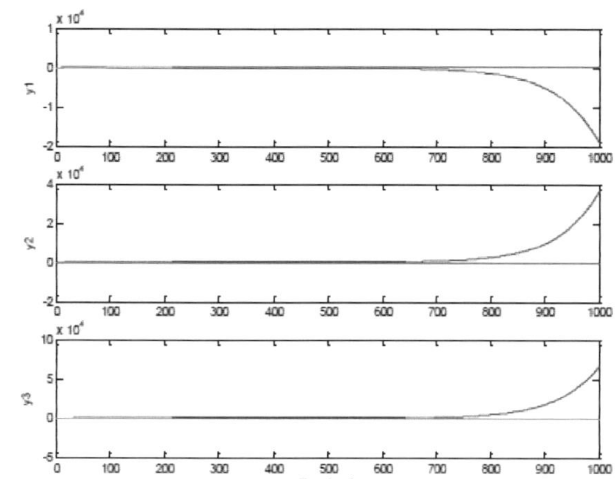

Figure 5. Simulation results for α = 0.25

References

[1] Marlin T E 2000 *Process Control* (New York: McGraw Hill)
[2] Firouzbahrami M and Nobakhti A 2013 Decentralized integral controllability of linear uncertain systems with correlated and independent uncertainties *Chem. Eng. Sci.* **94** pp 120–126
[3] Skogestad S and Hovd M 1995 Letters *J. Process Contr.* **5** pp 399–400
[4] Seki H and Naka Y 2006 A hierarchical controller design for a reactor/separator system with recycle *Ind. Eng. Chem. Res.* **45** pp 6518–24.
[5] Bristol E 1966 On a new measure of interaction for multivariable process control *IEEE T. Automat. Contr.* **11** pp 133–134
[6] Skogestad S and Postlethwaite I 1996 *Multivariable Feedback Control* (New York: Wiley)
[7] Chen D and Seborg D E 2002 Relative gain array analysis for uncertain process models *AICHE J.* **48** pp 302–310
[8] Kariwala V, Skogestad S and Forbes J F 2006 Relative gain array for norm-bounded uncertain systems *Ind. Eng. Chem. Res.* **45** pp 1751–57
[9] Agustriyanto R and Zhang J 2007 Obtaining the worst case RGA and RDGA for unccertain systems via optimization *Proc. of the 2007 American Control Conf.* (New York) pp 5360–65
[10] Ogunnaike B A, Lemaire J P, Morari M and Ray W.H 1983 Advanced multivariable control of a pilot-plant distillation column *AICHE J.* **29** pp 632–640
[11] Haggblom K 2008 Integral controllability and integrity for uncertain systems *Proc. of the 2008*

American Control Conference (Seattle) pp 5192–97

[12] Grosdidier P, Morari M and Holt B R 1985 Closed loop properties from steady state gain information *Ind. Eng. Chem. Fundamen.* **24** pp 221–235

An efficiency improvement in warehouse operation using simulation analysis

N Samattapapong

School of Industrial Engineering, Institute of Engineering, Suranaree University of Technology, Thailand

E-mail: nara@sut.ac.th

Abstract. In general, industry requires an efficient system for warehouse operation. There are many important factors that must be considered when designing an efficient warehouse system. The most important is an effective warehouse operation system that can help transfer raw material, reduce costs and support transportation. By all these factors, researchers are interested in studying about work systems and warehouse distribution. We start by collecting the important data for storage, such as the information on products, information on size and location, information on data collection and information on production, and all this information to build simulation model in Flexsim® simulation software. The result for simulation analysis found that the conveyor belt was a bottleneck in the warehouse operation. Therefore, many scenarios to improve that problem were generated and testing through simulation analysis process. The result showed that an average queuing time was reduced from 89.8% to 48.7% and the ability in transporting the product increased from 10.2% to 50.9%. Thus, it can be stated that this is the best method for increasing efficiency in the warehouse operation.

1. Introduction

Presently, warehouse maintenance is considered as one of the main factors in industry production. It can measure the efficiency of a factory, which is expressed in terms of economy and management. A factory with good ability to order and manage their warehouse is a factory with good management. There are a lot of warehouse problems in small factories due to the lack of ability in management and that causes a huge effect to the factory in terms of storage and unnecessary costs. Therefore, researchers are interested in warehouse management to increase storage efficiency and reduce costs in the factory.

2. Literature review

This study on using simulation in warehouse management has two main objectives. The first is having enough products in sufficient quantities to service the customers and to increase sales to maintain effective market penetration. The second is to reduce investment in product to decrease the costs of production, such as by keeping inventory as low as possible [1].

Warehouse management requires five main activities:

- The product (Receiving) is the process of obtaining the products.
- Storage (Put-Away) is the system that examines the size and area of storage.

- Converting units (Let-Down) is the unit conversion to facilitate distribution or stock management for increased convenience.
- Release of goods (Picking) is the process that allows products to be released.
- Warehouse count (Counting) is the way to check the amount of actual products in the stock [2]. Recently, the factory has increased value of products by making packaging for marketing and making the codes to be more flexible for transportation and counting [3].

At present, the factory must focus more on warehouse management due to the fact that the warehouse is one of the most influential factors to costs. Warehouse management can be either advantageous or disadvantageous, depending on how factories manage their warehouse. The objective is to deliver quality products in an optimal time with efficient service to the customers, while achieving the lowest costs in order to remain competitive with other providers [4]. Improvement and development in the warehouse means finding a way to optimize the warehouse by standardizing time in the transport of products from production to warehouse. Otherwise, bottlenecks can be created that lead to problems that will directly affect the costs of production [5].

The layout of the warehouse can be the main factor. By improving the positioning in the warehouse, a factory can maximize the use of space and have a better management in the system [6].

3. Research methods

3.1. Sources of information
According to the study on product transportation regarding data collection on product movement and interviewing staff about storage management, results showed that a significant amount of time is locked up in the transport of products. Thus, simulations were created to attempt to solve that problem.

3.2. Data analysis
The Figure 1, 2, 3 and 4 below show the data of transportation regarding.

Data Characteristic	Value
Source file	<edited>
Observation type	Real valued
Number of observations	30
Minimum observation	1.12000
Maximum observation	5.47000
Mean	2.63067
Median	2.36000
Variance	1.01004
Coefficient of variation	0.38204
Skewness	1.04913

Figure 1. Statistical data summary of inter-arrival time.

IOP Conf. Series: Materials Science and Engineering **273** (2017) 012013 doi:10.1088/1757-899X/273/1/012013

Relative Evaluation of Candidate Models

Model	Relative Score	Parameters	
1 - Pearson Type VI	97.22	Location	0.00000
		Scale	1.00000
		Shape #1	27.77911
		Shape #2	11.55950
2 - Inverse Gaussian	92.59	Location	0.00000
		Scale	2.63067
		Shape	19.44826
3 - Pearson Type V	92.59	Location	0.00000
		Scale	19.01226
		Shape	8.20474

28 models are defined with scores between 0.00 and 97.22

Absolute Evaluation of Model 1 - Pearson Type VI

Evaluation: Good

Suggestion: Additional evaluations using Comparisons Tab might be informative.

Additional Information about Model 1 - Pearson Type VI

"Error" in the model mean relative to the sample mean -5.6801e-5 = 0.00%

Figure 2. Analysis of inter-arrival time data.

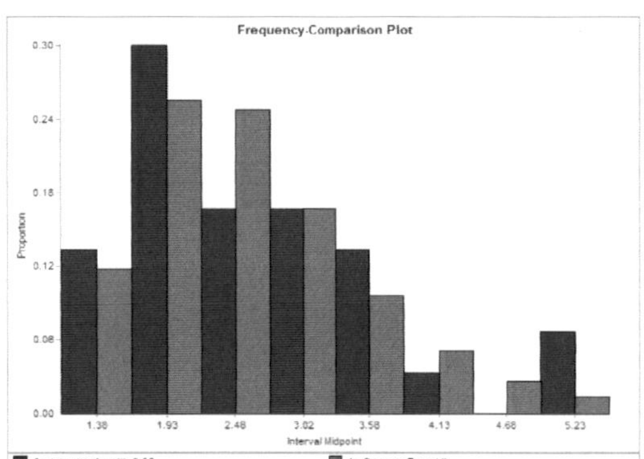

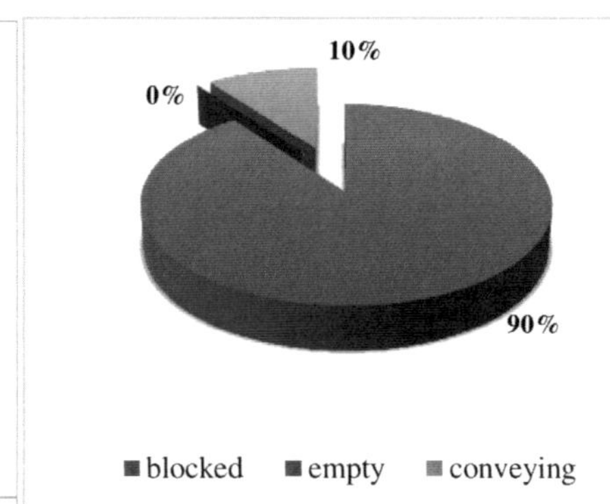

Figure 3. Probability of service time.

Figure 4. Working process in belt before improvement.

3.3. Modeling simulation

This research studied the unloading and storage of products using Flexsim® to solve the bottlenecking problem by focusing on the transportation of goods. The researchers offered three solutions. These solutions look at adding an additional worker in the transportation of products, adding a transportation belt to reduce queuing during transportation and adding both an additional worker and a transportation belt. In the warehouse, additional shelving can be added for additional space for storage. For transportation, the researcher used the principle of FIFO, in which the previously produced products will be sent to the client first. The simulation framework is shown in Figure 5.

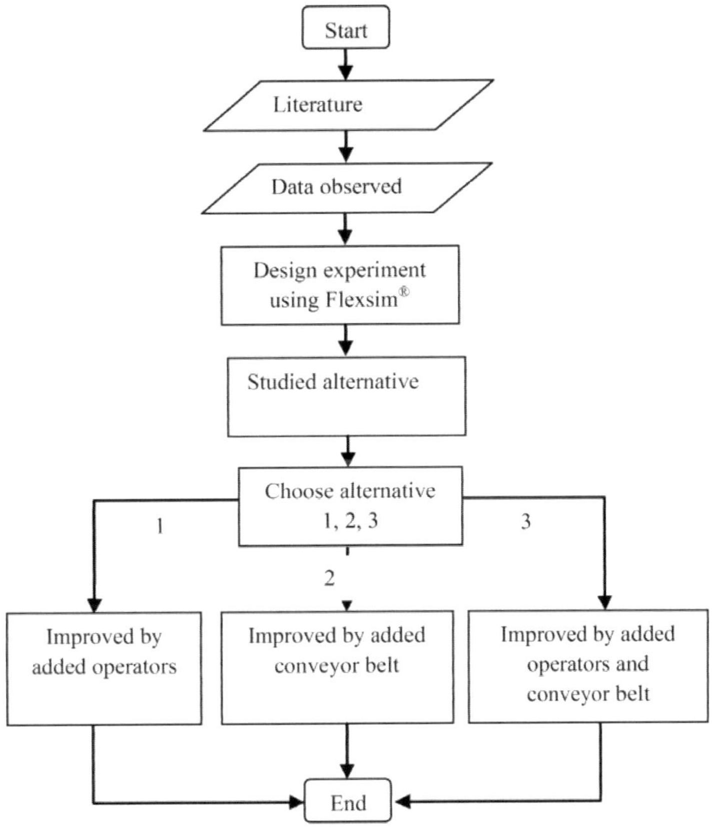

Figure 5. Working process.

The layout before improvement can be seen in Figure 6.

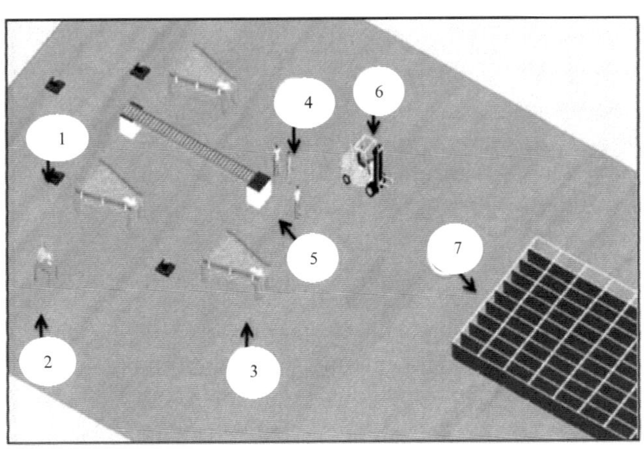

Figure 6. Modeling before improvement.

The data used in the models are explained in Table 1.

Table 1. Explanation of the simulation objects in Figure 6.

Number	Description
1	Released sacks of animal feed ready to be fed into the conveyor system
2	The dispatcher managing the work of the operator
3	Separator, which distributes sacks of animal feed onto the conveyor belt
4	The group releasing the pallets of sacks
5	The combiner, who combines the sacks of animal feed and the pallet for easy transportation
6	The conveyor, which transports sacks of animal feed
7	The conveyor, which transports sacks of animal feed
8	The transporter, who is responsible for transporting sacks of animal feed, which had been prepared onto a pallet, ready to be listed into the warehouse storage system
9	The operator who lifts the sacks of animal feed from the conveyor belt to the pallet

3.4. Instruments used in the study

The workers were interviewed by recording, taking note of the information from cited sources, including statistical information, documents, books, journals and research.

3.5. Scope of research

- To study warehouse management to optimize transportation.
- To improve warehouse efficiency.
- To reduce the time and cost of transportation.
- To study and analyze the possibility of improving shelving in the warehouse.

4. Results

The results of the simulation modeling analysis in each alternative are explained below.

4.1. Solution by the addition of an extra worker

To reduce queuing of product in between transportation, the addition of more staff was considered. However, the continuous and quick arrival of product was not consistent with the ability of the worker in transportation as shown in Table 2.

4.2. Solution by the addition of a conveyor belt

The addition of a conveyor belt adds increased distribution of the product to reduce queuing time. However, the ability of workers was not consistent with the ability of the added conveyor belt as shown in Table 2.

4.3. Solution by the addition of both an additional staff and conveyor belt

This simulation aimed to add both an additional worker and a conveyor belt to increase the workforce and capability of transport, as well as reduce queuing time through improved distribution of products on the conveyor belt. Results for this simulation can be seen in Table 2.

Table 2. Value of the operation from conveyor belt

Model Characteristics		Blocked	Empty	Waiting for transporter	Conveying
Original		89.8%	0.0%	-	10.2%
Addition of worker		90.3%	0.0%	-	9.7%
Addition of belt	Belt1	93.9%	0.3%	0.1%	5.8%
	Belt2	93.8%	0.3%	0.0%	5.9%
Addition of belt and worker	Belt1	50.2%	0.3%	0.2%	49.4%
	Belt2	47.2%	0.3%	0.2%	52.4%

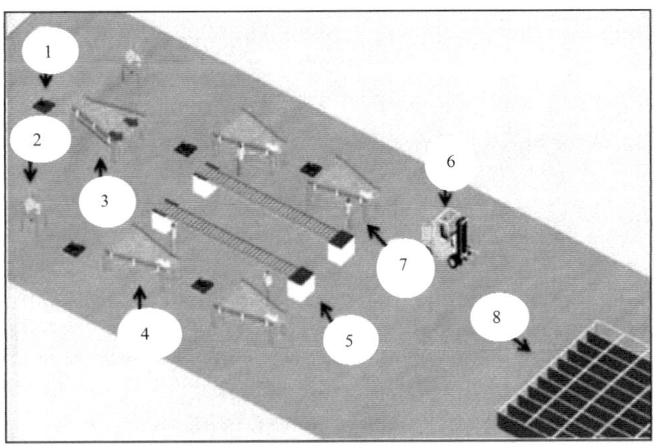

Figure 7. Modeling after improvement

5. Conclusion

From modeling the simulation by Flexsim®, it was found that the best simulation on improvement in the warehouse is achieved by adding both an additional worker and a conveyor belt.Improving the conveyor belt system makes it more flexible and reduces queuing from 89.8% to 48.7%. Using this model, the ability of transporting products increases from 10.2% to 50.9%.The improvement makes the warehouse management more efficient.

6. References

[1] Jitjong W 2009 *Logistics Thailand* **79** pp 34–38
[2] Thanathiporn J and Kanjanakumnerd A 2009 *Logistics Thailand* **87** pp 10–26
[3] Thanwongwan S 2008 *Logistics Thailand* pp 41–50
[4] Tharanon J 2005 Enhancements the warehouse management by system radio frequency identification in the A company Ltd (Master Thesis) (Bangkok: The University of the Thai Chamber of Commerce)
[5] Niebel B and Freivalds A 2003 *Methods Standards and Work Design* (New York: McGraw Hill)
[6] Tompkins J A, White J A, Tanchoco J M A and Bozer Y A 2003 *Facilities Planning* (USA: Wiley)

Modeling of the minimum variable blank holder force based on forming limit diagram (FLD) in deep drawing process

S Candra[1], I M L Batan[2], W Berata[2] and A S Pramono[2]

[1] Department of Manufacturing Engineering, University of Surabaya, Raya Kalirungkut, Surabaya 60293, Indonesia
[2] Mechanical Engineering, Institute of Technology Sepuluh Nopember (ITS), Surabaya, Indonesia

E-mail: susila_c@staff.ubaya.ac.id

Abstract. This paper presents the mathematical approach of minimum blank holder force to prevent wrinkling in deep drawing process of the cylindrical cup. Based on the maximum of minor-major strain ratio, the slab method was applied to determine the modeling of minimum variable blank holder force (VBHF) and it compared to FE simulation. The Tin steel sheet of T4-CA grade, with the thickness of 0.2 mm was used in this study. The modeling of minimum VBHF can be used as a simple reference to prevent wrinkling in deep drawing.

1. Introduction

The wrinkle is a type of product defect that often occurs in deep drawing. One thing that can be pursued to prevent product defects is determining the magnitude of the blank holder force (BHF) accurately. To solve this problem, the magnitude of BHF should be set by safe value. Some research related to the determination of blank holder force for prevention product defect is still widely performed. Reddy et al. [1], Agrawal et al. [2], Wang et al. [3], and Qin et al. [4] conducted the research to obtain the linear BHF distribution magnitude for reducing wrinkling. Based on restraining energy calculating and buckling criterion, Agrawal et al. [2], Wang et al. [3] and Qin et al. [4] developed the mathematical modeling of minimum constant BHF to avoid wrinkling. The wrinkling would occur when the buckling strain as the tangential strain is excessive. Furthermore, they have developed the buckling analysis to predict the onset of wrinkling and determined the BHF magnitude by using the theory of strain energy.

Kadkhodaya et al. [5], Chu et al. [6], Correia et al. [7], and Shafaat et al. [8, 9] have proposed the mathematical model of wrinkling by using the approach of bifurcation function and Tresca criterion. Furthermore, these models are used to predict the minimum constant BHF without wrinkling, during the punch stroke. Candra et al. [10] continued these researches to estimate the variable blank holder force by using the maximum gap criterion.

Based on the researches above, this paper develops a different approach to determine the minimum VBHF by using the forming limit diagram (FLD) criterion to prevent wrinkling. Modeling of the minimum VBHF would also be verified by FE simulation.

2. Mathematical modeling of minimum variable blank holder force

Based on the equilibrium force diagrams in a small element and considering the friction on the flange surface, as shown in Figure 1, the tangential stress equation is obtained as follows [10, 11].

$$\sigma_{t,i} = \sigma_{r,i} - \sqrt{\frac{2(\bar{R}_n+1)}{1+2\bar{R}_n}}(\sigma_f) \quad (1)$$

And

$$\sigma_{t,i} = \left\{\sqrt{\frac{2(\bar{R}_n+1)}{1+2\bar{R}_n}}\left(\sigma_{fmp1-2,i}\right)\left(\ln\frac{d_{1,i}}{d_m}\right)\right\}+\left\{\frac{2\mu F_{bh,i}}{\pi d_{1,i} s_0}\right\}-\sqrt{\frac{2(\bar{R}_n+1)}{1+2\bar{R}_n}}(\sigma_f) \quad (2)$$

Where $\sigma_{r,i}$ is the stress in radial direction, $\bar{R}_n$ is average of material anisotropic value, $\sigma_{fmp1-2,i}$ is the mean flow stress on flange over punch stroke, $d_{1,i}$ is the outer local diameter of flange - function of punch stroke, d_m is the average radius of cylindrical cup deep drawing, μ is the coefficient of friction, $F_{bh,i}$ is the variable blank holder force (VBHF), s_o is the initial thickness of material and σ_f is the flow stress a moment on *flange*.

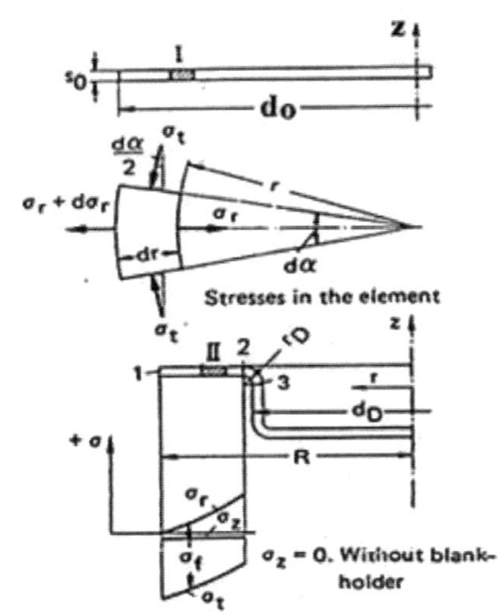

Figure 1. The equilibrium force diagram in a small element on flange [10, 11].

The mean flow stress on flange over punch stroke ($\sigma_{fm,p1-2,i}$) can be calculated by Eq. (3) [10].

$$\sigma_{fm,p1\text{-}2,i}=\left(\frac{1}{2}\right)\left(\sqrt{\frac{2(\bar{R}_n+1)}{1+2\bar{R}_n}}\right)^n K\left[\left(\ln\frac{d_o}{d_{1,i}}\right)^n+\left(\ln\frac{d_{int,i}}{d_D+2r_d}\right)^n\right] \quad (3)$$

Where d_D is the die diameter and r_d is the die edge radius and $d_{int,i}$ is the inside diameter.
To determine the changing of diameter should be calculated under the condition a constant volume. The outer local diameter of flange and inside diameter each step could be determined by equation as follows [10, 11].

$$d_{1,i} = f(d_0, d_D, r_p, r_d, i{=}h) = \sqrt{(d_0)^2 + 4d_D\{h - (0{,}43r_p - 0{,}43r_d)\}} \quad (4)$$

Where r_p is the punch edge radius, i=h is the punch stroke and $d_{int,i}$ can be calculated by Eq. (5),

$$d_{int,i} = f(d_0, d_D, r_p, h, d_{1,i}) = \sqrt{(d_0)^2 + (d_D + 2r_d)^2 - (d_{1,i})^2} \quad (5)$$

The deformation of deep drawing process can be represented by a minor strain (strain tangential) and major (tensile strain). The safe value of the radial and tangential strain ratio in the flange area could be determined by the forming limit diagram (FLD) as shown in Fig. 2. The force and stress tangential equation each punch stoke with considering the friction and no differences of the cross section area (see Figure 3) can be determined by Eqs. $F_{t,i} = F_{t\,minor,i} - F_{friction,i}$ and $\sigma_{t,i} = \sigma_{t\,minor,i} - \sigma_{friction,i}$, respectively. Where, $F_{t,i}$ is the tangential force, $F_{t\,minor,i}$ is the ideal tangential force, $F_{friction,i}$ is the tangential friction force, $\sigma_{t\,minor,i}$ is the ideal tangential stress and $\sigma_{friction,i}$ is the tangential friction stress.

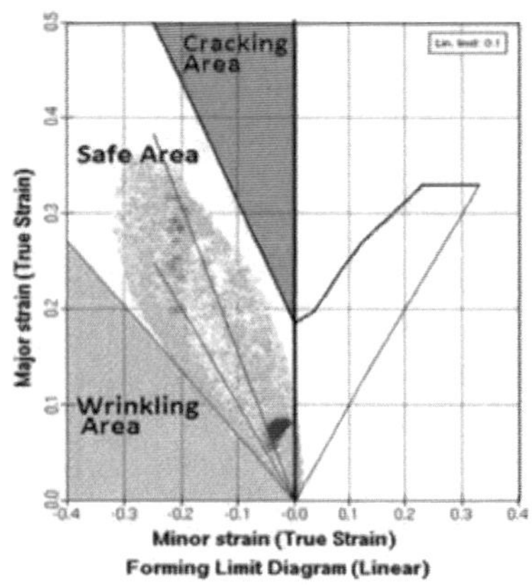

Figure 2. Forming limit diagram [19].

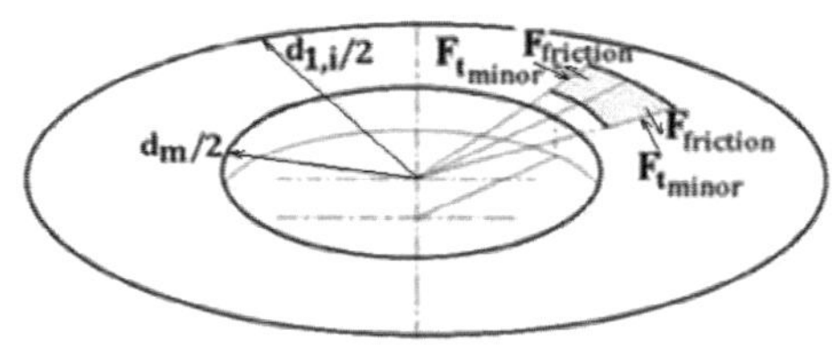

Figure 3. Equilibrium of the force diagram in tangential direction.

Based on defect criterion and the forming limit diagram (FLD), the wrinkling will be predicted when the ratio of minor and major strain around 1.3. In order to achieve these conditions, the equation of tangential stress of each punch stroke also becomes,

$$\sigma_{t,i}=\left(\frac{1}{2}\right)\left(\left(1.3\sqrt{\frac{2(\bar{R}_n+1)}{1+2\bar{R}_n}}\right)^{n} K\left(\left(\ln\frac{d_0}{d_{1,i}}\right)^{n}+\left(\ln\frac{d_{int,i}}{d_m}\right)^{n}\right)\right)-\left(\frac{2\mu\, F_{bh,i}}{s_0\left(\frac{d_{1,i}-(d_D+r_d)}{2}\right)}\right) \quad (6)$$

Combining Eqs. (2) and (8), the minimum VBHF ($F_{bh,i}$) could be determined with the equation,

$$F_{bh,i}=\text{VBHF}=\frac{\left(\frac{1}{2}\right)\left(\left(1.3\sqrt{\frac{2(\bar{R}_n+1)}{1+2\bar{R}_n}}\right)^{n} K\left(\left(\ln\frac{d_0}{d_{1,i}}\right)^{n}+\left(\ln\frac{d_{int,i}}{d_m}\right)^{n}\right)\right)-\left(\left\{\sqrt{\frac{2(\bar{R}_n+1)}{1+2\bar{R}_n}}(\sigma_{fmp1-2,i})\left(\ln\frac{d_{1,i}}{d_m}\right)\right\}-\left(\sqrt{\frac{2(\bar{R}_n+1)}{1+2\bar{R}_n}}\sigma_f\right)\right)}{\left((2\mu/\pi d_{1,i}s_0)+\left(2\mu/s_0\left(\frac{\left(\frac{d_{1,i}+d_{int,i}}{2}\right)-\left(\frac{d_D+r_d}{2}\right)}{2}\right)\right)\right)} \quad (7)$$

3. FE modeling

In this study, the analytical solution provided was compared to FE Modeling. Forming simulation used the implicit Auto form FE solver as shown a virtual image in Fig. 4. The FE simulation was utilized element formation using elastic-plastic deformation approach and ignore the effects of spring back. The number and shape of meshing were done automatically by FE software. The number of meshing was determined based on deformation conditions. Meanwhile, the shape of meshing is defined as a triangle, which is often used for sheet metal forming simulations.

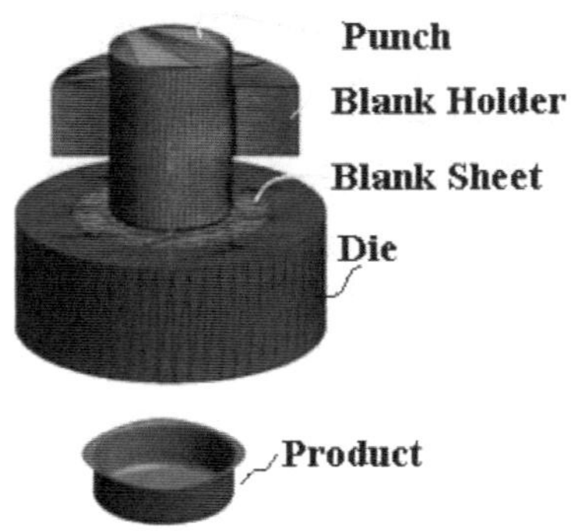

Figure 4. Finite Element modeling.

The analytical approach and FE simulation used the detail value of the punch-die dimension, the steel sheet and process parameters, as shown in Table1.

Table 1. Dimension of tools, mechanical properties and variables of processes.

Dimension of tools		Mech. properties Tin plate T4 CA		Processes parameters	
d_0	64 mm	K	559 N/mm^2	μ	0,45 (dry)
d_D	40 mm	n	0.176		
d_p	39.576 mm	$\overline{R}$	2	BHF	Not constant
Clearance	0.288 mm	UTS	355 N/mm^2		
s_0	0.2 mm	c	1.125		
r_d	1 mm				
r_p	2 mm				
H	15 mm				

4. Results and discussion: analytical and FE simulation of minimum VBHF

Table 2 shows the magnitude of tangential-radial stress and minimum VBHF by using the equations 1 to 7, based on the minor and major limiting strain ratio, during the punch stroke. A minimum VBHF application under maintaining the strain ratio was effective for controlling the equilibrium of stress deformation. The profile of minimum VBHF increases with an average of gradient slopes around 39.6 to prevent wrinkling.

Table 2. Minimum $F_{bh,i}$(VBHF) from analytical based on the limiting strain ratio.

i=h Punch stroke (mm)	The absolute of minor strain and major strain ratio	$\sigma_{t,i}$	$\sigma_{r,i}$	Min. $F_{bh,i}$ (MinimumVBHF)
6	1.3	418.35	399.47	682.40
8	1.3	444.43	424.37	739.38
10	1.3	464.92	443.94	788.80
12	1.3	482.21	460.45	831.00
14	1.3	497.55	475.10	865.13
Etc.				

Table 3 shows the details of the FE simulation result, related to tangential stress, radial stress and minimum blank holder force, based on the minor and major strain ratio and the forming limit diagram (FLD). The simulation was conducted to present an application of the minimum VBHF, should be applied slightly increase. The virtual image of minimum VBHF application leads to the major strain and minor strain at punch stroke 6, 10 and 14 mm, as shown in Fig. 5. In addition, the minor and major strain ratio can be maintained around 1.3 without excessive wrinkling product. FE simulations prove that the modeling of minimum VBHF could be applied as a reference in deep drawing process without excessive wrinkling. Results of FE simulation have no different significantly compared to the analytical results before.

Table 3. Minimum $F_{bh,i}$ (VBHF) from FE simulation.

i=h Punch stroke (mm)	The absolute of minor strain and major strain ratio	$\sigma_{t,i}$	$\sigma_{l,i}$	Minimum $F_{bh,i}$ (MinimumVBHF)
6	1.23	381.3	402.2	546.4
8	1.25	406.9	423.2	577.9
10	1.35	425.5	438.7	631.6
12	1.35	439.8	450.7	719.3
14	1.40	455.3	464.5	852.8
Etc.				

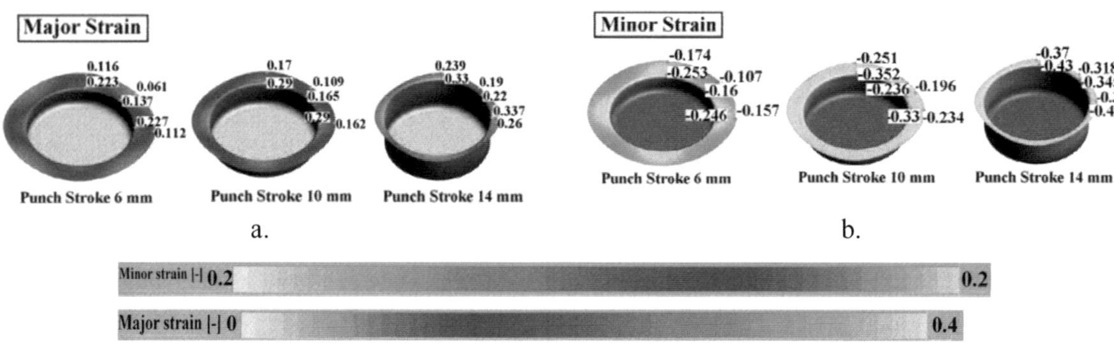

Figure 5. a. Major Strain, b. Minor Strain.

The final results of calculation and FE simulation have been displaying the minimum VBHF each punch stroke, as shown Fig. 6. To prevent wrinkling at each punch increment, the BHF should be maintained slightly above the minimum VBHF line. These conditions will keep the value of minor and major strain ratio around 1.3 and control the magnitude of the tangential stress. Finally, the process can be operated properly without excessive wrinkling product.

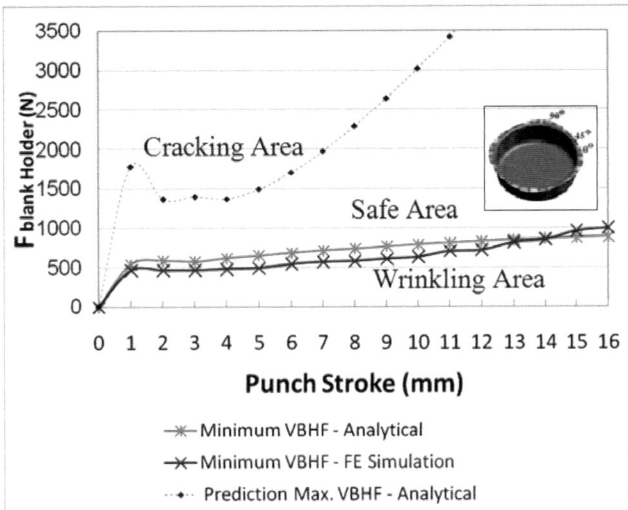

Figure 6. Minimum VBHF vs Punch Stroke.

5. Conclusion

The results of the minimum VBHF calculation are not much different from FE simulation. VBHF could be effective for preventing the occurrence of wrinkling, and also for improving the permeability of the process. VBHF modeling indicated a similar trend compared to FE simulation results. Mathematical modeling of VBHF can be used as a simple approach for estimating the magnitude of minimum blank holder force in every punch stroke. The minimum VBHF has provided the information about a safe area on cylindrical cup deep drawing. Therefore, the safe area must be maintained above the minimum VBHF. Modeling minimum VBHF based on the ratio of minor and major strain and FLD can be used as a reference in the process of deep drawing.

Acknowledgments

The authors would like to acknowledge to Prof. Dr. Dedi Priadi, DEA (Dean of Engineering Faculty - University of Indonesia) for allowing using the AUTOFORM ver. 3.1, and for helpful discussion.

References

[1] Reddy R V, Reddy T A J and Reddy G C M 2012 Effect of various parameters on the wrinkling deep drawing cylindrical cups *Int. J. Engineering Trends and Technology* **315** pp 53–58
[2] Agrawal A, Reddy N V and Dixit P M 2007 Determination of optimum process parameters for wrinkle free products in deep drawing process *J. Mater. Process. Tech.* **191** pp 51–54
[3] Wang X and Cao J 2000 Analytical prediction of flange wrinkling in sheet metal forming *J. Manuf. Process.* **2** 100–106
[4] Qin S-J, Xiong B-Q, Lu H and Zhang T-T 2012 Critical blank-holder force in axisymmetric deep drawing *T. Nonferr. Metal Soc.* **22** pp s239–s246
[5] Kadkhodaya M and Moayyedian F 2011 Analytical elastic-plastic study on flange wrinkling in deep drawing process *J. of Scientia Iranica Transaction B: Mechanical Engineering* **18** pp 250–260
[6] Chu E and Yu X 2001 An elastoplastic analysis of flange wrinkling in deep drawing process *Int. J. Mech. Sci.* **43** pp 1421–40
[7] Correia J P M and Ferron G 2002 Wrinkling prediction in the deep-drawing process of anisotropic metal sheets *J. Mater. Process. Tech.* **128** pp 178–190
[8] Shafaat M A, Abbasi M and Ketabchi M 2011 Investigation into wall wrinkling in deep drawing process of conical cups *J. Mater. Process. Tech.* **211** pp 1783–95

[9] Shafaat M A, Abbasi M and Ketabchi M 2011 Effect of different yield criteria on prediction of wrinkling initiation of interstitial-free galvanized steel sheet *J. of Material and Design* **32** pp 3370–76
[10] Candra S, Batan I M L, Berata W and Pramono A S 2015 Analytical study and FEM simulation of the maximum VBHF to prevent cracking on cylindrical cup deep drawing *Procedia CiRP* **26** pp 548–553
[11] Lange K 1985 *Handbook of Metal Forming* (Berlin: McGraw-Hill) chapter 20 pp 1–65

IOP Conf. Series: Materials Science and Engineering **273** (2017) 012015 doi:10.1088/1757-899X/273/1/012015

Single-tier city logistics model for single product

N I Saragih[1,2], S Nur Bahagia[1], Suprayogi[1] and I Syabri[3]

[1] Faculty of Industrial Technology, Bandung Institute of Technology, Bandung, Indonesia
[2] Department of Industrial Engineering, Widyatama University, Bandung, Indonesia
[3] School of Architecture, Planning, and Policy Development, Bandung Institute of Technology, Bandung, Indonesia

E-mail: nova.indah@widyatama.ac.id

Abstract. This research develops single-tier city logistics model which consists of suppliers, UCCs, and retailers. The problem that will be answered in this research is how to determine the location of UCCs, to allocate retailers to opened UCCs, to assign suppliers to opened UCCs, to control inventory in the three entities involved, and to determine the route of the vehicles from opened UCCs to retailers. This model has never been developed before. All the decisions will be simultaneously optimized. Characteristic of the demand is probabilistic following a normal distribution, and the number of product is single.

1. Introduction

This research continues the research of [1] to develop mathematical model for single-tier city logistics system. Single-tier city logistics system consists of three entities, that is, point of supplies, logistics facilities, and point of demands [2]. The location of point of supplies and logistics facilities is outside of the city and the location of point of demands is inside of the city. In this research, point of supplies are then called suppliers, logistics facilities are called UCCs (urban consolidation centers), and point of demands are called retailers. The problems that will be answered in this research are how to determine the location of UCCs, to allocate retailers to opened UCCs, to assign suppliers to opened UCCs, to control inventory in the three entities involved, and to determine the route of the vehicles from opened UCCs to retailers. All the decisions will be simultaneously optimized for city logistics is the process for totally optimizing the logistics and transport activities [3]. City logistics model that can answer the described problems simultaneously in single-tier city logistics system has never been developed before [4].Characteristics of the demand is the probabilistic one, following a normal distribution, and the number of product is single.

2. Methodology

This research used [5] as reference model to develop the single-tier city logistics model. The entities involved in [5] were single supplier, multi depot, and multi retailers, while the entities in single-tier city logistics model that would be developed in this research were multi suppliers, multi UCCs, and multi retailers. The illustration of entities involved in this research can be seen at Figure 1.

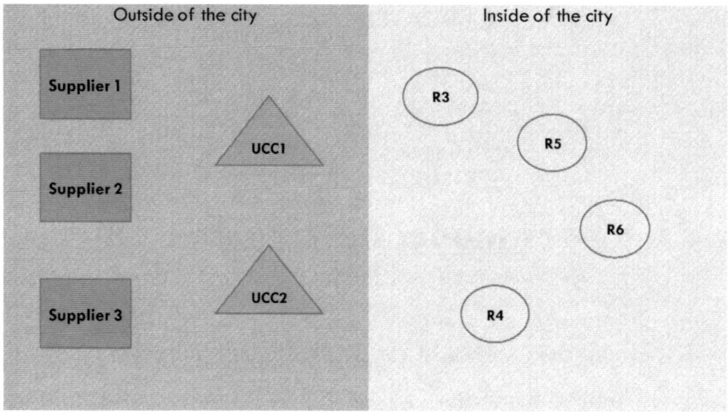

Figure 1. Illustration of entities involved in single-tier city logistics system.

3. Mathematical model

Mathematical model for single-tier city logistics model for single product that will be developed in this research is described as follows.

3.1. Index sets

K set of retailers
J set of potential UCCs
N_j set of capacity levels available to UCC $(j \in J)$
I set of suppliers
V set of vehicles
M merged set of retailers and potential UCCs, i.e. $(K \cup J)$

3.2. Parameters and notations

At retailer

μ_k mean of demand at retailer *k* (Unit/day) $(\forall k \in K)$
σ_k^2 variance of demand at retailer *k* (Unit/day) $(\forall k \in K)$
h_k inventory holding cost at retailer *k* (Rp/unit/day) $(\forall k \in K)$
a_k ordering cost at retailer *k* (Rp/order) $(\forall k \in K)$
lt_k lead time of retailer *k* (Day) $(\forall k \in K)$
s_k shortage cost at retailer *k* (Rp/unit) $(\forall k \in K)$
α fill rate
z_α left α -percentile of standard normal random variable z

At UCC

f_j^n fixed cost for opening and operating UCC *j* with capacity level *n* (Rp/day) $(\forall j \in J, \forall n \in N_j)$
b_j^n capacity with level *n* for UCC*j* (Unit/day) $(\forall j \in J, \forall n \in N_j)$
d_{kl} transportation cost between node *k* and node *l* (Rp/delivery) $(\forall k, l \in M)$
h_j inventory holding cost at UCC*j* (Rp/unit/day) $(\forall j \in J)$

a_j ordering cost at UCCj (Rp/order) $(\forall j \in J)$
lt_j lead time of UCCj (Day) $(\forall j \in J)$
vc capacity of vehicle (Unit)

At supplier
h_i inventory holding cost at supplier i (Rp/unit/day) $(\forall i \in I)$
a_i ordering cost at supplier i (Rp/order) $(\forall i \in I)$
lt_i lead time of supplier i (Day) $(\forall i \in I)$
b_i capacity for supplier i (Unit/day) $(\forall i \in I)$

At system
w transportation cost of truck (Rp/truck)
pp capacity of truck (Unit)
B number of retailers contained in set K, i.e. $B = |K|$

3.3. Decision variables

At retailer
NP_k order frequency of retailer $k (\forall k \in K)$
E order frequency of every retailer
Q_k lot size of retailer k (Unit) $(\forall k \in K)$
MK_k number of shortage at retailer k (Unit) $(\forall k \in K)$
RK_k reorder point at retailer k (Unit) $(\forall k \in K)$
SS_k safety stock at retailer k (Unit) $(\forall k \in K)$

At UCC
U_j^n 1 if UCCj is opened with capacity level n, 0 if otherwise $(\forall j \in J, \forall n \in N_j)$
D_j demand of UCC j (Unit/day) $(\forall j \in J)$
Y_{jk} 1 if retailer k is assigned to UCC j, 0 if otherwise $(\forall j \in J, \forall k \in K)$
NP_j order frequency of UCC$j (\forall j \in J)$
Z order frequency of every UCC
Q_j lot size of UCCj (Unit) $(\forall j \in J)$
RK_j reorder point at UCCj (Unit) $(\forall j \in J)$
R_{klv} 1 if k precedes l in route of vehicle v, 0 if otherwise $(\forall k, l \in M, \forall v \in V)$
M_{kv} auxiliary variable defined for retailer k for subtour elimination in route of vehicle v $(\forall k \in K, \forall v \in V)$
X_j number of truck at UCCj (Truck) $(\forall j \in J)$

At supplier

D_i demand of supplier (Unit/day) $(\forall i \in I)$

Q_i lot size of supplier *i* (Unit) $(\forall i \in I)$

RK_i reorder point at supplier *i* (Unit) $(\forall i \in I)$

X_i number of truck at supplier *i* (Truck) $(\forall i \in I)$

G_{ij} 1 if supplier *i* supplies UCC*j*, 0 if otherwise $(\forall i \in I, \forall j \in J)$

V_{ij} amount of demand of UCC*j* supplied by supplier $i(\forall i \in I, \forall j \in J)$

At system

TC total cost (Rp/day)

T single cycle time (Day)

3.4. The model

Mathematical model of single-tier city logistics model for single product is given below.

Objective function:

$$\min TC = \sum_{j\in J}\sum_{n\in N_j} f_j^n U_j^n + \frac{E}{T}\sum_{v\in V}\sum_{k\in M}\sum_{l\in M} d_{kl}R_{klv} + \sum_{k\in K}\left[\frac{a_k\mu_k}{Q_k} + h_k\left(\frac{Q_k}{2} + SS_k\right) + s_k MK_k\left(\frac{\mu_k}{Q_k}\right)\right] +$$
$$\sum_{j\in J}\left[\frac{a_jD_j}{Q_j} + h_j\left(\frac{Q_j}{2} + \sum_k (lt_k\mu_k + SS_k)Y_{jk}\right) + wX_j\frac{Z}{T}\right] + \quad (1)$$
$$\sum_{i\in I}\left[\frac{a_iD_i}{Q_i} + h_i\left(\frac{Q_i}{2} + \left(\sum_j\sum_k ((lt_j + lt_k)\mu_k + SS_k)Y_{jk}\right)G_{ij}\right) + wX_i\frac{1}{T}\right]$$

Subject to:

$$\sum_{v\in V}\sum_{l\in M} R_{klv} = 1, \forall k \in K \quad (2)$$

$$\sum_{l\in K}\sum_{k\in M} Q_l R_{klv} \le vc, \forall v \in V \quad (3)$$

$$M_{kv} - M_{lv} + (B \times R_{klv}) \le B - 1, \forall k, l \in K, \forall v \in V \quad (4)$$

$$\sum_{l\in M} R_{klv} - \sum_{l\in M} R_{lkv} = 0, \forall k \in M, \forall v \in V \quad (5)$$

$$\sum_{j\in J}\sum_{k\in K} R_{jkv} \le 1, \forall v \in V \quad (6)$$

$$\sum_{l\in M} R_{klv} + \sum_{l\in M} R_{jlv} - Y_{jk} \le 1, \forall j \in J, \forall k \in K, \forall v \in V \quad (7)$$

$$\sum_{n\in N_j} U_j^n \le 1, \forall j \in J \quad (8)$$

$$\sum_{k\in K} \mu_k Y_{jk} \le \sum_{n\in N_j} b_j^n U_j^n, \forall j \in J \quad (9)$$

$$\sum_{k\in K} \mu_k Y_{jk} = \sum_{n\in N_j} D_j U_j^n, \forall j \in J \quad (10)$$

$$\sum_{j\in J} Y_{jk} = 1, \forall k \in K \tag{11}$$

$$\sum_{j\in J} G_{ij} \geq 1, \forall i \in I \tag{12}$$

$$\sum_{n\in N_j} D_j U_j^n = \sum_{i\in I} V_{ij} G_{ij}, \forall j \in J \tag{13}$$

$$\sum_{j\in J} V_{ij} = D_i, \forall i \in I \tag{14}$$

$$\sum_{j\in J} V_{ij} \leq b_i, \forall i \in I \tag{15}$$

$$SS_k = z_\alpha \sqrt{lt_k \sigma_k^2}, \forall k \in K \tag{16}$$

$$MK_k = \sqrt{lt_k \sigma_k^2}\left[f(z_\alpha) - z_\alpha \psi(z_\alpha)\right], \forall k \in K \tag{17}$$

$$RK_k = lt_k \mu_k + SS_k, \forall k \in K \tag{18}$$

$$RK_j = \left(\sum_k (lt_j + lt_k)\mu_k + SS_k\right) Y_{jk}, \forall j \in J \tag{19}$$

$$RK_i = \left(\sum_k lt_i \mu_k + \left(\sum_j \sum_k (lt_j + lt_k)\mu_k + SS_k\right)\right) Y_{jk}, \forall i \in I \tag{20}$$

$$T = \frac{Q_i}{D_i} = \frac{NP_j Q_j}{D_j} = \frac{NP_j NP_k Q_k}{\mu_k} \tag{21}$$

$$NP_k = E, \forall k \in K \tag{22}$$

$$NP_j = Z, \forall j \in J \tag{23}$$

$$X_j = \left\lceil \frac{Q_j}{pp} \right\rceil, \forall j \in J \tag{24}$$

$$X_i = \left\lceil \frac{Q_i}{pp} \right\rceil, \forall i \in I \tag{25}$$

$$U_j^n \in \{0,1\}, \forall j \in J, \forall n \in N_j \tag{26}$$

$$Y_{jk} \in \{0,1\}, \forall j \in J, \forall k \in J \tag{27}$$

$$R_{klv} \in \{0,1\}, \forall k,l \in M, \forall v \in V \tag{28}$$

$$G_{ij} \in \{0,1\}, \forall i \in I, \forall j \in J \tag{29}$$

$$M_{kv} \geq 0, \forall k \in K, \forall v \in V \tag{30}$$

$$Q_i, Q_j, Q_k \geq 0, \forall i \in I, \forall j \in J, \forall k \in K \tag{31}$$

$$T > 0 \tag{32}$$

$$N, N_j, N_k \geq 1, N, N_j, N_k \in \text{int}, \forall j \in J, \forall k \in K \tag{33}$$

The difference between single-tier city logistics model that has been developed with the model of [5] is at constraints (12)–(15) where constraint (12) ensures that a supplier can supply more than one UCCs, constraint (13) ensures that demand at opened UCC must be fulfilled, constraint (14) ensures that demand at supplier is sum of the amount of demand of UCCs supplied by that supplier, and constraint (15) ensures that amount of demand of UCCs supplied by a supplier does not exceed its capacity. The explanation of other constraints can be seen at [5].

4. Numerical example

The model was then solved with Lingo 12.0 for some numerical examples. The data used were generated. Figure 2 gives the illustration for numerical example of 3 suppliers, 2 UCCs, and 4 retailers.

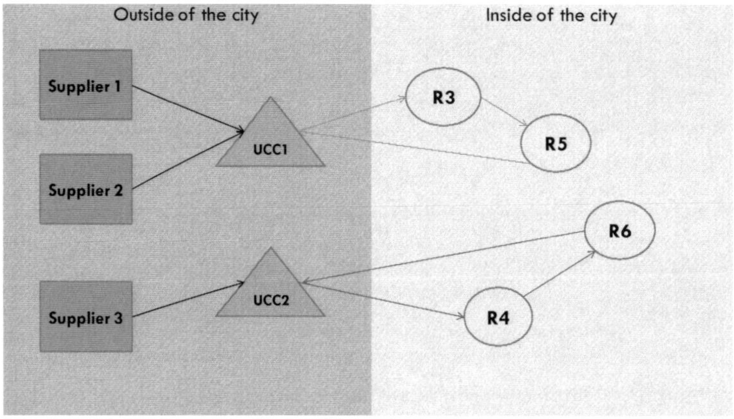

Figure 2. Illustration for numerical example of 3 suppliers, 2 UCCs, and 4 retailers.

From Figure 2 it was found that both UCC1 and UCC2 were opened. UCC1 was supplied by Supplier1 and supplier2 in accordance with UCC1's lot size. UCC2 was supplied by Supplier1 also in accordance with UCC2's lot size. UCC1 supplied R3 and R5 in accordance with their lot size and to serve R3 and R5 there was a route from UCC1, that is, UCC1-R3-R5-UCC1. UCC2 supplied R4 and R6 also in accordance with their lot size. To serve R3 and R6, there was also a route that started from UCC2, that is, UCC2-R4-R6-UCC2. All decisions were simultaneously optimized.

5. Conclusion

This research has developed single-tier city logistics model for single product which has never been developed before. All the decisions are simultaneously optimized. The decisions are location-allocation of UCCs retailers, assignment of suppliers that supply opened UCCs, inventory control at all entities involved, and vehicle routes to serve retailers. This research also gives challenges for future studies to develop heuristic method to solve the model in large scale for the model itself belongs to NP-hard problems [6].Considering multi product also can be a challenge for future studies.

6. References

[1] Saragih N I, Bahagia S N, Suprayogi and Syabri I 2015 City logistics for mega city: a conceptual model (case study: DKI Jakarta) *Joint Int. Conf. on Electric Vehicular Technology and Industrial, Mechanical, Electrical and Chemical Engineering (Surakarta)* (Surakarta: Sebelas Maret University) pp 4C2-1:1–5

[2] Benjelloun A and Crainic T G 2008 Trends, challenges, and perspectives in city logistics *Transportation And Land Use Interaction Proc. TRANSLU (Bucharest)* (Bucharest: Editura Politecnica Press) pp 269–284

[3] Taniguchi E, Thompson R G, Yamada T and van Duin R 2001 *City Logistics: Network Modelling and Intelligent Transport Systems* (Oxford: Pergamon)

[4] Saragih N I, Bahagia S N, Suprayogi and Syabri I 2016 A survey on location-routing-inventory problem *Proc. of the 8th Widyatama Int. Seminar on Sustainability (Bandung)* (Bandung: Widyatama University) pp 139–141

[5] Saragih N I, Bahagia S N, Suprayogi and Syabri I 2017 Integration model of location, vehicle route, and inventory control decision on three echelon supply chain system *The J. Teknik Industri* **19** pp 1–10
[6] Javid A H and Azad N 2010 Incorporating location, routing and inventory decisions in supply chain network design *Transport. Res. E-Log.* **46** pp 582–597

Inventory model optimization for supplier-manufacturer-retailer system with rework and waste disposal

A R Dwicahyani, E Kholisoh, W A Jauhari, C N Rosyidi and P W Laksono

Department of Industrial Engineering, Sebelas Market University, Surakarta, Indonesia

E-mail: anindyard94@gmail.com

Abstract. This study developed a model for a CLSC inventory system which consisted of a supplier, a manufacturer and a retailer. We applied single remanufacturing cycle and multiple manufacturing cycle, $(1,P)$, policy and performed a comparison to the previous study. We conducted an investigation of an imperfect manufacturing process whose defective items were being reworked. Other considerations were quality dependent return rate and waste disposal activity, for any returned items that did not exceed the acceptable quality level. Some parameters including demand, proportion of defect, waste and refurbished item were assumed deterministic and constant. We proposed a solution procedure and presented a numerical example to illustrate the application of the model. The results showed that $(1,P)$ policy allowed higher profit for the system than $(R,1)$ policy presented in the previous study.

Keywords: Inventory model, reverse logistics, closed loop supply chain, waste disposal, imperfect production, rework, quality dependent return rate

1. Introduction

Managing end-of-life (EoL) products returned from the market is a key to succeed a sustainable supply chain [1]. In addition, it also guarantees some benefits including reducing costs, maintaining customer loyalty, improving customer services and increasing market share [2, 3]. According to [4], activities related to EoL product management may include collection, inspection, cleaning, disassembly or product recovery that in the end will be resold to the market. Many manufacturing companies such as electronic products and automotive parts [5, 6], have broadly applied the concept of EoL product management. According to [7], since it has never been an easy way to manage EoL products from the market, some typical industries should do alliance between firms that later be called joint ventures collaboration. A supply chain that implements the concept of return item management is then classified as closed-loop-supply-chain (CLSC). By performing joint venture collaboration, some operations of the EoL product management are being performed by the help of certain firms in the joint venture [8]. Therefore, costs can be reduced and inventories within the supply chain can be managed more efficiently.

The model developed in this paper is an inventory model that aims to maximize joint total profit of all firms that collaborate in the supply chain. Formerly, some studies have initiated the development of inventory model in the system with product returns, including [9–11]. Some researchers have also extended the model by including some aspects such as collaborations, stochastic demand and returns, dependent return rate, imperfect production and reworks [12–16]. Yuan and Gao [14] developed a cost minimization model of the CLSC inventory system and determined the optimal solutions for each party. The model was an extended study of [21], in which they investigated both (R,1) and (1,P) policies. The decision to choose the right number of manufacturing and remanufacturing cycles is surely a challenging effort in the system of CLSC, since each operation is performed sequentially due to the lack of production equipment and machinery. Teunter [15], [14] and [16] are some studies that investigated multiple cycle policy of remanufacturing and manufacturing operations in the system of reversed logistics.

However, most previous studies have not yet considered the collaboration in the supply chain and imperfect production process all at once. In this model, we aim to develop a model of CLSC inventory system that considers the imperfect manufacturing process, reworks, quality dependent return rate, as well as waste disposal activity. The model is intended to determine the optimal solution, including number of manufacturing cycle, number of shipment between parties, as well as retailer ordering cycle time, that maximizes joint total profit for all parties. Here, we also give a comparison of (R,1) and (1,P) policies to help the managers in choosing the most profitable and efficient policy of manufacturing and remanufacturing.

2. Notations

Some parameters used in the model development are denoted by these following notations.

D: annual demand rate (units/year);
A_r: ordering cost for retailer (\$/order);
F_r: holding cost fraction for retailer's stock;
A_R: remanufacturing set up cost (\$/batch);
A_{mw} : raw material ordering cost (\$/order);
F_M: holding cost fraction for serviceable stock;
F_{mw} : holding cost fraction for raw material stock;
F_{rw}: holding cost fraction for recoverable stock;
P_{rm}: raw material purchasing price (\$/unit);
P_u: used item purchasing price (\$/unit);
P_f: refurbished item selling price (\$/unit);
T: planning period of the model (year)
T_R: remanufacturing cycle time (year);
T_P: manufacturing cycle time (year);
T_w: setup time for reworking process (year);
T_{P2} : reworking process period (year);
P: number of manufacturing cycle per T;
C_{iscp}: cost to inspect and sort (manufactured items) (\$/unit);
A_s: supplier fixed cost for each order (\$/order);
F_s: holding cost fraction ($0<F_s<1$);
P_c : retail price (\$/unit);
P_r: wholesale price (\$/unit);
A_m: manufacturing set up cost (\$/batch);
N : total number of shipment per T;
M : fixed cost to process all demand (\$);
r : remanufacturing rate ($r>D$, unit/year);
β : multiplying factor for p ($0<\beta<1$);
γ: multiplying factor for r rate ($0<\gamma<1$);
f : refurbishable item proportion ($0<f<1$);
q: recoverable item proportion ($0<q<1$);
k : defective item proportion ($0<k<1$);
C_d : defective item holding cost (\$/unit);
C : annual return rate (unit/year);
C_P: cost to manufacture (\$/unit);
C_R : cost to remanufacture (\$/unit);
C_f: cost to refurbish (\$/unit);
C_w : cost of waste disposal (\$/unit);
C_{isc}: cost to inspect and sort (return items) (\$/unit);
P_s: supplier purchasing cost (\$/unit);
L : fixed cost to process all orders (\$);

3. Problem description, mathematical model and solution procedure

The development of the model began by illustrating the relevant system. The proposed model concerns with a closed-loop-supply-chain (CLSC) inventory system considering three parties: a retailer, a manufacturer, and a supplier. This study develops the model belongs to [17] to a single-remanufacturing-multiple-manufacturing cycle policy (1,P) model. The results from this model are then compared to the results from [17]. The CLSC inventory system is depicted in Figure 1.

In the system, manufacturer delivers products to retailer in n shipments during manufacturing process and m shipments during remanufacturing process. Manufacturer orders raw material from suppliers and receives the orders in l shipments per one cycle T_P. The problem here lies on how many shipments manufacturer and supplier should determine, hence ordering cost, holding cost, production cost, purchasing cost and other related costs could be minimized to gain the maximum joint total profit. Here, refurbishing refers to a process to increase the quality of a portion of recoverable used items (f) that cannot be remanufactured, as its quality does not exceed the acceptable quality level. Refurbished items are then sold to the secondary market at a lower price, with $P_f<P_r$. Reworking process is needed, since manufacturing processes conducted by the manufacturer produce any defective items needed to be reworked immediately. The proportion of defective items (k) is assumed constant and known. In this study, we consider that amount of product returns (C) depends on quality level, q, where q is the portion of recoverable returned items. Hence, (1-q) is the portion of items that disposed from the system. Return rate, C, is a portion of demand, D, returns to the system, then $0<C(q)/D<1$. The function $C(q)=Dbe^{-\varphi q}$ is adopted from [18], with $0<b<1$ and $\varphi>1$.

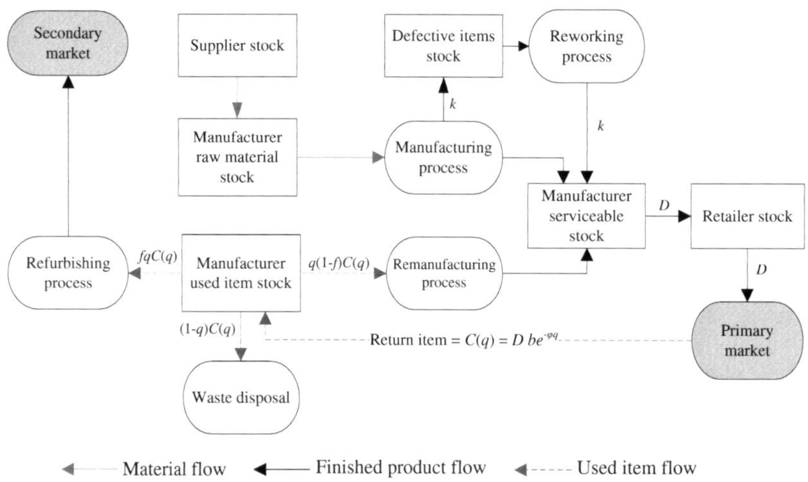

Figure 1. The proposed closed-loop supply chain inventory system.

This model aims to maximize the annual joint total profit (JTP) for manufacturer, retailer and supplier by determining the optimal retailer cycle time, T_r^*, manufacturer optimal number of shipment, m^* and n^*, supplier optimal number of shipment, l^*, and optimal number of manufacturing cycle, P^*. The inventory profile of the system is shown in Figure 2.

3.1 *Retailer profit* (TP_r)

Inventory cost to the retailer consists of ordering cost (OC_r), purchasing cost (PC_r) and holding cost (HC_r). Retailer profit (TP_r) is obtained by subtracting total revenue (TR_r) with total costs (TIC_r) as follows.

$$TP_r = TR_r - TIC_r = DP_c - \frac{A_r}{T_r} - DP_r - \frac{DF_r P_r T_r}{2} \quad (1)$$

3.2 *Manufacturer profit* (TP_m)

As shown in Figure 2, manufacturer has four types of stock, including raw material, serviceable items, defective items and recoverable items. Each kind of stock has different components of inventory cost. Inventory cost for raw material stock includes ordering cost (OC_{rm}), purchasing cost (PC_{rm}), and holding cost (HC_{rm}). Manufacturer serviceable inventory cost consists of production setup cost (SC_m), manufacturing cost (PC_m), reworking cost (rC_m), inspection cost (IC_m), remanufacturing cost (RC_m) and holding cost (HC_m). Manufacturer recoverable inventory cost (TIC_{rw}) consists of purchasing cost (PC_{rw}), inspection and sorting cost (ISC_{rw}), refurbishing cost (TC_f), waste disposal cost (WDC_{rw}) and holding cost (HC_{rw}). Manufacturer defective items inventory cost (TIC_d) is an incurred cost to hold stocks of defective items, the function of TIC_d is formulated below. Subsequently, we obtain the following equation for manufacturer total profit (TP_m).

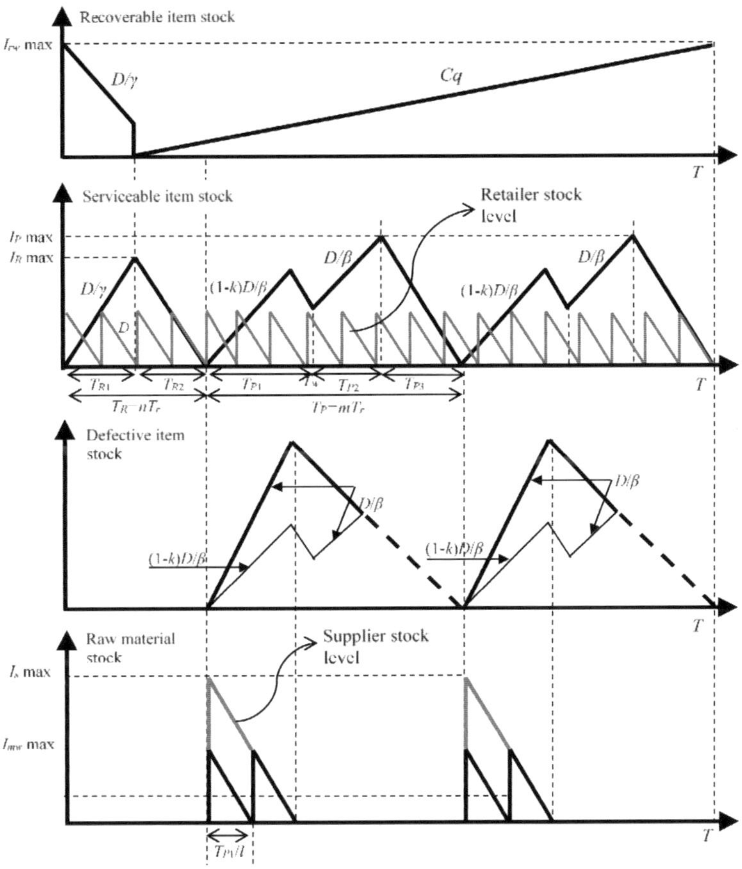

Figure 2. Retailer, manufacturer, and supplier stocks level (l=2, P=2, n=4, m=7, N=18).

$$TP_m = TR_m - TIC_{rm} - TIC_m - TIC_{rw} - TIC_d$$

$$= P_r D + P_f\left(\frac{nfD}{(1-f)(n+mP)}\right) - \left[\frac{A_{mw}\,P\,l}{T_r(n+Pm)} + \frac{mDP_{rm}P}{(n+Pm)} + F_{mw}P_{rm}\left(\frac{P\,DT_r\,\beta\,m^2}{2\,l\,(n+Pm)}\right)\right]$$

$$-\left[\begin{array}{l}\frac{A_R+A_mP+A_wP+M(n+mP)}{T_r(n+mP)} + \frac{DC_pmP}{n+mP} + \frac{DC_pkmP}{n+mP} + \frac{DC_{iscp}mP}{n+mP} + \frac{DC_Rn}{n+mP} \\ +F_M\left(\frac{P_un+mP_{rm}P}{n+mP}\right)\left(\frac{n\left(DT_r\gamma + \frac{Cq\,T_r(1-f)\,(1-\gamma)\,(n+mP)}{2}\right)}{n+mP} + \frac{mDP\left(-2kT_w+2T_r\beta-T_r\left(\left(1+k+k^2\right)\beta-1\right)n^2\right)}{2(n+mP)} - \frac{DT_r}{2}\right)\end{array}\right]$$

$$-\left[\frac{P_unD}{(1-f)q(n+mP)} + \frac{C_{isc}nD}{(1-f)q(n+mP)} + \frac{C_f\,nfD}{(1-f)(n+mP)} + \frac{C_wnD(1-q)}{(1-f)q(n+mP)} + F_{rw}P_u\left(2Cfq\gamma n + DT_r\gamma n^2 + CqT_r\left((\gamma-1)n-mP\right)^2\right)\right]$$

$$-\left[\frac{C_dkmPD\,(2T_w+(1+k)T_r\beta m)}{2(n+mP)}\right] \qquad (2)$$

3.3 *Supplier profit* (TP_s)

Inventory cost to the supplier consists of delivery cost (DC_s), purchasing cost (PC_s) and holding cost (HC_s). Supplier profit (TP_s) is obtained by subtracting total revenue (TR_s) with total costs (TIC_s).

$$TP_s = TR_s - TIC_s = \frac{P_{rm}mPD}{(n+mP)} - \left(\frac{P(A_s+lL)}{T_r(n+mP)} + \frac{P_s mPD}{(n+mP)} + \frac{F_s P_s D T_r \beta m^2 P(l-1)}{2l(n+mP)}\right) \quad (3)$$

Finally, the joint total profit of the CLSC system (*JTP*) is formulated by equation (4) below.

$$JTP(T_r, n, m, P, l) = TP_r + TP_m + TP_s \quad (4)$$

3.4 *Solution procedure*

The optimal values of T_r, P, m, n and l that maximize $JTP(T_r, P, m, n, l)$ are determined by the following solution procedure. Here, we only guarantee local optimal solution. By substituting ($N=n+mP$)to ($nDT_r = Cq\ (1-f)T_r\,(n+mP)$), we obtain $n = \frac{N\,C\,q(1-f)}{D}$ and $m = \frac{N(1-f)(1-\frac{Cq}{D})}{P}$. The function of n and m are then substituted into equation (4). Hence, the function of JTP in equation (4) becomes $JTP(T_r, P, N, l)$.

To obtain the optimal value of T_r, let the derivative of $JTP(T_r, P, N, l)$ with respect to (for short w.r.t.) T_r equal to zero, then we get $T_r^{\#}(P, N, l)$.The optimal value of N and l can be obtained by letting the derivative of equation (4) w.r.t. N and l equal to zero, then substituting the optimal value of $T_r^{\#}(P, N, l)$to those equation respectively. Hence, we obtain the values of $N^{\#}(P, l)$ and $l^{\#}(P)$.The value of $T_r^{\#}$, $N^{\#}$ and $l^{\#}$ are used as approximations to find the optimal solution. The value of P^*, N^*,m^*, n^*,l^* and T_r^* that maximize $JTP(T_r, P, m, n, l)$ can be derived by the following procedure

Step 1 Given $C(q)=Dbe^{-\varphi q}$,calculate the value of $C(q)$
Step 2 Set $P^{(0)}=1$, then find the value of $l^{\#}$,$N^{\#}$and $T_r^{\#}$
Step 3 Determine the optimal value of N which satisfies:

$$JTP(N^*-1,P^*,l^*,T_r^*) \le JTP(N^*,P^*,l^*,T_r^*) \ge JTP(N^*+1,P^*,l^*,T_r^*) \quad (5)$$

Step 4 Repeat step 2 to 5 for $P^{(a+1)}=P^{(a)}+1$ to find the optimal value of P which satisfies:

$$JTP(N^*,P^*-1,l^*,T_r^*) \le JTP(N^*,P^*,l^*,T_r^*) \ge JTP(N^*,P^*+1,l^*,T_r^*) \quad (6)$$

Step 5 Derive the optimal value of l which satisfies:

$$JTP(N^*, P^*,l^*-1,T_r^*) \le JTP(N^*, P^*,l^*,T_r^*) \ge JTP(N^*, P^*,l^*+1,T_r^*) \quad (7)$$

Step 6 Find the combination value of P^*, N^*, T_r^* and l^* that gives the maximum value of JTP
Step 7 Given a known N^* and P^*, calculate the value of $n^* = \frac{N\,C\,q(1-f)}{D}$ and $m^* = \frac{N(1-f)(1-\frac{Cq}{D})}{P}$

4. Numerical example and analysis

We present a numerical example to illustrate the application of the proposed P(1,*P*) model. The following parameters are adopted from [14]: D=2000 unit/year, A_r=\$100/order, A_{rw}=\$350/order, A_{mw}=\$350/collection, A_s=\$200/order, A_m=\$2000/batch, A_R=\$2500/batch, P_c=\$180/unit, P_r=\$160/unit, P_{rm}=\$100/unit, P_u=\$60/unit, P_s=\$80/unit, P_f=\$120/unit, F_r=0.3/unit/year, F_M=0.5/unit/year, F_{mw}=0.4/unit/year, F_{rw}=0.25/unit/year, F_s=0.2/unit/year, M=\$350/batch, L=\$150/delivery, C_f=\$20/unit, C_d=\$30/unit, C_w=\$5/unit, C_{isc}=\$1/unit, C_{iscp}=\$1/unit, C_p=\$23/unit, C_r=\$15/unit, γ=0.4, β=0.6, f=0.3, k=0.1, q=0.83, b=0.95 and ω=1.5. Using the proposed solution procedure explained before, we obtain the optimal results as shown in Table 1.

Table 1. Optimization results of the proposed P(1,P) model.

P (cycles)	l (shipments)	N (shipments)	T_r (year)	TP_r ($/year)	TP_s ($/year)	TP_m ($/year)	JTP ($/year)
1	2	16	0.065	$35,349	$27,351	$59,080	$121,780
		17	0.062	$35,417	$27,321	$58,799	$121,537
		18	0.059	$35,471	$27,292	$58,516	$121,279
		19	0.057	$35,514	$27,264	$58,232	$121,010
2*	3*	16	0.095	$34,374	$27,692	$60,556	$122,622
		17*	**0.091***	**$34,536***	**$27,682***	**$60,411***	**$122,628***
		18	0.087	$34,676	$27,671	$60,260	$122,607
		19	0.083	$34,799	$27,659	$60,105	$122,563
⋮	⋮	⋮	⋮	⋮	⋮	⋮	⋮
10	5	16	0.219	$29,025	$28,071	$51,249	$108,345
		17	0.208	$29,492	$28,096	$51,353	$108,941
		18	0.199	$29,911	$28,118	$51,434	$109,463
		19	0.191	$30,291	$28,139	$51,495	$109,925

The number of returned items calculated using equation (1) is 548 units. Table 1 shows the optimal solutions of the case. The comparison of the proposed P(1,P) model with its prior P(R,1) model belongs to [17] is shown in Table 2.

Table 2. Comparison of the optimization results of P(R,1) and P(1,P) policy.

Policy	R^* (cycles)	P^* (cycles)	l^* (shipments)	m^* (shipments)	n^* (shipments)	T_r^*	JTP^*
P(R,1)	1	1	2	5.4	4.6	32 days	$121,979
P(1,P)	1	2	3	4.6	7.8	33 days	$122,628

We conclude that (1,P) policy gives higher profit to the system rather than (R,1) policy. Hence, the system should choose only one remanufacturing cycle and at least one manufacturing cycle to gain maximum profit. However, the decision may be different if the values for each parameter are changed.

5. Conclusion and further research

This study has developed an inventory model of a CLSC system that implements (1,P) policy and considers inspection, sorting, waste disposal, imperfect manufacturing and reworks. From the proposed model and solution procedure, the system might maximize its annual joint total profit among three parties, i.e. retailer, manufacturer, and supplier, by determining the optimal value of number of shipments for each party, number of manufacturing cycles and retailer cycle time. From the comparison results to the model of [17], the (1,P) policy is better than (R,1) policy since (1,P) policy gives higher value of JTP than (R,1) policy. For further research, this model may be extended by adding some more considerations such as leadtime, shortage, stochastic demand and returns and environmental effects.

References

[1] Kleindorfer P R, Singhal K and Van Wassenhove L N 2005 Sustainable operations management *Prod. Oper. Manag.* **14** pp 482–492

[2] Chaves G d L D and Martins R S 2005 Diagnóstico da logísticareversanacadeia de suprimentos de alimentosprocessados no oesteparanaense *VIII Simpósio de Administração da Produção, Logísticae OperaçõesInternacionais (SIMPOI)* (São Paulo)

[3] Andrade R, Vieira J M, Lucato W and Vanalle R 2014 Review of the relationship between reverse logistics and competitiveness *Adv. Mat. Res.* **845** pp 614–617

[4] de Brito M P and Dekker R 2004 A framework for reverse logistics *Reverse Logistics: Quantitative Models for Closed-Loop Supply Chains* eds Dekker R, Fleischmann M, Interfurth K and Van Wassenhove L N (Berlin: Springer) pp 3–27
[5] Xia W-H, Jia D-Y and He Y-Y 2011 The remanufacturing reverse logistics management based on closed-loop supply chain management processes *Procedia Environmental Science* **11** pp 351–354
[6] Akçalı E and Çetinkaya S 2011 Quantitative models for inventory and production planning in closed-loop supply chains *Int. J. Prod. Res.* **49** pp 2373–2407
[7] Dwicahyani A R, Jauhari W A and Jonrinaldi 2017 A regular production-remanufacturing inventory model for a two-echelon system with price-dependent return rate and environmental effects investigation *J. Phys. Conf. Ser.* **855**
[8] Kasper A C, Berselli G B T, Freitas B D, Tenório J A S, Bernardes A M and Veit H M 2011 Printed wiring boards for mobile phones: Characterization and recycling of copper *Waste Management* **31** pp 2536–45
[9] Shcrady D A 1967 A deterministic inventory model for repairable items *Nav. Res. Logist. Q.* **14** pp 57–71
[10] Nahmias N and Rivera H 1979 A deterministic model for repairable item inventory system with a finite repair rate *Int. J. Prod. Res.* **17** pp 215–221
[11] Teunter, R H 2004 Lot-sizing for inventory systems with product recovery *Comput. Ind. Eng.* **46** pp 431–441
[12] Haji A, Haji R and Sajadifar S M 2008 Lot sizing with non-zero setup times for rework *J. Syst. Sci. Syst. Eng.* **17** pp 230–240
[13] Chung S L, Wee H M and Yang P C 2008 Optimal policy for a closed-loop supply chain inventory system with remanufacturing *Math. Comput. Model.* **48** pp 867–881
[14] Yuan K F and Gao Y 2010 Inventory decision-making models for a closed-loop supply chain system *Int. J. Prod. Res.* **48** pp 6155–87
[15] Bazan E, Jaber M Y and Zanoni S 2017 Carbon emissions and energy effects on a two-level manufacturer-retailer closed-loop supply chain model with remanufacturing subject to different coordination mechanisms *Int. J. Prod. Econ.* **183** pp 394–408
[16] Jauhari W A, Dwicahyani A R and Kurdhi N A 2017 Lot sizing decisions in a closed-loop supply chain system with remanufacturing *Int. J. Proc. Manag.* **10** pp 381–409
[17] Kholisoh E, Jauhari W A and Rosyidi C N 2016 A closed-loop supply chain model for supplier-manufacturer-retailer system with rework and waste disposal *Proc. 2nd. Int. Conf. Industrial, Mechanical, Electrical, Chemical Engineering* (Yogyakarta: Universitas Sebelas Maret)
[18] El Saadany A M A and Jaber M Y 2010 A production/remanufacturing inventory model with price and quality dependent return rate *Comput. Ind. Eng.* **58** pp 352–362

A periodic review integrated inventory model with controllable setup cost, imperfect items, and inspection errors under service level constraint

R S Saga, W A Jauhari and P W Laksono

Department of Industrial Engineering, Sebelas Maret University, Surakarta, Indonesia

E-mail: rendysuryass@gmail.com

Abstract. This paper presents an integrated inventory model which consists of single vendor and buyer. The buyer managed its inventory periodically and orders products from the vendor to satisfy the end customer's demand, where the annual demand and the ordering cost were in the fuzzy environment. The buyer used a service level constraint instead of the stock-out cost term, so that the stock-out level per cycle was bounded. Then, the vendor produced and delivered products to the buyer. The vendor had a choice to commit an investment to reduce the setup cost. However, the vendor's production process was imperfect, thus the lot delivered contained some defective products. Moreover, the buyer's inspection process was not error-free since the inspector could be mistaken in categorizing the product's quality. The objective was to find the optimum value for the review period, the setup cost, and the number of deliveries in one production cycle which might minimize the joint total cost. Furthermore, the algorithm and numerical example were provided to illustrate the application of the model.

Keywords: fuzzy demand; imperfect inspection; inventory model; periodic review; service level constraint

1. Introduction

Along with the changing of times, many companies compete with each other in satisfying the customer's requirement by producing better products and services in a minimum cost. During the recent years, most inventory problem that focuses on the integration between vendor-buyer has become more important because the vendor and the buyer want to increase their mutual benefit.

The integrated vendor-buyer inventory model was introduced for the first time by Goyal [1]. His research has encouraged other researchers to develop various types of integrated inventory system. Banerjee [2] generalized Goyal [1] model and proposed an integrated model with a joint economic-lot-size where the vendor produces for buyer's order on a lot-by-lot basis. From that time on, there are more researchers developing the integrated inventory system, such as Goyal and Nebebe [3], Ouyang, Wu, and Ho [4], and many more.

Even at this time, the literature about the integrated inventory system with a periodic review policy is rather rare compared to the continuous review policy. However, this periodic review model is considered easier to be applied for many cases. Montgomery et al. [5] studied deterministic and stochastic models under backorders and lost sales. Donaldson [6] developed a periodic review inventory model with normal demand, and obtained the optimal order-up-to level and review period. Ouyang and Chuang [7]) proposed a periodic review inventory with variable lead time. Furthermore, Lin [8]) proposed a periodic review inventory model with fuzzy expected demand short and backorder rate.

In the earlier literature, many researchers have assumed that the process from beginning to the end is error free. Whereas in reality, neither the production process nor inspection process are perfect due to unreliable process or human error. Lin [9] developed periodic review integrated model with imperfect production process and backorder discount. Salamah [10] developed an integrated inventory model with stochastic demand, imperfect quality, and inspection errors. Jauhari et al. [11] developed Lin [9] by considering adjustable production rate and variable lead time. Jauhari [12] also developed integrated inventory model with defective items, inspection error, and stochastic demand.

In real situation, everything including inventory management will face an uncertainty. This can cause difficulties in building an inventory model as it would be more complex. Hence, we can use fuzzy number to resolve this problem. Yao and Wu [13] developed an inventory model with triangular fuzzy number approach. Kurdhi et al. [14] developed an inventory model considering fuzzy demand, transportation cost, setup cost reduction, and partially backorder. Furthermore, Jauhari et al. [15] developed an inventory model with fuzzy demand, imperfect production process, and inspection error.

Researches concerning stochastic integrated inventory model in a single vendor-buyer can be divided into two broad groups, the full cost model, and the service level approach model. Service level approach model is more popular than full cost model. This approach introduces a service level constraint (SLC) instead of the shortage cost term in the model. It is because the determination of the stock-out cost is considered more difficult. Ouyang and Wu [16], Chen and Krass [17] replaced the stock-out cost with SLC in order to prevent unacceptable stock-out. Lin and Lin [18] established an integrated inventory model with inspection process, ordering cost reduction, and SLC. Priyan and Uthayakumar [19] developed an inventory model with variable lead time and SLC in a fuzzy cost environment.

To summarize the above description, an integrated periodic review inventory model has been much developed, but none has considered fuzzy annual demand and ordering cost, setup cost reduction, imperfect items, and inspection error under service level constraint, which is proposed in this study. Therefore, an algorithm and numerical example are provided to establish the model's optimal solution.

2. Notations and assumptions

Notations

T	review period (year) (decision variable)
S	setup cost (\$/setup) (decision variable)
n	number of deliveries (decision variable)
D	demand rate (units/year)
V	transportation cost (\$/shipment)
$I(S)$	the capital investment required to reduce the setup cost per setup
c_p	unit purchasing cost (\$/unit)
S_0	original setup cost (\$/setup)
g	transportation time (year)
A	ordering cost (\$/order)
P	production rate (units/year)
a_1	fixed production cost (\$)
a_2	variable production cost (\$/unit)
c_{dV}	post-sale defective item cost for vendor (\$/unit)
c_{dB}	post-sale defective item cost for buyer (\$/unit)
c_w	warranty cost (\$/unit)
c_i	inspection cost (\$/unit)
h_B	buyer's holding cost (\$/unit/year)
c_r	false rejection of non-defective item cost (\$/unit)
h_V	vendor's holding cost (\$/unit/year)
e_1	inspection error type I probability
e_2	inspection error type II probability
γ	probability of defect
X	the protection interval demand
R	buyer's target level
ξ	proportion of demands which are not met from stock, i.e. $1-\xi$ is the service level

2.1. Assumptions

1. This periodic review inventory model considers one vendor-buyer with a single-product.
2. The demand in buyer side is normally distributed with mean D and standard deviation σ.

3. There is a choice to do an investment in reducing the setup cost. The investment required to reduce the setup cost from initial setup cost S_0 to a target level S is denoted by $I(S)$.
4. To satisfy the end customer demand's, the buyer orders DT units each time and vendor produces nDT items at one set-up with a finite production rate P and delivers each lot in n equally batch sized shipment. The transportation cost V is charged to the vendor for each delivery of DT units.
5. The demand during lead-time follows normal distribution with mean $D(T+L)$ and standard deviation $\sigma\sqrt{T+L}$, where the length of lead-time L is depending on production time DT/P and transportation time g.
6. A production process is imperfect in a way that each time the shipment lot has arrived at the buyer, and it contains some defective products with rate γ. All products will be inspected by the buyer, but the inspection process is also imperfect. The inspector may classify the product's quality incorrectly.
7. Instead of considering stock-out cost, the buyer sets the target service level corresponding to the proportion of demands which are not met from stock. This service level constraint sets the limit of the stock-out level per cycle, where it should not exceed the value of ξ.

3. Model development

The vendor's total cost per year is composed of production cost, setup cost, investment in setup cost reduction, holding cost, cost for rejecting non-defective item (type I error cost), post-sale failure cost (type II error), and warranty cost. The formulation of production cost adopted from Khouja and Mehrez [20] is $\left(\frac{a_1}{P} + a_2 P\right) D$ and the setup cost is $\frac{S}{nT}$.

Furthermore, we consider the logarithmic investment function $I(S)$, as used in Lin [21], which is

$$I(S) = d \ln \frac{S_0}{S}, \qquad for\ 0 < S_0 \leq S \tag{1}$$

where $d = \frac{1}{\delta}$ and δ is a percentage decrease in setup cost S per dollar increase in $I(S)$. Thus, the investment cost per unit per year is $\tau I(S)$, where τ is fractional opportunity cost of capital per unit time. The vendor's holding cost can be expressed as

$$HoldC_V = h_V \frac{DT}{2}\left[\frac{D(2-n)}{P} + (n-1)\right] \tag{2}$$

As explained in the previous section, the model developed in this study considers that the production process and the inspection process are imperfect. Therefore, the vendor bears the type I error cost $c_r(1-\gamma)De_1$ that occurs because the inspector is incorrectly categorizing non-defective products as defective products, and the type II error cost $c_{dV}\gamma De_2$ that occurs because the inspector is incorrectly categorizing defective products as non-defective products, and warranty cost $c_w\gamma D$ that occurs because of the vendor's consequences for producing defective products.

Therefore, the total cost of vendor can be expressed as:

$$TC_V = \left(\frac{a_1}{P} + a_2 P\right) D + \frac{S}{nT} + \frac{\tau}{\delta} ln \frac{S_0}{S} + h_V \frac{DT}{2}\left[\frac{D(2-n)}{P} + (n-1)\right] + c_r(1-\gamma)De_1 + c_{dV}\gamma De_2 + c_w\gamma D \tag{3}$$

The buyer's total cost per year is composed of ordering cost, purchasing cost, transportation cost, inspection cost, cost of type II error, and holding cost where the formulation can be expressed as ordering cost $\frac{A}{T}$, purchasing cost $c_p D$, and transportation cost $\frac{V}{T}$. As mentioned before, an inspection process $c_i D$ will be charged by the buyer for every time the shipment lot arrived at its place where the lot always contains γ defective products. Hence, the buyer also bears the type II error cost $c_{dB}\gamma De_2$.

The formulation of buyer's holding cost developed in this study is adopted from Lin [9] and Hsu and Hsu [22] which consists of holding cost for not only defective and non-defective items found by inspector but also the returned products from end customer. The buyer's holding cost formulation is

$$HoldC_B = h_B\left(\frac{DT}{2} + DT(1-\gamma)e_1 + DT\gamma(1-e_2) + \frac{DT\gamma e_2}{2}\right) \tag{4}$$

We assume that the demand during protection period X follows a normal distribution with the mean $D(T+L)$ and the standard deviation $\sigma\sqrt{T+L}$. The value of lead time L is $\frac{DT}{P}+g$ and the target level is

$R=\left(T+\frac{DT}{P}+g\right)+k\sigma\sqrt{T+\frac{DT}{P}+g}$. Thus, the expected of demand shortage at the end of the cycle is

$$E(X-R)^{+}=\int_{R}^{\infty}(X-R)f_{x}(x)dx=\sigma\psi(k)\sqrt{T+\frac{DT}{P}+g} \quad (5)$$

where $\psi(k)=fs(k)-k[1-Fs(k)]$.

As mentioned earlier, we do not use a stock-out cost term but a service level constraint in the objective function in this study. Thus, the service level constraint can be set as

$$\frac{\sigma\psi(k)\sqrt{T+\frac{DT}{P}+g}}{D\left(T+\frac{DT}{P}+g\right)(1-\gamma)(1-e_1)}\leq\xi \quad (6)$$

Hence, the total cost of buyer can be expressed by the equation below:

$$TC_B=\frac{A}{T}+\frac{F}{T}+c_pD+c_iD+c_{dB}\gamma De_2+h_B\left(\frac{DT}{2}+DT(1-\gamma)e_1+DT\gamma(1-e_2)+\frac{DT\gamma e_2}{2}\right)$$

subject to $\frac{\sigma\psi(k)\sqrt{T+\frac{DT}{P}+g}}{D\left(T+\frac{DT}{P}+g\right)(1-\gamma)(1-e_1)}\leq\xi$ (7)

Joint total cost for this system is established by adding by the vendor's total cost and buyer's total cost. While D and A is fuzzified to be $\tilde{D}$ and $\tilde{A}$, the joint total cost for the system is expressed by

$$\widetilde{JTC}=\left(\frac{a_1}{P}+a_2P\right)\tilde{D}+\frac{S}{nT}+\frac{\tau}{\delta}ln\frac{S_0}{S}+h_V\frac{\tilde{D}T}{2}\left[\frac{\tilde{D}(2-n)}{P}+(n-1)\right]+c_r(1-\gamma)\tilde{D}e_1+c_{dV}\gamma\tilde{D}e_2+c_w\gamma\tilde{D}+\frac{\tilde{A}}{T}+\frac{V}{T}+c_p\tilde{D}+c_i\tilde{D}+c_{dB}\gamma\tilde{D}e_2+h_B\left(\frac{\tilde{D}T}{2}+\tilde{D}T(1-\gamma)e_1+\tilde{D}T\gamma(1-e_2)+\frac{\tilde{D}T\gamma e_2}{2}\right) \quad (8)$$

We defuzzify *JTC* by using the signed distance method. The signed distance of *JTC* is expressed by

$$d(\widetilde{JTC},\tilde{0})=\left(\frac{a_1}{P}+a_2P\right)(\tilde{D},\tilde{0})+\frac{S}{nT}+\frac{\tau}{\delta}ln\frac{S_0}{S}+h_V\frac{(\tilde{D},\tilde{0})T}{2}\left[\frac{(\tilde{D},\tilde{0})(2-n)}{P}+(n-1)\right]+c_r(1-\gamma)(\tilde{D},\tilde{0})e_1+c_{dV}\gamma(\tilde{D},\tilde{0})e_2+c_w\gamma(\tilde{D},\tilde{0})+\frac{\tilde{A}}{T}+\frac{V}{T}+c_p(\tilde{D},\tilde{0})+c_i(\tilde{D},\tilde{0})+c_{dB}\gamma(\tilde{D},\tilde{0})e_2+h_B\left(\frac{(\tilde{D},\tilde{0})T}{2}+(\tilde{D},\tilde{0})T(1-\gamma)e_1+(\tilde{D},\tilde{0})T\gamma(1-e_2)+\frac{(\tilde{D},\tilde{0})T\gamma e_2}{2}\right) \quad (9)$$

As adopted from Jauhari et al. [15], for any a and $0\in R$, the signed distance between the real numbers a and 0 is $d_0(a,0)=a$. The signed distance of $\tilde{A}$ to $\tilde{0}$ is

$$d(\tilde{A},\tilde{0})=\frac{1}{4}(a+2b+c)=\frac{1}{4}[(A-p_1)+2A+(A+p_2)]=A+\frac{1}{4}(p_2-p_1) \quad (10)$$

The joint total cost for the integrated vendor and buyer system can be rewritten as

$$d(\widetilde{JTC},\tilde{0})=\left(\frac{a_1}{P}+a_2P\right)\left(D+\frac{1}{4}(q_2-q_1)\right)+\frac{S}{nT}+\frac{\tau}{\delta}ln\frac{S_0}{S}+h_V\frac{\left(D+\frac{1}{4}(q_2-q_1)\right)T}{2}\left[\frac{\left(D+\frac{1}{4}(q_2-q_1)\right)(2-n)}{P}+(n-1)\right]+c_r(1-\gamma)\left(D+\frac{1}{4}(q_2-q_1)\right)e_1+c_{dV}\gamma(D+\frac{1}{4}(q_2-q_1))\,e_2+c_w\gamma\Big(D+$$

$$\frac{1}{4}(q_2 - q_1)\Big) + \frac{\left(A+\frac{1}{4}(p_2-p_1)\right)}{T} + \frac{V}{T} + c_p\left(D + \frac{1}{4}(q_2 - q_1)\right) + c_i\left(D + \frac{1}{4}(q_2 - q_1)\right) +$$
$$c_{dB}\gamma\left(D + \frac{1}{4}(q_2 - q_1)\right)e_2 + h_B\left(\frac{\left(D+\frac{1}{4}(q_2-q_1)\right)T}{2} + \left(D + \frac{1}{4}(q_2 - q_1)\right)T(1-\gamma)e_1 +\right.$$
$$\left.\left(D + \frac{1}{4}(q_2 - q_1)\right)T\gamma(1-e_2) + \frac{\left(D+\frac{1}{4}(q_2-q_1)\right)T\gamma e_2}{2}\right)$$

subject to
$$\frac{\sigma\psi(k)\sqrt{T+\frac{\left(D+\frac{1}{4}(q_2-q_1)\right)T}{P}+g}}{\left(D+\frac{1}{4}(q_2-q_1)\right)\left(T+\frac{\left(D+\frac{1}{4}(q_2-q_1)\right)T}{P}+g\right)(1-\gamma)(1-e_1)} \leq \xi \tag{11}$$

4. Solution methodology

To find the optimal solution of problem (11), we temporarily ignore the service level constraint and consider a fixed value of n, then find the first-order partial derivatives of $JTC(T, S, n)$ with respect to T and S. Next, by setting the equations to zero, rearranging and simplifying we obtain

$$T^* = \sqrt{\frac{2P\left(\left(A+\frac{1}{4}(p_2-p_1)\right)n+Vn+S\right)}{\left(D+\frac{1}{4}(q_2-q_1)\right)n\left(\left(D+\frac{1}{4}(q_2-q_1)\right)h_V(2-n)+P\left(h_Vn-h_V+h_B(2e_1(1-\gamma)+(2-e_2)\gamma+1)\right)\right)}} \tag{12}$$

and

$$S^* = \frac{nT\tau}{\delta} \tag{13}$$

Therefore, we can develop an iterative procedure to solve the proposed problem and obtain the optimal value of all decision variables. The proposed algorithm is as follows

1. Fuzzify D and A to be a triangular fuzzy number, $\widetilde{D} = (D\text{-}q_1, D, D+q_2)$ and $\tilde{A} = (A\text{-}p_1, A, A+p_2)$, where p_i and q_i, for $i = 1,2$ are decided by decision makers.
2. Use the signed distance method to defuzzify the fuzzy number $\widetilde{D}$ and $\tilde{A}$.
3. Set $n = 1$ and JTC($T_{n\text{-}1}$, $S_{n\text{-}1}$, n-1) = ∞.
4. Perform (i)-(v)
   (i) Start with $S = S_0$.
   (ii) Substituting the value of S into Equation (12) evaluates T.
   (iii) Using T in Equation (13) and determines S.
   (iv) Repeat (ii) and (iii) until no change occurs in the values of T and S.
5. Compare S and S_0
   (i) If $S < S_0$, S is feasible, denote the solution by (T^* and S^*) then go to Step 6.
   (ii) If $S > S_0$, S is not feasible. Set $S = S_0$ and calculate the corresponding value of T from Equation (12), then go to Step 6.
6. Compute JTC(T_n^*, S_n^*, n^*) from Equation (11).
7. For a given safety factor k, check the Equation (6), if it satisfied, the solutions satisfy the service level constraint, and the solution JTC(T_n^*, S_n^*, n^*) is the optimal solution for fixed n. Otherwise, find another solution until it satisfies the service level constraint. If all solutions do not satisfy the service level constraint, this inventory model has no feasible solution.
8. If JTC(T_n^*, S_n^*, n^*) ≤ JTC($T_{n\text{-}1}^*$, $S_{n\text{-}1}^*$, $n\text{-}1^*$) and the service level constraint is satisfied, repeat Steps 3-7 with $n = n + 1$, otherwise go to Step 9.

9. Set JTC(T_n^*, S_n^*, n^*) = JTC(T_{n-1}^*, S_{n-1}^*, $n\text{-}1^*$), then T_n^*, S_n^*, and n^* are the solutions of the model.

5. Numerical example

The numerical example given below is for illustrating the feasibility of the above solution procedure. Let us consider an integrated inventory model with the following data: D = 1,000 units/year, q_1 = 100, q_2 = 50, A = \$100/order, p_1 = 20, p_2 = 50, S_0 = \$1,500/setup, V = \$25/shipment, c_p = \$65/unit, g = 1/12 year, P= 2500 units/year, a_1= \$2,000/batch, a_2=\$1/2,000/unit, c_{dB}= \$200/unit, c_{dV}= \$270/unit, c_i= \$1/unit, c_r= \$100/unit, c_w= \$100/unit, h_B= \$8/unit/year, h_V=\$6/unit/year,e_1 = 0.04, e_2 = 0.04, the service level 1-ξ = 99.8% (i.e. ξ = 0.02), γ = 0.005,σ = 10 units/week, τ = 0.25 per dollar per year, δ = 0.01%, we take an assumptions that one year = 52 weeks. Using the solution procedure explained above, the optimal value for each variable can be obtained.

Therefore, the optimal value of n^*, T^*, and S^*, satisfying the proposed solution procedure are 5 deliveries, 0.1850 years, and \$1,500/setup which can minimize the value of JTC at \$174,650.

6. Conclusion

This paper proposed an integrated periodic review inventory model involving a vendor and a buyer under fuzzy annual demand and ordering cost, setup cost reduction, imperfect items, inspection error, and service level constraint. The review period T, the setup cost S, and the number of deliveries n were considered as decision variables. An algorithm was developed to determine the model's optimal solutions. The numerical example's results showed us that the higher the service level, the higher joint total cost. This paper can help managers to decide whether investing in reducing the setup cost in this inventory model is necessary or not.

References

[1] Goyal S K 1976 An integrated inventory model for a single supplier-single customer problem *Int. J. Prod. Res.* **15** pp 107–111
[2] Banerjee A 1986 A joint economic-lot-size model for purchaser and vendor *Decision Sci.* **17** pp 292–311
[3] Goyal S K and Nebebe F 2000 Determination of economic production-shipment policy for a single-vendor-single buyer system *Eur. J. Oper. Res.* **121** pp 175–178
[4] Ouyang L Y, Wu K S and Ho C H 2004 Integrated vendor buyer cooperative models with stochastic demand in controllable lead time *Int. J. Prod. Econ.* **92** pp 255–266
[5] Montgomery D C, Bazaraa M S and Keswani A K 1973 Inventory models with a mixture of backorders and lost sales *Nav. Res. Log.* **20** pp 255–263
[6] Donaldson W A 1984 An equation for the optimal value of T, the inventory replenishment review period when demand is normal *J. Opl. Res. Soc.* **35** pp 137–139
[7] Ouyang L Y and Chuang B R 2001 A periodic review inventory-control system with variable lead time *Information and Management Sciences* **12** pp 1–13
[8] Lin Y J 2008 A periodic review inventory model involving fuzzy expected demand short and fuzzy backorder rate *Comput. Ind. Eng.* **54** pp 666-676
[9] Lin Y J 2010 A stochastic periodic review integrated inventory model involving defective items, backorder price discount, and variable lead time *4OR–Q. J. Oper. Res.* **8** pp 281–297
[10] Al-Salamah M 2012 Economic order quantity with stochastic demand, imperfect quality, and inspection errors *J. Engineering and Computer Sciences* **5** pp 131–145
[11] Jauhari W A, Sejati N P and Rosyidi C N 2016 A collaborative supply chain inventory model with defective items, adjusted production rate and variable lead time *Int. J. Proc. Manag.* **9** pp 733–750
[12] Jauhari W A 2016 Integrated vendor-buyer model with defective items, inspection error and stochastic demand *Int. J. Mathematics in Operational Research* **8** pp 342–359
[13] Yao J S and Wu K 2000 Rangking fuzzy numbers based on decomposition principle and signed

distance *Fuzzy Set. Syst.* **116** pp 275–288

[14] Khurdi N A, Lestari S M P and Susanti Y 2015 A fuzzy collaborative supply chain inventory model with controllable setup cost and service level constraint for imperfect items *Int. J. of Applied Management Science* **7** pp 93–122

[15] Jauhari W A, Mayangsari S, Kurdhi N A and Wong K Y 2017 A fuzzy periodic review integrated inventory model involving stochastic demand, imperfect production process and inspection errors *Cogent Engineering* **4** pp 1–24

[16] Ouyang L Y and Wu K S 1997 Mixture inventory model involving variable lead time with a service level constraint *Comput. Oper. Res.* **24** pp 875–882

[17] Chen F Y and Krass D 2001 Inventory models with minimal service level contraints *Eur. J. Oper. Res.* **134** pp 120–140

[18] Lin Y J and Lin H J 2016 Optimal ordering and recovery policy in a periodic review integrated inventory model *Int. J. Syst. Sci.* **3** pp 200–210

[19] Priyan S and Uthayakumar R 2016 Economic design of multi-echelon inventory system with variable lead time and service level constraint in a fuzzy cost environment *Fuzzy Information and Engineering* **8** pp 465–511

[20] Khouja M and Mehrez A 1994 Economic prodction lot size model with variable production rate and imperfect quality *The J. of the Operational Research Society* **45** pp 1405-17

[21] Lin H J 2012 Effective investment to reduce setup cost in a mixture inventory model involving controllable backorder rate and variable lead time with a service level constraint *Math. Probl. Eng.* **2012** pp 1–15

[22] Hsu J T and Hsu L F 2012 An integrated single-vendor single buyer production-inventory model for items with imperfect quality and inspection errors *Int. J .of Industrial Engineering Computations* **3** pp 703–720

A joint economic lot-sizing problem with fuzzy demand, defective items and environmental impacts

W A Jauhari[1] and P W Laksono[2]

[1,2] Department of Industrial Engineering, Sebelas Maret University, Surakarta, Indonesia

E-mail: wachid_aj@yahoo.com

Abstract. In this paper, a joint economic lot-sizing problem consisting of a vendor and a buyer was proposed. A buyer ordered products from a vendor to fulfill end customer's demand. A produced a batch of products, and delivered it to the buyer. The production process in the vendor was imperfect and produced a number of defective products. Production rate was assumed to be adjustable to control the output of vendor's production. A continuous review policy was adopted by the buyer to manage his inventory level. In addition, an average annual demand was considered to be fuzzy rather than constant. The proposed model contributed to the current inventory literature by allowing the inclusion of fuzzy annual demand, imperfect production emission cost, and adjustable production rate. The proposed model also considered carbon emission cost which was resulted from the transportation activity. A mathematical model was developed for obtaining the optimal ordering quantity, safety factor and the number of deliveries so the joint total cost was minimized. Furthermore, an iterative procedure was suggested to determine the optimal solutions.

Keywords: Inventory; continuous review; fuzzy demand; adjustable production rate; joint total cost.

1. Introduction

In the last few decades, new technologies have grown exponentially and are changing the world trade. The globalization of market and increased competition will definitely shape the future of supply chain logistics. Integrating vendor–buyer inventory system is critical to lower the joint total cost of supply chain and gain competitive advantages [1]. In inventory literature, the problem concerning with production and inventory decisions is known as joint economic lot-sizing problem (JELP). Goyal [2] was the first researcher who introduced JELP involving a vendor and a buyer under deterministic demand. Banerjee [3] relaxed the assumption of infinite production rate used by Goyal and proposed that the vendor has a finite production. Furthermore, the model of Goyal [2] was then extended into various cases, i.e. discount [4], defective items [5], inspection errors [6] and learning [7].

Previous research on JELP focused on the production shipment schedule in terms of the number and quantity of batches delivered from vendor to buyer. However, most of these models assumed that the demand is deterministic rather than stochastic. In reality, the demand changes over time assuming it to be deterministic was too restrictive assumption. Ben-Daya and Hariga [8] proposed a JELP consisting of a vendor and a buyer with stochastic demand and variable lead time. Hsiao [9] proposed two kinds of the reorder point in a continuous review JELP under stochastic demand. Glock [10] analyzed the lead time reduction strategies where the transportation time can be reduced using a crashing cost. Jauhari and Winingsih [11] developed JELP involving of a vendor and a buyer with freight and quantity discount, and considered a quantity discount.

In most of the earlier literature dealing with stochastic JELP, researchers have assumed that the production process was100% defect-free. In reality, the production process is imperfect due to an unreliable process. Lin [12] developed periodic review integrated vendor-buyer model with imperfect production and backorder price discount. Jauhari et al. [13] then developed the model of Lin [12] by proposing adjustable production rate and variable lead time. Furthermore, Lin and Lin [14] studied the impact of vendor's recovery process and the investment in reducing ordering cost in a JELP.

One major drawback of published stochastic JELP was that most of the authors assumed that the annual demand was treated as a constant parameter rather than fuzzy. In reality, it is usually difficult for managers to set demand as crisp values due to lack of data, e.g., new product introduction and single period setting. Thus, fuzzy set theory is reasonable to be used to address inventory problems. In addition, in most of published stochastic JELP, researchers have neglected environmental issues, i.e. carbon emission and energy, in developing the inventory models. Environmental issues are becoming more of a concern to supply chain stakeholders, and investigating them in the modeling and design of supply chains is necessary [15].

Based on the above description, a stochastic JELP has been discussed in the past, but none has considered the inclusion of fuzzy demand, carbon emission cost and imperfect production. Therefore, we propose a JELP consisting of a vendor and a buyer under situation in which the annual demand is treated as fuzzy, the production rate is adjustable and the production process is imperfect. In addition, the carbon emission resulted from transportation activity is also considered in the model.

2. Problem statement

Let us consider a supply system consisting of a vendor and a buyer. The buyer faces demand from end customers which is assumed to follow normal distribution with demand rate D and standard deviation σ. A continuous review policy is adopted by the buyer to manage his inventory level. The buyer orders nQ units to the vendor to satisfy the end customer's demand. An ordering cost A is charged to the buyer for each order. The demand that cannot be satisfied by the buyer is assumed to be fully backordered. The vendor produces a batch of product nQ with a finite production rate P ($P>D$). The vendor incurs a setup cost K for each production run. The vendor then delivers a batch of product over n times in each production run. The demand during lead time of the first shipment is assumed to follow normal distribution with mean $DL(Q)$ and standard deviation $\sigma\sqrt{L(Q)}$.

A production process is not 100% perfect, so it will produce a certain number of defective products with a known probability. Consequently, the lot received by the buyer may contain defective products. Thus, the buyer will inspect the incoming lot to categorize the quality of the product. A screening cost s is charged to the buyer for each unit product. After finishing the screening process, the defective items found by the inspector will be returned to the vendor. For each item returned by the buyer, the vendor will incur warranty cost w. In addition, the production rate is adjustable between P_{min} and P_{max}.

The objective function of the model is to minimize the joint total cost for the supply chain system and determine the decision variables, which are the size of equal shipments from the vendor to the buyer (Q), safety factor of first shipment (k_1) and the number of deliveries (n). Whereas the parameters used in the proposed model are denoted by these following notations:

D	:	demand rate in units per unit time;	S_b	fixed carbon emission cost for buyer;
σ	:	standard deviation of demand per unit time;	S_v	fixed carbon emission cost for vendor;
A	:	ordering cost per order;	z_b	buyer's variable carbon emission cost;
F	:	transportation cost per shipment;	z_v	vendor's variable carbon emission cost;
h_b	:	holding cost per unit per unit time for	a_1	fixed production cost;

		buyer;		
h_v	:	holding cost per unit per unit time for vendor;	a_2	variable production cost;
π	:	backorder cost per unit backordered;	P	production rate in units per unit time;
Y	:	defect rate;	K	setup cost;
w	:	warranty cost per unit item;	$L(Q)$	lead time as a function of shipment size;
s	:	screening cost per unit item;	T_w	setup and transportation time;
x	:	screening rate;	T_s	transportation time;

3. Model development

Suppose that the demand during lead time of the first shipment is normally distributed with mean $DL(Q)$ and standard deviation $\sigma\sqrt{L(Q)}$. As we assumed in previous section that the incoming lot contains some defective items with a percentage Y and has the probability density function $f(y)$. Then, the average inventory level of the buyer is can be formulated as $E\left(\frac{Y}{1-Y}\right)\frac{QD}{x}+\frac{[1-E(Y)]Q}{2}$ and the safety stock is $k_1\sigma\sqrt{Q/_P+T_w}$. The expected holding cost for the buyer per year is given by

$$h_b\left(E\left(\frac{Y}{1-Y}\right)\frac{QD}{x}+\frac{[1-E(Y)]Q}{2}+k_1\sigma\sqrt{Q/_P+T_w}\right) \tag{1}$$

The expected of the number of backorders per year can be calculated by considering the number of cycles per year$E\left(\frac{Y}{1-Y}\right)\frac{D\sigma}{nQ}$, number of backorder for first shipment $\sqrt{Q/_P+T_w}\,\psi(k_1)$ and the number of backorder of next shipments $(n-1)\sqrt{T_s}\,\psi(k_2)$, which is

$$\left(E\left(\frac{1}{1-Y}\right)\frac{\pi D\sigma}{nQ}\right)\left(\sqrt{Q/_P+T_w}\,\psi(k_1)+(n-1)\sqrt{T_s}\,\psi(k_2)\right) \tag{2}$$

Here, we also consider the carbon emission cost resulted from transportation activity. The expected fixed carbon emission cost for buyer per year is $\mathrm{E}\left(\frac{1}{1-\mathrm{Y}}\right)\frac{\mathrm{DS_b}}{\mathrm{Q}}$and the expected variable carbon emission cost for buyer is $\mathrm{E}\left(\frac{1}{1-\mathrm{Y}}\right)\mathrm{z_b D}$. Thus, the expected carbon emission cost incurred by the buyer is given by

$$E_b=E\left(\frac{1}{1-Y}\right)\frac{DS_b}{Q}+E\left(\frac{Y}{1-Y}\right)z_bD \tag{3}$$

Thus, the expected total cost for the buyer composing of ordering cost, transportation cost, inspection cost, carbon emission cost, holding cost and backorder cost, is given by

$$\begin{aligned}ETC_B=&E\left(\frac{1}{1-Y}\right)\frac{D}{nQ}(A+nF)+E\left(\frac{1}{1-Y}\right)sD+E\left(\frac{1}{1-Y}\right)\frac{DS_b}{Q}+E\left(\frac{Y}{1-Y}\right)z_bD\\&h_b\left(E\left(\frac{Y}{1-Y}\right)\frac{QD}{x}+\frac{[1-E(Y)]Q}{2}+k_1\sigma\sqrt{Q/_P+T_w}\right)\\&\left(E\left(\frac{1}{1-Y}\right)\frac{\pi D\sigma}{nQ}\right)\left(\sqrt{Q/_P+T_w}\,\psi(k_1)+(n-1)\sqrt{T_s}\,\psi(k_2)\right)\end{aligned} \tag{4}$$

The expected vendor's inventory level can be calculated by subtracting the accumulated buyer's consumption from accumulated vendor's production. The expected vendor's holding cost per year is expressed by

$$h_v\frac{Q}{2}\left(n\left[1-E\left(\frac{1}{1-Y}\right)\frac{D}{P}\right]-1+E\left(\frac{1}{1-Y}\right)\frac{2D}{P}\right) \tag{5}$$

The expected carbon emission cost resulted from delivering defective items from buyer to vendor is given by

$$E_v=E\left(\frac{1}{1-Y}\right)\frac{DS_v}{Q}+E\left(\frac{Y}{1-Y}\right)z_vD \tag{6}$$

The expected vendor's cost composing of setup cost, warranty cost, carbon emission cost, holding cost and production cost, is given by the following expression

$$ETC_v = E\left(\frac{1}{1-Y}\right)\frac{DK}{Q} + E\left(\frac{Y}{1-Y}\right)wD + E\left(\frac{Y}{1-Y}\right)z_v D + E\left(\frac{1}{1-Y}\right)\frac{DS_v}{Q}$$
$$h_v\frac{Q}{2}\left(n\left[1 - E\left(\frac{1}{1-Y}\right)\frac{D}{P}\right] - 1 + E\left(\frac{1}{1-Y}\right)\frac{2D}{P}\right) + E\left(\frac{1}{1-Y}\right)\left(\frac{a_1}{P} + a_2 P\right)D \quad (7)$$

By using equation $M = E\left(\frac{1}{1-Y}\right)$ and D is fuzzified to be $\widetilde{D}$,the expected total cost for supply chain is given by

$$\widetilde{EJTC} = \frac{M\widetilde{D}}{nQ}(A + nF) + Ms\widetilde{D} + \frac{M\widetilde{D}S_b}{Q} + Mz_b\widetilde{D} - z_b\widetilde{D}$$
$$+ h_b\left(\frac{MQ\widetilde{D}}{x} - \frac{Q\widetilde{D}}{x} + \frac{[1 - E(Y)]Q}{2} + k_1\sigma\sqrt{Q/_P + T_w}\right)$$

$$+\left(\frac{M\pi\widetilde{D}\sigma}{nQ}\right)\left(\sqrt{Q/_P + T_w}\,\psi(k_1) + (n-1)\sqrt{T_s}\,\psi(k_2)\right) + \frac{M\widetilde{D}K}{Q} + Mw\widetilde{D} - w\widetilde{D} + Mz_v\widetilde{D} - z_v\widetilde{D}$$
$$+\frac{M\widetilde{D}S_v}{Q} + h_v\frac{Q}{2}\left(n\left[1 - \frac{M\widetilde{D}}{P}\right] - 1 + \frac{2M\widetilde{D}}{P}\right) + \left(\frac{a_1}{P} + a_2 P\right)M\widetilde{D} \quad (8)$$

Next, we defuzzify $\widetilde{EJTC}$ by using the signed distance method. The signed distance of $\widetilde{EJTC}$ is given by

$$d(\widetilde{EJTC},\tilde{0}) = \frac{M(\widetilde{D},\tilde{0})}{nQ}(A + nF) + Ms(\widetilde{D},\tilde{0}) + \frac{M(\widetilde{D},\tilde{0})S_b}{Q} + Mz_b(\widetilde{D},\tilde{0}) - z_b(\widetilde{D},\tilde{0}) + Mz_v(\widetilde{D},\tilde{0}) - z_v(\widetilde{D},\tilde{0})$$
$$+ h_b\left(\frac{MQ(\widetilde{D},\tilde{0})}{x} - \frac{Q(\widetilde{D},\tilde{0})}{x} + \frac{[1 - E(Y)]Q}{2} + k_1\sigma\sqrt{Q/_P + T_w}\right)$$
$$+\left(\frac{M\pi(\widetilde{D},\tilde{0})\sigma}{nQ}\right)\left(\sqrt{Q/_P + T_w}\,\psi(k_1) + (n-1)\sqrt{T_s}\,\psi(k_2)\right) + \frac{M(\widetilde{D},\tilde{0})K}{Q} + Mw(\widetilde{D},\tilde{0})$$
$$- w(\widetilde{D},\tilde{0})$$
$$+\frac{M(\widetilde{D},\tilde{0})S_v}{Q} + h_v\frac{Q}{2}\left(n\left[1 - \frac{M(\widetilde{D},\tilde{0})}{P}\right] - 1 + \frac{2M(\widetilde{D},\tilde{0})}{P}\right) + \left(\frac{a_1}{P} + a_2 P\right)M(\widetilde{D},\tilde{0}) \quad (9)$$

Let Ψ be the family of all fuzzy sets $\widetilde{D}$ defined in$R$ for which the α-cut$D(\alpha) = [D_L(\alpha), D_U(\alpha)]$ exists for every $\alpha \in [0,1]$. Both $D_L(\alpha)$ and $D_U(\alpha)$ are continuous functions in $\alpha \in [0,1]$. Then it can be said that for any $\widetilde{D} \in \Psi$, we have $\widetilde{D} = \cup_{0\le\alpha\le1}[D_L(\alpha)_\alpha, D_U(\alpha)_\alpha]$. Hence, for $\widetilde{D} \in \Psi$ the signed distance from $\widetilde{D}$ to $\tilde{0}$ can be defined as:

$$d(\widetilde{D},\tilde{0}) = \frac{1}{2}\int_0^1 (D_L(\alpha) + D_U(\alpha))\,d\alpha.$$

For a triangular fuzzy number $\widetilde{D} = (a,b,c)$, the α-cut of $\widetilde{D}$ is $D(\alpha) = [D_L(\alpha), D_U(\alpha)]$ for $\alpha \in [0,1]$, where $D_L(\alpha) = a + (b-a)\alpha$ and $D_U(\alpha) = c - (c-b)\alpha$. The signed distance from $\widetilde{D}$ to $\tilde{0}$ is:

$$d(\widetilde{D},\tilde{0}) = \frac{1}{4}(a + 2b + c).$$
$$= \frac{1}{4}[(D\text{-}z_1) + 2D + (D + z_2)] = D + \frac{1}{4}(z_2 - z_1) \quad (10)$$

Therefore, the expected total cost for supply chain is can be rewritten as follows

$$d(\widetilde{EJTC},\tilde{0}) = \frac{M\left(D + \frac{1}{4}[z_2 - z_1]\right)}{nQ}(A + nF) + Ms\left(D + \frac{1}{4}[z_2 - z_1]\right) + \frac{M\left(D + \frac{1}{4}[z_2 - z_1]\right)S_b}{Q}$$
$$+ Mz_b\left(D + \frac{1}{4}[z_2 - z_1]\right) - z_b\left(D + \frac{1}{4}[z_2 - z_1]\right) + Mz_v\left(D + \frac{1}{4}[z_2 - z_1]\right)$$
$$- z_v\left(D + \frac{1}{4}[z_2 - z_1]\right) + \left(\frac{a_1}{P} + a_2 P\right)M\left(D + \frac{1}{4}[z_2 - z_1]\right) + \frac{M\left(D + \frac{1}{4}[z_2 - z_1]\right)K}{Q}$$
$$+ h_b\left(\frac{MQ\left(D + \frac{1}{4}[z_2 - z_1]\right)}{x} - \frac{Q\left(D + \frac{1}{4}[z_2 - z_1]\right)}{x} + \frac{[1 - E(Y)]Q}{2} + k_1\sigma\sqrt{Q/_P + T_w}\right)$$

$$+\left(\frac{M\pi\left(D+\frac{1}{4}[z_2-z_1]\right)\sigma}{nQ}\right)\left(\sqrt{Q/_P+T_w}\,\psi(k_1)+(n-1)\sqrt{T_s}\,\psi(k_2)\right)$$
$$+\frac{M\left(D+\frac{1}{4}[z_2-z_1]\right)S_v}{Q}+h_v\frac{Q}{2}\left(n\left[1-\frac{M\left(D+\frac{1}{4}[z_2-z_1]\right)}{P}\right]-1+\frac{2M\left(D+\frac{1}{4}[z_2-z_1]\right)}{P}\right)$$
$$+Mw\left(D+\frac{1}{4}[z_2-z_1]\right)-w\left(D+\frac{1}{4}[z_2-z_1]\right) \quad (11)$$

4. Solution methodology

By considering a fixed value of n, the minimum value of the above expected total cost for supply chain occurs at the point (n,Q,k_1,P) which satisfies $\frac{\partial EJTC(n,Q,k_1,P)}{\partial k_1}=0$, $\frac{\partial EJTC(n,Q,k_1,P)}{\partial Q}=0$ and $\frac{\partial EJTC(n,Q,k_1,P)}{\partial P}=0$ simultaneously. To find the solution of the problem, we first find the first partial derivative of $EJTC(n,Q,k_1,P)$ with respect to and k_1, Q and P, respectively and set the equations to zero, so we will obtain the following expressions:

$$1-F_s(k_1)+(n-1)\left[1-F_s\left(k_1\sqrt{\frac{Q/_P+T_w}{T_s}}\right)\right]=\frac{h_b nQ}{M\pi\left(D+\frac{1}{4}[z_2-z_1]\right)} \quad (12)$$

$$Q=\sqrt{\frac{2M\left(D+\frac{1}{4}[z_2-z_1]\right)\left\{\left[\frac{A+K}{n}+F\right]+S_b+S_v+\frac{\pi\sigma\left(\sqrt{Q/_P+T_w}\,\psi(k_1)+(n-1)\sqrt{T_s}\,\psi(k_2)\right)}{n}\right\}}{h_v\left\{n\left(1-\frac{M\left(D+\frac{1}{4}[z_2-z_1]\right)}{P}\right)-1+\frac{2M\left(D+\frac{1}{4}[z_2-z_1]\right)}{P}\right\}+h_b[1-E(Y)]+\frac{2h_bM\left(D+\frac{1}{4}[z_2-z_1]\right)}{x}-\frac{2h_b\left(D+\frac{1}{4}[z_2-z_1]\right)}{x}+\frac{h_bk_1\sigma}{P\sqrt{Q/_P+T_w}}+\frac{M\left(D+\frac{1}{4}[z_2-z_1]\right)\pi\sigma\,\psi(k_1)}{nPQ\sqrt{Q/_P+T_w}}}} \quad (13)$$

$$P=\sqrt{\frac{a_1M\left(D+\frac{1}{4}[z_2-z_1]\right)+MQh_v\left(D+\frac{1}{4}[z_2-z_1]\right)-\frac{MQh_vn\left(D+\frac{1}{4}[z_2-z_1]\right)}{2}+\frac{h_bk_1\sigma}{2\sqrt{Q/_P+T_w}}+\frac{\frac{M\left(D+\frac{1}{4}[z_2-z_1]\right)\sigma}{n}\psi(k_1)}{2\sqrt{Q/_P+T_w}}}{a_2M\left(D+\frac{1}{4}[z_2-z_1]\right)}} \quad (14)$$

Here, we propose an iterative procedure to obtain the optimal decision variables, which is

1. Fuzzify D to be a triangular fuzzy number, $\widetilde{D}=(D-z_1, D, D+z_2)$, where z_i, for i =1,2are decided by the decision makers. Then, defuzzify the fuzzy number $\widetilde{D}$ by using the signed distance method
2. Set n=1 and $EJTC(n-1,Q_{n-1},k_{1n-1},P_{n-1})=\infty$
3. Set the initial values of Q and P with equations (13) and (14) by setting the stochastic variables equal to zero
4. Compute k_1 by substituting Q into equation (12).
5. Compute P from equation (14).
6. For a given previous value of k_1 and P, compute Q from equation (13).
7. Compute k_1 by substituting Q into equation (12).
8. Repeat steps 3 - 6 until no change occurs in the values of Q, k_1 and P.
9. Set $Q_n= Q$, $k_{1n}=k_1$ and $P_n= P$ and compute $EJTC(n,Q_n,k_{1n},P_n)$ from equation (11).
10. If $EJTC(n,Q_n,k_{1n},P_n)\leq EJTC(n-1,Q_{n-1},k_{1n-1},P_{n-1})$ repeat steps 2-8 with $n=n+1$, otherwise go to step 11.

11. Compute $EJTC(n^*,Q^*,k_1^*,P^*)= EJTC(n-1,Q_{n-1},k_{1n-1},P_{n-1})$, then n^*,Q^*,k_1^*,P^* are the optimal solution.

5. Numerical example

Let us consider a supply chain inventory model consisting of a vendor and a buyer with the following data: D=1,000 units/year, z_1=100, z_2=50, σ=25 units/year, A=\$100/order, F=\$50/shipment, K=\$400/setup, h_v=\$1/unit/year, h_b=\$3/unit/year, π=\$50/unit, P_{min}=1,500 units/year, x=7,000 units/year, P_{max}=5,000 units/year, a_1=\$2,500, a_2=\$0.0004/unit, S_b=\$3/shipment, S_v=\$4/shipment, z_b=\$1/unit, z_v=\$1.25/unit, w=\$2/unit, s=\$0.5/unit, T_w=0.05, T_s=0.04. The probability density function for the percentage of defectives is taken to be

$$f(y) = \begin{cases} 25, & 0 \le y \le 0.04 \\ 0, & otherwise \end{cases}$$

Therefore, we will have

$$E(Y) = \int_0^{0.04} 25y dy = 0.02 \quad \text{and} \quad E\left(\frac{1}{1-Y}\right) = \int_0^{0.04} \frac{25}{1-y} dy = 1.02055.$$

The optimal values of n^*, k^*, Q^*, and P^*, satisfying the proposed solution procedure, are 7 deliveries, 2,0659, 191 units and 2,296 units which can minimize the value of JTC at \$4,944.6.

6. Conclusions

In this paper, an integrated inventory model for single-vendor single-buyer under fuzzy annual demand and imperfect items was proposed. The demand was assumed to be normally distributed and the lead time was taken to be variable and dependent upon ordering size and production rate. A simple algorithm was suggested to obtain the model's solutions. The proposed model can help the practitioners to keep the vendor-buyer inventories more efficiently. It can assist managers to decide the optimum ordering quantity, safety factor, number of deliveries and level of production rate. Furthermore, it may also give guidance to managers to reduce the risk of defective rate by altering the ordering quantity. Further research can look into considering many aspects, such as vendor's recovery process, inspection errors, multi-stage supply chain system and learning curve.

References

[1] Wangsa I D and Wee H M 2017 An integrated vendor–buyer inventory model with transportation cost and stochastic demand *Int. J. Syst. Sci.: Oper. & Logst.*

[2] Goyal S K 1976 An integrated inventory model for a single supplier-single customer problem *Int. J. Prod. Res.* **15** pp 107–111

[3] Banerjee A 1986 A joint economic-lot-size model for purchaser and vendor *Decision Sci.* **17** pp 292–311

[4] Jauhari W A, Fitriyani A and Aisyati A 2016 An integrated inventory model for single-vendor single-buyer system with freight rate discount and stochastic demand *Int. J. Oper. Res.* **25** pp 327–350

[5] Dey O and Giri B C 2014 Optimal vendor investment for reducing defect rate in a vendor-buyer integrated system with imperfect production process *Int. J. Prod. Econ.* **155** pp 222–228

[6] Khan M, Jaber M Y and Ahmad A R 2014 An integrated supply chain model with errors in quality inspection and learning in production *Omega* **42** pp 16–24

[7] Glock C H and Jaber M Y 2013 A multi-stage production-inventory model with learning and forgetting effects, rework and scrap *Comput. Ind. Eng.* **64** pp 708–720

[8] Ben-Daya M and Hariga M 2004 Integrated single vendor single buyer model with stochastic demand and variable lead time *Int. J. Prod. Econ.* **92** pp 75–80

[9] Hsiao Y C 2008 A note on integrated single vendor single buyer model with stochastic demand and variable lead time *Int. J. Prod. Econ.* **114** pp 294–297

[10] Glock C H 2012 Lead time reduction strategies in a single-vendor-single buyer integrated inventory model with lot size-dependent lead times and stochastic demand *Int. J. Prod. Econ.* **136** pp 37–44

[11] Jauhari W A and Winingsih K 2016 Cooperative inventory model for vendor-buyer system with quantity and freight discounts *Int. J. Serv. and Operations Manag.* **25** pp 508–530

[12] Lin Y J 2010 A stochastic periodic review integrated inventory model involving defective items, backorder price discount, and variable lead time *4OR–Q. J. Oper. Res.* **8** pp 281–297

[13] Jauhari W A, Sejati N P and Rosyidi C N 2016 A collaborative supply chain inventory model with defective items, adjusted production rate and variable lead time *Int. J. Proc. Manag.* **9** pp 733–750

[14] Lin Y J and Lin H J 2016 Optimal ordering and recovery policy in a periodic review integrated inventory model *Int. J. Syst. Sci.: Oper. & Logst.* **3** pp 200–210

[15] Bonney M and Jaber M Y 2011 Environmentally responsible inventory models: non-classical models for a non-classical era *Int. J. Prod. Econ.* **133** pp 43–53

Development of coordination system model on single-supplier multi-buyer for multi-item supply chain with probabilistic demand

G Olivia, A Santoso and D N Prayogo

Department of Industrial Engineering, University of Surabaya, Raya Kalirungkut, Surabaya 60293, Indonesia

E-mail: gloriaolivia13@gmail.com

Abstract. Nowadays, the level of competition between supply chains is getting tighter and a good coordination system between supply chains members is very crucial in solving the issue. This paper focused on a model development of coordination system between single supplier and buyers in a supply chain as a solution. Proposed optimization model was designed to determine the optimal number of deliveries from a supplier to buyers in order to minimize the total cost over a planning horizon. Components of the total supply chain cost consist of transportation costs, handling costs of supplier and buyers and also stock out costs. In the proposed optimization model, the supplier can supply various types of items to retailers whose item demand patterns are probabilistic. Sensitivity analysis of the proposed model was conducted to test the effect of changes in transport costs, handling costs and production capacities of the supplier. The results of the sensitivity analysis showed a significant influence on the changes in the transportation cost, handling costs and production capacity to the decisions of the optimal numbers of product delivery for each item to the buyers.

Keywords: single supplier-multi buyer; coordination system; supply chain; multi-item; probabilistic demand

1. Introduction

Due to the increasing challenge of cost efficiency in order to enhance organization competitive edge in the global market, optimization in all aspects of organization functions has been the main concern of every organization. One way to overcome this challenge is by creating a coordination system between suppliers and buyers in a supply chain. Good coordination system between suppliers and buyers can support optimal decision making in an integrated supply chain. Chang and Chou [1] has developed an optimization model for the coordination system between single supplier and multi buyer to determine the optimal number of deliveries with the objective function to minimize the total supply chain cost. Buyer demand patterns considered in their model are deterministic, and supplier might produce only a single item.

In Chang and Chou [1] model, the coordination model has considered several important factors. However, there are some aspects that need to be taken into account in the coordination system model. According to Tersine [6], a deterministic demand patternis rarely found in real conditions. Therefore, this study develops an optimization model of the coordinate system with probabilistic demand patterns. In addition, the development of the proposed model is intended for a supplier that can produce and supply various types of product to buyers.

Furthermore, this paper will be divided into five sections. Section 2 will discuss the literature review while Section 3 will describe the research method. The development as well as the discussion of the proposed coordination system model for a supplier and buyers with multi-item and probabilistic demand will be described in Section 4. Section 5 will describe the conclusions and further research.

2. Literature review

Several papers have discussed coordination systems of supplier and buyer. Table 1 shows a comparison of the characteristics of the coordination system model of previous studies and this research.

Table 1. A comparison of the characteristics of coordination system models.

Characteristic	(Hejazi et al.) [3]	(Zavanella and Zanoni) [7]	(Chang and Chou) [1]	(This research, 2014)
Demand pattern	Deterministic	Deterministic	Deterministic	Probabilistic
Product variation	Single-item	Single-item	Single-item	Multi-item
Number of entity	Single supplier-single buyer	Single supplier-multi buyer	Single supplier-multi buyer	Single supplier-multi buyer
Total costcomponent	Purchasing cost, order cost, handling cost in supplier and buyer	Set-up cost, order cost, handling cost in supplier and buyer	Set-up cost, transportation, order, handling in supplier and buyer, receiving	Transportation cost, handling cost in supplier and buyer, shortage cost

Hejazi et al. [3], Zavanella and Zanoni [7] as well as Chang and Chou [1] developed a model with deterministic demand data for single item supply chain. Hejazi et al. [3] and Zavanella and Zanoni [8] using annual demand data, while Chang and Chou [1] using monthly demand data.

The proposed optimization model in this paper is coordination between single supplier that can produce and supply various types of item with buyers with probabilistic demand pattern. The components of the total cost that is considered in this paper are transportation cost, handling cost in supplier and buyer as well as stock out cost.

3. Research methodology

The model development in this paper will be conducted in two stages. The first stage is to create some adjustment from mathematical modeling from Chang and Chou [1].The adjustments made in this stage are to remove ordering cost, receiving cost and set up cost from a component of total cost since these costs do not affect the decision in optimization model. Meanwhile, the second stage is to develop a proposed model by considering probabilistic demand pattern and multiple items. The completion of a mathematical model in this paper is using Lingo 11.0. Model validation and sensitivity analysis are also conducted for the proposed model in this paper.

4. Results and discussion

Some assumptions used in the initial model adjustment are: the supplier only produces one type of product, and demand data pattern is deterministic. Adjustments made for the initial model consist of the elimination of ordering cost, order receiving cost and set-up cost from total cost components. In a coordinated replenishment system, the order process is done only once in the beginning of planning horizon. Ordering cost formulation according to Chang and Chou [1] is $\sum_{j=1}^{n} Aj$, and setup cost is CxS in which there are no decision variables involved in that formulations. For order receiving cost, Chang and Chou [1] formulate it as $\sum_j [V_j(\sum_i Q_{ij}X_{ij})]$. As long as demand data is the same, total order receiving cost would be the same in a year, and it will not influence the decision of optimization model. Briefly, this initial model has objective function to minimize total cost. The Initial model also ensures that there is no shortage allowed, and the delivery quantity will not exceed supplier's stock level.

This proposed model is developing a mathematical model for the coordination system between a supplier and buyers. The supplier can produce and supply various types of item to multiple buyers who have probabilistic demand pattern. There are some assumptions used in this research, such as supplier's lead time is 0, meaning that the goods will be delivered immediately after the supplier finish their production and the delivery can be done simultaneously for all kind of goods. Another assumption is that a buyer is willing to accept all the delivery quantity from the supplier. The demand data in this research have normal distribution pattern, and there is no limit on warehouse and shipping capacity. In overall, the proposed model can be described as follows:

i= index of period; T = number of period;
j= index of buyer; B = number of buyer;
k= index of number of product type (item) P = number of product;

D_{jk} = Total annual demand of item- k for buyer-j
d_{ijk} = Demand of item-k in period-i for buyer-j
F_j = Fixed transportation cost for buyer-j per trip
h_{bjk} = Handling cost per unit of item-k for buyer-j
h_{sk} = Handling cost per unit of item-k for supplier
Y_{ik} = Stock position item-k in supplier after production process finished in period-i
$(Y_{ik} - \sum_j Q_{ijk})$ = Final stock supplier for item-k in period-i
Y_{ijk} = Stock position item-k inbuyer-j after receiving delivery in period-i
$(Y_{ijk} - d_{ijk})$ = Final stock buyer-j for item-k in period-i
R_{ik} = Production quantity of item-k in period-i
K = Production capacity
Q_{ijk} = Optimal shipping quantity item-k in period-i for buyer-j
x_{ijk} = Binary variable
x_{ijk}=1, if there is delivery of item-k in period-i for buyer-j
x_{ijk}= 0, otherwise
Z_{ijk} = Binary variable
Z_{ijk}=1, if there is stockout of item-$k$for buyer-j in period-i
Z_{ijk} = 0, otherwise
N_{ij} = Binary decision
N_{ij} = 1, if there is delivery activity from buyer-j in period-i
N_{ij} = 0, otherwise

Objective function

Minimize:

$$\sum_{k=1}^{P} h_{bk} \sum_{j=1}^{B} \sum_{i=i}^{T} (Y_{ijk} - d_{ijk})(1 - Z_{ijk}) + \sum_{k=1}^{P} h_{sk} \sum_{i=1}^{T} (Y_{ik} - \sum_{i=1}^{T} Q_{ijk}) + \sum_{j=1}^{B} F_j \sum_{i=1}^{T} N_{ij} + \sum_{j=1}^{B} \sum_{k=1}^{P} SO_{jk} \sum_{i=1}^{T} [(d_{ijk} - Y_{ijk})] Z_{ijk} \quad (1)$$

Constraints

Subject to:

$$\sum_{i=1}^{T} Q_{ijk} = D_{jk}\,, \quad i = 1, \ldots, T; j = 1, \ldots, B; k = 1, \ldots, P \quad (2)$$

$$Y_{ik} = (Y_{(i-1)k} - \sum_{j=1}^{B} Q_{(i-1)jk})) + R_{ik}, \quad i = 1, \ldots, T; j = 1, \ldots, B; k = 1, \ldots, P \quad (3)$$

$x_{1jk} = 1, j = 1, \ldots, B; k = 1, \ldots, P$ (4)

$Y_{ik} - \sum_{j=1}^{B} Q_{ijk} \geq 0, \quad i = 1, \ldots, T; j = 1, \ldots, B; k = 1, \ldots, P$ (5)

$Q_{ijk} \leq x_{ijk} * M, \quad i = 1, \ldots, T; j = 1, \ldots, B; k = 1, \ldots, P$ (6)

$Y_{ijk} = (Y_{(i-1)jk} - d_{ijk}) + Q_{ijk}, \quad i = 1, \ldots, T; j = 1, \ldots, B; k = 1, \ldots, P$ (7)

$Y_{ijk} - d_{ijk} \leq (1 - Z_{ijk}) * M, \quad i = 1, \ldots, T; j = 1, \ldots, B; k = 1, \ldots, P$ (8)

$d_{ijk} - Y_{ijk} \leq Z_{ijk} * M, \quad i = 1, \ldots, T; j = 1, \ldots, B; k = 1, \ldots, P$ (9)

$\sum_{k=1}^{P} x_{ijk} \leq N_{ij} * M, \ i = 1, \ldots, T; j = 1, \ldots, B; k = 1, \ldots, P$ (10)

$\sum_{k=1}^{P} R_{ik} \leq K, \ i = 1, \ldots, T; k = 1, \ldots, P$ (11)

$R_{ik} + (Y_{(i-1)k} - \sum_{j=1}^{B} Q_{(i-1)jk}) \geq \sum_{j=1}^{B} Q_{ijk}, \ i = 1, \ldots, T; j = 1, \ldots, B; k = 1, \ldots, P$ (12)

The objective function in this model is a total cost minimization which consists of transportation cost, handling cost in supplier and buyer, and also shortage cost, and it can be seen from equation (1). Equation (2) states that total delivery item-k to buyer-j in all of the period should be same with total demand item-k for buyer-j. The supplier's inventory position before delivering the goods is shown in equation (3), which is equal to the sum of the initial supplier's stock of item-k in period-$i$andthe production quantity of item- k in period- i. Equation (4) ensures that there is a delivery on period 1 for all buyers and all items. Equation (5) ensures that there is enough quantity of item-k in period-i for supplying item- k to all buyers. Binary variable x_{ijk} on equation (6) states whether there is delivery of item- k to buyer- j in period- i or not. x_{ijk}=1 means that there is a delivery of item-k for buyer- j in period- i, and otherwise. The amount of inventory position of item-k in buyer-j after receiving delivery in period- i is the sum of initial stock buyer-j for item-k and delivery quantity item-k for buyer-j in period-i which stated in equation (7). Equation (8) and (9) ensure that when Z_{ijk}=1, it means that there is stock out of item-k for buyer-j in period-i, and otherwise. N_{ij} in equation (10) is a binary variable that states whether there is delivery for buyer-j in period- i or not.N_{ij}=1 when there is a delivery for buyer-j in period-i, and otherwise when N_{ij}=0. Equation (11) ensures that the production quantity does not exceed production capacity of the supplier. Equation (12) ensures that sum of production quantity and initial stock of item-k in period-i should be able to meet delivery quantity of item-k to buyer-j in period-i.

Case study for the proposed mathematical model consists of one supplier and five buyers that can be applied for consumer goods manufacturer. The supplier produces three items: item A, item B and item C with a production capacity of 4200 items/month. For a case study in this proposed model, demand data is probabilistic, and safety stock should be calculated. After calculating safety stock, total demand from each buyer can be determined by summing up the demand and safety stock for each buyer.

Table 2. Demand buyer- $j$for item-$k$in period-i(unit).

d_{ijk}	*Buyer 1*			*Buyer 2*			*Buyer 3*			*Buyer 4*			*Buyer 5*		
	A	B	C	A	B	C	A	B	C	A	B	C	A	B	C
1	234	209	204	301	227	268	281	152	308	243	244	280	179	291	257
2	207	143	196	257	227	227	223	229	269	231	191	254	203	286	252
3	238	194	243	270	186	256	287	249	255	191	279	353	186	263	201
4	198	216	293	248	161	284	152	253	211	173	241	311	255	263	198
5	204	207	226	279	179	252	193	173	281	176	266	236	209	206	257
6	164	215	235	293	206	270	159	210	213	165	193	276	199	244	203
7	178	165	288	249	190	242	241	239	268	265	279	250	145	202	204
8	267	195	266	208	234	294	298	256	231	160	253	241	217	233	259

d_{ijk}	*Buyer* 1			*Buyer* 2			*Buyer* 3			*Buyer* 4			*Buyer* 5		
	A	B	C	A	B	C	A	B	C	A	B	C	A	B	C
9	223	184	270	305	185	266	179	273	313	253	263	268	199	242	243
10	220	200	260	251	192	271	210	214	353	230	274	261	207	273	267
11	208	198	369	285	183	193	266	172	371	216	240	203	138	163	193
12	202	184	162	339	156	341	200	208	271	278	234	181	229	305	208
Total	2543	2310	3012	3285	2326	3164	2689	2628	3344	2581	2957	3114	2366	2971	2742

Table 3. The handling cost in supplier and buyer for each items.

Item	Buyer's handling cost	Supplier's handling Cost
Item A	IDR 117	IDR90
Item B	IDR 157	IDR 121
Item C	IDR 196	IDR 151

Table 4. Fix transportation cost.

Buyer	Fix transportation cost (IDR/trip)
1.	600.000
2.	500.000
3.	450.000
4.	600.000
5.	450.000

Table 5. Optimal delivery quantity for item-k for buyer-j in period-i.

Q_{ijk}	*Buyer* 1			*Buyer* 2			*Buyer* 3			*Buyer* 4			*Buyer* 5		
	A	B	C	A	B	C	A	B	C	A	B	C	A	B	C
1	362	317	287	433	322	374	54	217	194	391	125	66	300	409	350
2	0	0	0	0	0	0	445	151	276	1036	1186	1106	0	0	0
3	279	280	320	400	332	370	219	222	263	0	0	0	156	978	382
4	270	221	418	550	326	569	353	212	278	0	0	0	546	0	457
5	0	0	0	0	0	0	93	81	60	1154	1646	1052	113	0	0
6	619	866	920	0	0	0	487	544	764	0	0	0	0	0	0
7	0	0	0	1116	853	486	0	0	0	0	0	0	254	700	791
8	0	0	0	0	0	0	617	372	1509	0	0	0	424	516	763
9	981	568	1067	0	0	1085	264	235	0	0	0	0	0	0	0
10	0	0	0	0	0	0	0	0	0	0	0	890	574	368	0
11	0	0	0	206	394	280	157	594	0	0	0	0	0	0	0
12	31	58	0	580	99	0	0	0	0	0	0	0	0	0	0
Σ	2543	2310	3012	3285	2326	3164	2689	2628	3344	2581	2957	3114	2366	2971	2742

Table 6. Production quantity of item-k in period-i.

	R_{ik}(unit)			TOTAL
i	Item A	Item B	Item C	
1	1.539	1.390	1.270	4.200
2	1.481	1.337	1.382	4.200
3	1.053	1.811	1.335	4.200
4	1.719	759	1.722	4.200
5	1.360	1.727	1.112	4.200
6	1.106	1.411	1.683	4.200
7	1.370	1.553	1.277	4.200
8	1.040	888	2.272	4.200
9	1.246	803	2.151	4.200
10	574	368	890	1.832
11	363	988	280	1.631
12	612	157	0	769

The result of this proposed model is a decision for optimal shipping quantity using the scenario that has been defined. The optimal shipping quantity can be seen in Table 5, while the production quantity is in Table 6.

5. Conclusion

This study developed amathematical model for coordination system between single supplier and multi buyer for multi item supply chain with probabilistic demand pattern. The objective function in this proposed model was to minimize total cost which consists of transportation cost, handling cost in supplier and buyers as well as stockout cost. Sensitivity analysis showed that the changing parameters have a significant influence on the value of decision variables and objective function. The transportation cost was inversely proportional to the total delivery frequency for one year, and the handling cost was directly proportional to the frequency of delivery within one year, and the production capacity was directly proportional to the handling cost.

References

[1] Chang C and Chou H 2013 A coordination system for seasonal demand problems in the supply chain *Appl. Math. Model.* **37** pp 3674–86
[2] Chopra S and Meindl P 2013 *Supply Chain Management: Strategy, Planning, and Operation* (Asia: Pearson Education) pp 13–216
[3] Hejazi S R, Arkan A and Rezvan T 2009 Coordination mechanism for two-level supply chain with one manufacturer and one buyer under credit period *Int. J. Management Science and Engineering Management* **4** pp 39–48
[4] Nahmias S 2009 *Production and Operation Analysis* (New York: McGraw-Hill) p 311
[5] Simchi-Levi D, Kaminsky P and Simchi-Levi E 2008 *Designing and Managing the Supply Chain: Concepts, Strategies, and Case Studies* (New York: McGraw-Hill Irwin) pp 161–162
[6] Tersine R J 1994 *Principles of Inventory and Materials Management* (New Jersey: Prentice-Hall) p 205
[7] Zavanella L and Zanoni S 2009 A one-vendor multi buyer integrated production-inventory model *Int. J. Prod. Econ.* **118** pp 225–232

IOP Conf. Series: Materials Science and Engineering **273** (2017) 012020 doi:10.1088/1757-899X/273/1/012020

Using genetic algorithm to determine the optimal order quantities for multi-item multi-period under warehouse capacity constraints in kitchenware manufacturing

D Saraswati, D K Sari and V Johan

Industrial Engineering Department, Trisakti University, Jakarta, Indonesia

E-mail: docki_saraswati@trisakti.ac.id

Abstract. The study was conducted on a manufacturer that produced various kinds of kitchenware with kitchen sink as the main product. There were four types of steel sheets selected as the raw materials of the kitchen sink. The problem was the manufacturer wanted to determine how much steel sheets to order from a single supplier to meet the production requirements in a way to minimize the total inventory cost. In this case, the economic order quantity (EOQ) model was developed using all-unit discount as the price of steel sheets and the warehouse capacity was limited. Genetic algorithm (GA) was used to find the minimum of the total inventory cost as a sum of purchasing cost, ordering cost, holding cost and penalty cost.

1. Introduction

Companies are often faced with two problems of inventory management. A policy requires inventory in large quantities, and as the consequences, they will require a large space in a warehouse to stock the materials. On the other hand, inventories in small amounts are at risk of running out, which causes back-order or lost-sales. The best known inventory model is the classic economic order quantity (EOQ). The application of EOQ is widely used in a real environment. The extension of this model is relaxing some of its assumptions [1]. Matsuyama [2] described the dependence of the purchase price on ordering quantity with a certain function in order to maximize the one day's average profit. Mendoza and Ventura [3] developed EOQ model with transportation cost in which all unit discount was included to find the optimal inventory policy. Taleizadeh and Pentico [4] developed the EOQ model with all unit discounts and partial backordering.

Genetic algorithm (GA) is a search procedure to identify optimal or near optimal solutions from a population. Stockton and Quinn [5] proposed GA to solve economic lot size in deterministic inventory model. Using GA, the other group of researchers, Ongkunaruk *et al.* [6] developed a multi-item replenishment as a group from a single supplier with shipment constraint, budget constraint and transportation constraint to minimize the total expected cost per unit time. An algorithm for multi-item replenishment from a single supplier was presented by Goyal [7] with 15 items in order to determine the optimum ordering frequency.

In this study, the inventory policy was applied in the environment with multi-item multi period in a kitchenware manufacturer that produced household appliances with steel sheets as the main raw materials. Another factor to consider was the limited capacity of the warehouse and the discounts rate for ordering raw materials in a certain amount. Following that, this paper presented the genetic algorithm as an approach to model the problem and determined how much to order based on minimizing the total inventory cost.

2. Problem definition

This study was done in a kitchenware manufacturer with kitchen sink as the main products (Figure 1). The raw material of kitchen sink is steel sheets. During the planning horizon, the manufacturer produced79 items for 18 product family of kitchen sink. The planning horizon in this research was four periods. Based on ABC classification there were only four types of steel sheets in class A, namely steel sheets; S/S 201: 0.3 mm x 1220 mm (item 1), S/S 430: 0.37 mm x 660 mm (item 2), S/S 201: 0.5 mm x 1220 mm (item 3) and S/S 430: 0.4 mm x 1220 mm (item 4). The demand for the four types of steel sheets is based on the planned order release in four periods (Table 1).Ordering cost is the cost of preparing an order from the supplier. Demand and ordering cost for each item are shown in Table 1. Holding cost consists of cost of capital, storage and handling cost, taxes, insurance, obsolescence, pilferage and deterioration.

Kitchen sink 80S Kitchen sink 46 Kitchen sink 81

Figure 1. Types of kitchen sinks.

The objective of this study was to determine the optimal order quantities in order to minimize the total inventory cost including ordering, holding and purchasing. The notations used in the inventory modeling are the same as Pasandideh *et al.* [8]. The notations are as follows:

B_i : batch size of item i (i = 1,2,3,4)
$V_{i,t}$: number of orders of item i in period t (t = 1,2,3,4)
$D_{i,t}$: demand of item i in period t (kg)
H_i : holding cost of item i per kg per period (Rp/kg/period)
A_i : ordering cost of item i per order (Rp/order)
P_i : purchase cost of item i (Rp/unit)
$q_{i,k}$: upper bound quantity of the k-th price break of item i (kg)
$P_{i,k}$: purchase cost of item i in period t determined by the discount schedule of $Q_{i,t}$ (Rp/kg)
$m_{i,k}$: discount rate of purchasing item i at price break k ($0<m_{i,k}<1$)
$Q_{i,t}$: ordering quantity of item i in period t (kg)
$X_{i,t}$: beginning inventory of item i in period $t = 0$, beginning inventory for all items = 0 (kg)
TC : total inventory cost for all items in all period. (Rp.)
S : available warehouse capacity (kg)
S_i : storage space needed per unit of item i (kg/unit)
$U_{i,t,k}$: binary variable, equals to 1 if item i is purchased at price break k in period t, and 0 otherwise.
$W_{i,t}$: binary variable, equals to 1 if an order of item i is placed in period t, and 0 otherwise.

Table 1. Demand and ordering cost for 4 items in 4 periods.

t \ i	$D_{i,t}$ (kg) 1	2	3	4
1	8576.79	17737.64	12016.15	9065.65
2	3658.45	8160.12	6140.44	2740.93
3	5564.63	17039.91	9252.2	19322.71
4	7189.96	5055.39	2821.97	5815.29
A_i(Rp/order)	926,000	935,000	875,000	888,000

Table 2 shows the discount rate for the unit price of 4 types of stainless sheets. The price includes shipping costs from supplier located in Surabaya to Jakarta. The supplier's quantity discount applies all-unit discount (AUD) method. The unit of holding cost is defined 25% of the unit price. Stainless steel is sold by suppliers in the form of batch. Each batch weighs 5 kg. Therefore, the purchase of stainless steel should be a multiplication of 5.

Table 2. Quantity discount for 4 items.

Quantity (kg)	P_i(Rp/kg) 1	2	3	4
$0 \leq Q_i < 5000$	26,900	25,800	22,575	21,450
$5000 \leq Q_i < 10000$	26,400	25,300	22,075	20,950
$Q_i \geq 10000$	25,900	24,800	21,575	20,450

Each pallet for stainless steel roll cannot exceed 3500 kg because of the company's forklift capacity, while the warehouse capacity is enough for 15 pallets. Therefore, the available capacity for holding the 4 types of stainless steel rolls are 15 times 3500 kg equals to 52,500 kg.

3. Problem formulation

The beginning inventory of an item i in period t, $X_{i,t}$ ($i = 1,2,3,4$ and $t = 1,2,3,4$) is as follows (Figure 2);

$$X_{i,t} = X_{i,t-1} + Q_{i,t-1} + D_{i,t-1} \tag{1}$$

$$Q_{i,t} = B_{i,t} \times V_{i,t} \tag{2}$$

Since $Q_{i,t} + X_{i,t} \geq D_{i,t}$, therefore $B_{i,t} \times V_{i,t} + X_{i,t} \geq D_{i,t}$.

The purchasing cost per unit with all-units quantity discounts (AUD) and 3 price break points is defined as follows:

$$P_i = \begin{cases} P_{i,1}\,;\, 0 \leq Q_{i,t} < q_{i,2} \\ P_{i,2}\,;\, q_{i,2} \leq Q_{i,t} < q_{i,3} \\ P_{i,3}\,;\, q_{i,3} \leq Q_{i,t} \end{cases} \tag{3}$$

The mathematical model for finding the minimum total cost:

$$Min\,TC = \sum_{i=1}^{4} \sum_{t=1}^{4} A_i + \sum_{i=1}^{4} \sum_{t=1}^{3} \left(\frac{X_{i,t} + X_{i,t+1}}{2} H_i \right) + \sum_{i=1}^{4} \sum_{t=1}^{4} \sum_{k=1}^{3} U_{i,t,k} \left(P_{i,k} \times Q_{i,t} \right) \tag{4}$$

Subject to

$$X_{i,t} = X_{i,t-1} + Q_{i,t-1} - D_{i,t-1} \tag{5}$$

$$X_{i,1} = 0 \tag{6}$$

$$\sum_{i=1}^{4} S_i X_{i,t} \leq S \tag{7}$$

$$Q_{i,t} = B_{i,t} \times V_{i,t} \quad (8)$$

$$\sum_{k=1}^{K} U_{i,t,k} = 1 ; U_{i,t,k} \in \{0,1\} \quad (9)$$

$$i = 1,2,3,4 ; t = 1,2,3,4 ; k = 1,2,3$$

Equation (4) explains the minimum total cost as the sum of total ordering cost, total inventory holding cost, and total purchasing cost. Equation (5) describes the inventory at period t is the inventory of item i at period *t-1* plus ordering quantity of item *i* at period *t-1* minus demand of item *i* at period *t-1*.Inventory of item *i* in period *t* is $X_{i,t}$ as illustrated in Figure 2. Equation (6) assumes there is no inventory at the beginning of period 1 for all items $X_{1,1} = X_{2,1} = X_{3,1} = X_{4,1} = 0$. Equation (7) requires the space for item *i* at period *t* should be less or equal to the warehouse capacity. Equation (8) calculates the batch size times the number of ordering of item *i* at period *t* equal to ordering quantity for item *i* at period *t*. Equation (9) is a binary variable. The value equals 1 if item *i* is purchased at price break *k* in period *t* and 0 otherwise.

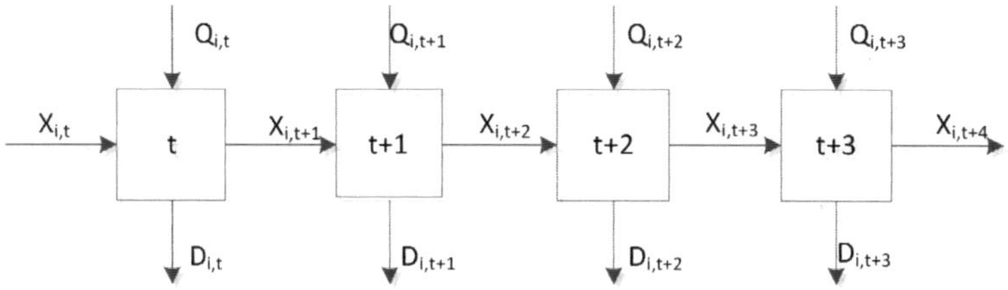

Figure 2. Inventory position for item *i* at each period.

Warehouse capacity

Chromosomes that do not meet the specified constraints will get penalty. It happens if the storage space used exceeds the warehouse capacity. The penalty value (PV) is derived from the sum of squared space used to store the 4 items in the same period minus the available warehouse capacity. If the total inventory of four items in each period is less than the capacity of the warehouse, there is no penalty or penalty value = 0.

$$\text{if } \sum_{i=1}^{4} X_{i,t} + Q_{i,t} \geq S \text{ ; PV} = \left\{ \sum_{i=1}^{4} \left(X_{i,t} + Q_{i,t} \right) - S \right\}^2 \text{ and if } \sum_{i=1}^{4} X_{i,t} \leq S \text{ ; PV} = 0 \quad (10)$$

$$\text{Total penalty value (TPV)} = \sum_{t=1}^{4} \left\{ \sum_{i=1}^{4} \left(X_{i,t} + Q_{i,t} \right) - S \right\}^2 \quad (11)$$

4. Genetic algorithms

GA begins with an initial set of solutions while the optimal ordering quantity for each item from a single supplier in each period is called a population. Each ordering quantity in the population is called a chromosome. Each chromosome is composed set of gens. The chromosome forms a matrix (*m* x *n*) with *m* rows for items and *n* columns for periods [8]. Figure 3 shows the matrix of gene for *i-th* row and the *j-th* column. This matrix indicates the ordering quantity of item *i* in period *t*.

$$\mathbf{Q} = \begin{array}{c|cccc} i \backslash t & 1 & 2 & 3 & 4 \\ \hline 1 & Q_{1,1} & Q_{2,1} & Q_{3,1} & Q_{4,1} \\ 2 & Q_{1,2} & Q_{2,2} & Q_{3,2} & Q_{4,2} \\ 3 & Q_{1,3} & Q_{2,3} & Q_{3,3} & Q_{4,3} \\ 4 & Q_{1,4} & Q_{2,4} & Q_{3,4} & Q_{4,4} \end{array}$$

Figure3. Matrix of chromosome Q

There are 20 chromosomes in the population and each chromosome forms a 4×4 matrix. All genes are generated by the random number generator in the form of a 4×80 matrix with a range of 500-3500 (Figure 4). Range 500-3500 is used since the minimum order quantity fixed by the supplier equals to 500 kg and the maximum weight that can be transported by the company's material handling is 3500 kg.

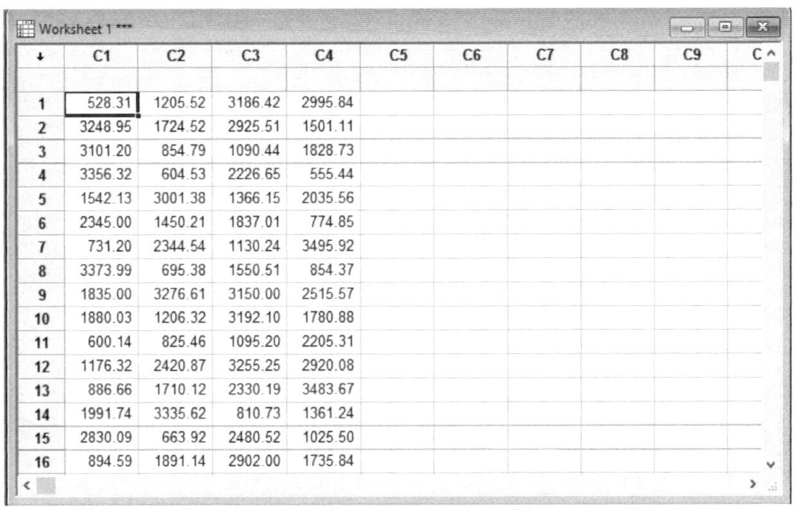

Worksheet 1 ***

↓	C1	C2	C3	C4	C5	C6	C7	C8	C9	C
1	528.31	1205.52	3186.42	2995.84						
2	3248.95	1724.52	2925.51	1501.11						
3	3101.20	854.79	1090.44	1828.73						
4	3356.32	604.53	2226.65	555.44						
5	1542.13	3001.38	1366.15	2035.56						
6	2345.00	1450.21	1837.01	774.85						
7	731.20	2344.54	1130.24	3495.92						
8	3373.99	695.38	1550.51	854.37						
9	1835.00	3276.61	3150.00	2515.57						
10	1880.03	1206.32	3192.10	1780.88						
11	600.14	825.46	1095.20	2205.31						
12	1176.32	2420.87	3255.25	2920.08						
13	886.66	1710.12	2330.19	3483.67						
14	1991.74	3335.62	810.73	1361.24						
15	2830.09	663.92	2480.52	1025.50						
16	894.59	1891.14	2902.00	1735.84						

Figure 4. Random number generators with range 500-3500.

The generated random number becomes the initial population. Since the supplier has determined the weight of each batch is 5 kg, then the initial population was adjusted to a multiple of 5 kg per batch. For example in cell (1,C1) the value of 528.31 became530. In this case, there were 20 chromosomes in the population with 80 genes. Each gene represented the batch size of each item in each period ($B_{i,t}$).As shown at Table 3, after obtaining the value of $B_{i,t}$ for 20 chromosomes and calculating the ordering quantity ($Q_{i,t}$) , the inventory began($X_{i,t}$).

GA parameters

Since there were 20 chromosomes in population, the population size (pop-size) was20. The crossover probability (P_c) = 0.7 indicated that as many as 0.7 of the total population would be crossover. The mutation probability (P_m) = 0.05 was set for only one gene mutated per chromosome. The mutation probability in this study was 0.05 obtained from 1 divided by 20. The number of generations (GEN) was not determined because iteration was complete when the average fitness generation declined.

Table 3. Batch size, order quantities and beginning inventory.

Chromosome	i	$B_{i,t}$				$Q_{i,t}$				$X_{i,t}$				
		t=1	t=2	t=3	t=4	t=1	t=2	t=3	t=4	t=1	t=2	t=3	t=4	t=5
1	1	530	1205	3185	2995	9010	18075	12740	8985	0	433.2	770.6	1494.4	1413.7
	2	3250	1725	2925	1500	6500	6900	5850	1500	0	2841.6	1581.4	1290.9	50.1
	3	3100	855	1090	1830	6200	17100	8720	20130	0	635.4	695.5	163.3	970.6
	4	3355	605	2225	555	10065	2420	4450	4440	0	2875.1	239.7	1867.7	492.4
2	5	1540	3000	1365	2035	9240	18000	12285	8140	0	663.2	925.6	1194.4	268.8
	6	2345	1450	1835	775	4690	7250	7340	1550	0	1031.6	121.4	1320.9	130.1
	7	730	2345	1130	3495	5840	18760	7910	20970	0	275.4	1995.4	653.3	2300.5
	8	3375	695	1550	855	10125	2780	3100	5130	0	2935.1	659.7	937.7	252.4
:	:	:	:	:	:	:	:	:	:	0	:	:	:	:
20	77	2355	2245	1595	3455	9420	17960	11165	10365	0	843.2	1065.5	214.4	1513.7
	78	2735	1010	2675	760	5470	7070	8025	760	0	1811.6	721.4	2606.0	625.1
	79	2475	1230	2985	2510	7425	15990	8955	20080	0	1860.4	810.5	513.3	1270.5
	80	2855	2055	3360	610	8565	4110	3360	4880	0	1375.1	429.7	967.7	32.4

Chromosomes

The fitness value of each chromosome was determined by evaluating the suitable objective function, in which the objective function was the summation of purchase cost, ordering cost and holding cost.

$$\text{Total ordering cost (TOC)} = \sum_{i=1}^{4}\sum_{t=1}^{4} A_i W_{i,t} \tag{12}$$

$$\text{Total holding cost (THC)} = \sum_{i=1}^{4}\sum_{t=1}^{4}\left(\frac{X_{i,t} + X_{i,t+1} + Q_{i,t}}{2}\right)H_i \tag{13}$$

$$\text{Total purchasing cost (TPC)} = \sum_{i=1}^{4}\sum_{t=1}^{4}\sum_{k=1}^{3} U_{i,t,k} \times \left(P_{i,k} \times Q_{i,t}\right) \tag{14}$$

$$\text{Total inventory cost (TC)} = \text{TOC} + \text{THC} + \text{TPC}$$

The total cost or objective function of 20 chromosomes in the first generation was IDR3,450,434,433.

Fitness and chromosome selection

The purpose of this study was to obtain inventory planning with minimum total cost. The selection process was done by making the chromosome with a small objective function had a high selected probability. This selection used fitness value equal to 1/objective function. As an example, the fitness value of the first chromosome in the first generation was0.00000000029.Next, the fitness probability was calculated for each chromosome with the formula = fitness value 1/total fitness value. For example, the probability of the fitness value for chromosome 1 was 0.0502.

Selection of chromosomes was carried out using a strategy roulette wheel, where the selection was done randomly. In the selection process, it was necessary to calculate the cumulative probability. Roulette wheel was rotated according to the number of chromosomes that was as much as 20 times to generate random numbers R with a range of 0-1. In each round, a chromosome was selected for the new population. If R<C[1] then chromosome 1would be selected as a parent. Otherwise, the *i*-th chromosome should be selected as the parent for C[i-1]<R<C[i]. For example, in the generation of random numbers in the first round was obtained a number of 0.7978. As shown in Table 4, the first random number R[1] was larger than C [15] but smaller than C[16] then chromosome 16 should be selected as the chromosome in the new population.

Table 4. Chromosome selection.

Chromo-some	Objective function	Fitness Value	Fitness proportion (p)	Cumulative p	Random number	Selected chromosome
1		0.000000000	0.0502	0.0502	0.7978	16
2		0.000000000	0.0503	0.1005	0.5345	11
3		0.000000000	0.0494	0.1499	0.2258	5
:	:	:	:	:	:	:
20		0.000000000	0.05	1	0.4383	9
	Total	0.000000005				

Crossover

Crossover operations were carried out in two stages: selection and operation. In the selection process, individuals were randomly selected for crossing with a probability of 0.7 (P_c = 0.7). If the generated random number was smaller than the specified P_c value, crossover was performed. Crossover would not be accomplished if the random number was greater than the value of P_c and the value of the parent became offspring. The selected chromosomes would be paired with each other. If the number was odd, then one of the chromosomes was not selected for crossover process. The Chromosome was 4×4 matrix, and crossover operation was executed by selecting the number of rows and columns randomly (Figure 5).

Mutation

The number of chromosomes that performed the gene mutation process was determined by the mutation probability parameter (P_m). The number of gen had the value of 16 because the matrix shape was 4×4 with 16 cells. Total genes = number of genes in chromosome x number of chromosomes in the population = 16×20 = 320 genes. The cell was mutated because it had a random number smaller than the mutation probability (P_m). In the first generation, there were 21 genes mutated.

Chromosome pair		Random number Σ rows	Random number Σ columns
6	4	3	3
8	3	2	1
11	20	3	2

Chromosome 8

$$Q_8 = \begin{pmatrix} \mathbf{2890} & 2300 & 3290 & 1145 \\ \mathbf{2380} & 2260 & 995 & 2785 \\ 2260 & 2205 & 2680 & 2695 \\ 1070 & 3225 & 2920 & 575 \end{pmatrix}$$

Chromosome 3

$$Q_3 = \begin{pmatrix} \mathbf{2820} & 2600 & 3205 & 1345 \\ \mathbf{1475} & 3240 & 2415 & 2965 \\ 800 & 990 & 3360 & 2790 \\ 900 & 1115 & 1090 & 1970 \end{pmatrix}$$

Offspring 8

$$Q_8 = \begin{pmatrix} \mathbf{2820} & 2300 & 3290 & 1145 \\ \mathbf{1475} & 2260 & 995 & 2785 \\ 2260 & 2205 & 2680 & 2695 \\ 1070 & 3225 & 2920 & 575 \end{pmatrix}$$

Offspring 3

$$Q_3 = \begin{pmatrix} \mathbf{2890} & 2600 & 3205 & 1345 \\ \mathbf{2380} & 3240 & 2415 & 2965 \\ 800 & 990 & 3360 & 2790 \\ 900 & 1115 & 1090 & 1970 \end{pmatrix}$$

Chromosome 6

$$Q_6 = \begin{pmatrix} 1740 & \mathbf{945} & \mathbf{3465} & \mathbf{915} \\ 580 & \mathbf{2490} & \mathbf{1020} & \mathbf{980} \\ 1470 & \mathbf{2340} & \mathbf{1945} & \mathbf{960} \\ 1825 & 1250 & 3105 & 975 \end{pmatrix}$$

Chromosome 4

$$Q_4 = \begin{pmatrix} 1835 & \mathbf{3275} & \mathbf{3150} & \mathbf{2515} \\ 1880 & \mathbf{1205} & \mathbf{3190} & \mathbf{1780} \\ 600 & \mathbf{825} & \mathbf{1095} & \mathbf{2205} \\ 1175 & 2420 & 3255 & 2920 \end{pmatrix}$$

Offspring 6

$$Q_6 = \begin{pmatrix} 1740 & \mathbf{3275} & \mathbf{3150} & \mathbf{2515} \\ 580 & \mathbf{1205} & \mathbf{3190} & \mathbf{1780} \\ 1470 & \mathbf{825} & \mathbf{1095} & \mathbf{2205} \\ 1825 & 1250 & 3105 & 975 \end{pmatrix}$$

Offspring 4

$$Q_4 = \begin{pmatrix} 1835 & \mathbf{945} & \mathbf{3465} & \mathbf{915} \\ 1880 & \mathbf{2490} & \mathbf{1020} & \mathbf{980} \\ 600 & \mathbf{2340} & \mathbf{1945} & \mathbf{960} \\ 1175 & 2420 & 3255 & 2920 \end{pmatrix}$$

Chromosome 11

$$Q_{11} = \begin{pmatrix} 2720 & 2555 & 3115 & 525 \\ 1000 & 3290 & \mathbf{3475} & \mathbf{970} \\ 2725 & 1400 & \mathbf{555} & \mathbf{685} \\ 1365 & 2635 & \mathbf{1160} & \mathbf{1405} \end{pmatrix}$$

Chromosome 20

$$Q_{20} = \begin{pmatrix} 625 & 1490 & 1580 & 1975 \\ 2485 & 2360 & \mathbf{2535} & \mathbf{2420} \\ 2975 & 2505 & \mathbf{990} & \mathbf{1610} \\ 2495 & 3160 & \mathbf{1440} & \mathbf{1235} \end{pmatrix}$$

Offspring 11

$$Q_{11} = \begin{pmatrix} 2720 & 2555 & 3290 & 1145 \\ 1000 & 3290 & \mathbf{2535} & \mathbf{2420} \\ 2725 & 1400 & \mathbf{990} & \mathbf{1610} \\ 1365 & 2635 & \mathbf{1440} & \mathbf{1235} \end{pmatrix}$$

Offspring 20

$$Q_{20} = \begin{pmatrix} 625 & 1490 & 1580 & 1975 \\ 2485 & 2360 & \mathbf{3475} & \mathbf{970} \\ 2975 & 2505 & \mathbf{555} & \mathbf{685} \\ 2495 & 3160 & \mathbf{1160} & \mathbf{1405} \end{pmatrix}$$

Figure 5. Crossover operations.

New generation

Iteration continues to produce a new generation. That is, the results of the first generation will become the initial population for the second generation, and so on. Iteration will be terminated if the stopping criteria have been fulfilled by obtaining the best chromosome with a minimum objective function value of the whole generations. As shown in Table 5, the best value of objective function is given by chromosome #7 with the total cost of IDR 3,374,322,432.

Table 5. Ordering quantity ($Q_{i,j}$).

Raw materials	Item (i)	Period (j) j=1	j=2	j=3	j=4
SS 201 0.30 mm x 1220 mm x coil (kg)	1	8700	17955	13400	7350
SS 201 0.50 mm x 1220 mm x coil (kg)	2	4235	7810	7620	1080
SS 430 0.37 mm x 660 mm x coil (kg)	3	7080	16060	8780	19390
SS 430 0.40 mm x 1220 mm x coil (kg)	4	8320	5220	2810	4550

5. Conclusion

This article addressed the inventory problem for multi-item multi period by considering the warehouse capacity. Moreover, this inventory model was implemented in the kitchenware manufacturer whose ordering of the raw materials was from a single supplier. Stainless steel is the main raw materials for kitchenware products. This article used metaheuristic approach, genetic algorithm (GA), to determine the ordering lot size of stainless steel based on minimizing total inventory cost. Moreover, this work can be explored for different value of GA's parameters such as population size or probability of crossover (P_c) or probability of mutation (P_m).

References

[1] Axsater S 2006 *Inventory Control* 2nd (Lund, Sweden: Springer)
[2] Matsuyama K 2001 The EOQ-models modified by introducing discount of purchase price or increase of setup cost *Int. J. Prod. Econ.* **73** pp 83–99
[3] Mendoza A and Ventura J A 2008 Incorporating quantity discount to the EOQ model with transportation cost *Int. J. Prod. Econ.* **113** pp 754–756
[4] Taleizadeh A A and Pentico D W 2014 An economic order quantity model with partial backordering and all units discounts *Int. J. Prod. Econ.* **155** pp 172–184
[5] Stockton D J and Quinn L 1993 Identifying economic order quantities using genetic algorithms *Int. J. Oper. Prod. Mgmt.* **13** pp 92–103
[6] Ongkunaruk P, Wahab M I M and Chen Y 2016 A genetic algorithm for a joint replenishment problem with resource and shipment constraints and defective items *Int. J. Prod. Econ.* **175** pp 142–152
[7] Goyal S K 1974 Optimum ordering policy for a multi item single supplier system *J. Opl. Res. Soc.* **25** pp 293–298
[8] Pasandideh S H R, Naiki S T A and Mosavi S M 2013 Two metaheuristic to solve a multi-item multiperiod inventory control problem under storage constraint and discounts *Int. J. Adv. Manuf. Tech.* **69** pp 1671–84

From ISO 9001:2008 to ISO 9001:2015: Significant changes and their impacts to aspiring organizations

Y Sari, E Wibisono, R D Wahyudi and Y Lio

Department of Industrial Engineering, University of Surabaya, Raya Kalirungkut, Surabaya 60293, Indonesia

E-mail: ysari@staff.ubaya.ac.id

Abstract. ISO 9001:2015 is the latest version of ISO Quality Management System standard that has been updated recently from ISO 9001:2008. It is necessary for all organizations that have implemented and been certified with ISO 9001:2008 to prepare the transition and upgrade their Quality Management System because the certification will expire by September 2018. This paper attempts to provide knowledge on the significant changes from ISO 9001:2008 to ISO 9001:2015, what new requirements are added, and how they would impact the organizations. An exploratory and applied research was chosen as the research approach and aimed to explore what transition designs are needed to anticipate the changes as well as their impacts. The research applied a methodology of Plan-Do-Check-Action (PDCA) cycle into four organizations and their results were compared and discussed to explain the transition designs. Some qualitative methods such as observation and interview were used to collect the data. By addressing the new requirements, three transition designs that should be prepared are: (i) identifying needs from interested parties, (ii) analyzing internal and external factors of the organizations to formulate relevant strategies and quality objectives, and (iii) registering risks associated to business processes as well as organizational strategies.

Keywords: Standardization, ISO 9001: 2015, Quality Management System

1. Introduction

ISO 9001 is an International Standard of Quality Management System (QMS). This Standard describes the requirements for organizations to help them promote continual improvements and achieve customer satisfaction. This Standard has been revised several times [1, 2]. The first version which was published in 1987 and known as ISO 9001:1987 was formulated based on the concept of quality assurance. It comprised three classifications i.e. ISO 9001:1987 as a model for organizations with the creation of new products, ISO 9002:1987 for those without the creation of new products, and ISO 9003:1987 as a model for final inspection and testing purposes. The Standard was later developed into the second version i.e. ISO 9001:1994 and was improved based on the concept of preventive action. The third version, ISO 9001:2000, was formulated based on the concept of quality management to help organizations improve their business process. The fourth version was ISO 9001:2008 which underwent minor changes (merely to improve and make a better version of the standards) from the previous version.

ISO 9001 Quality Management System Standard is now transforming itself into a new brand with major changes from the earlier 2008 version. This new version was published in 2015 and known as ISO 9001:2015. The changes include more clauses and use a new concept and approach. The comparison of the changes between the old and new versions was well explained by [3]. If ISO 9001:2008 emphasizes on continual improvement and customer satisfaction, ISO 9001:2015 puts more focus on risk-based thinking. Risk-based thinking, as the concept and approach added in the new version, requires organizations to identify and analyze potential risks that could arise both from inside and outside of the organizations. Thus, organizations can formulate strategies to prevent any impact of the risks and they can be expectantly more resilient and sustainable by accommodating the risks. Other changes in the new version are the consideration of the organizational stakeholders' needs, the importance of knowledge management and less emphasis on documentation [4, 5]. The new version of ISO 9001 has 10 clauses and a brief comparison of clauses between both versions is presented in Table 1.

Table 1. The comparison of clauses between two latest versions of ISO 9001.

ISO 9001:2008	ISO 9001:2015
Clause 1: Scope	Clause 1: Scope
Clause 2: Normative References	Clause 2: Normative References
Clause 3: Terms and Definitions	Clause 3: Terms and Definitions
Clause 4: Quality Management System	Clause 4: Organizational context.
Clause 5: Management Responsibility	Clause 5: Leadership
Clause 6: Resource Management	Clause 6: Planning
Clause 7: Product Realizations	Clause 7: Support
	Clause 8: Operation
Clause 8: Measurement, Analysis and Improvements	Clause 9: Performance Evaluation Clause 10: Improvement

Previous related studies have been consistently conducted in the last four years. The period between 2014 and 2016 showed that among 219 internal researches on the design and application of various ISO conducted by Industrial Engineering, University of Surabaya – Indonesia [6], five of them were case study researches on service and manufacturing companies producing a generic methodology using Plan-Do-Check-Action (PDCA) approach. A study to improve the effectiveness and efficiency of ISO implementation was done by integrating the lean concept [7, 8] and its result was known as Lean Quality Management System (LQMS). The study was then continued with the application of simplified LQMS on some micro and small enterprises through QMS certification program [9]. Roy & Ghose [10] revealed the stress sensation of the QMS certified companies and their urgencies to adjust their QMS. Cochran [3] explained theoretically the necessary preparations in term of the upgrading process; however, practical experiences on designing and applying new version of ISO 9001 is needed to see the compatibility. The increasing demands from the QMS certified organizations to prepare the transition by September 2018 (otherwise their certifications will expire) called for this research. Hence, this research aims to: (i) identify the gap between existing QMS and the requirement of ISO 9001:2015, (ii) formulate action plans for bridging the gap, and (iii) derive the suggested transition designs and their impacts on organizations to cover the significant changes of ISO 9001:2015.

2. PDCA approach for research methodology

This research was an applied-exploratory research; it tried to explore the practical application of ISO 9001:2015 in some QMS certified organizations. The mixed qualitative and quantitative methods, displayed in Table 2, such as structured interview, questionnaire and sampling, literature review, focus group discussion, statistical analysis were used during data collection and analysis [10]. Figure 1 describes briefly the framework of Plan-Do-Check-Action (PDCA) approach that was used.

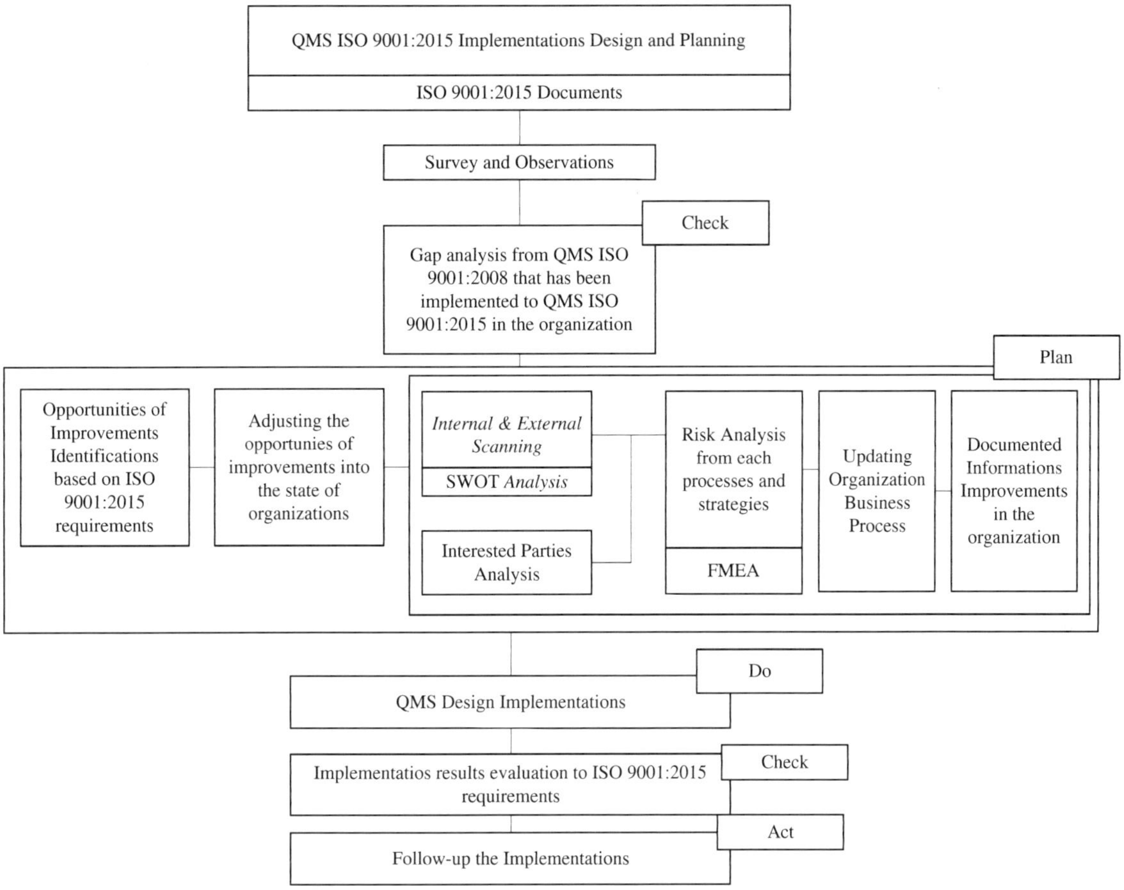

Figure 1. Research steps: PDCA cycle approach.

a. Gap identification and analysis (Check)

The main purpose of gap identification was to investigate how the current QMS complies with ISO 9001:2015 requirements. The gap identification was done through [6]: (i) a questionnaire comprising a list of questions derived from the requirements of the Standard, (ii) observation and survey, guided by the questionnaire, to collect the information on how the current QMS complies with the requirements. The gap identification, followed by its analysis, was aimed to present the percentage of conformity and nonconformity of current QMS towards the Standard. It also helped to identify opportunities for improvements in order to upgrade organizations' QMS.

b. Transition designs (Plan)

Based on the previous gap analysis, there would be several opportunities for improvement for transition designs, especially to take into consideration the changes in ISO 9001:2015. The transition designs covered the usage of concepts such as: (i) Strength Weakness Opportunity & Threat (SWOT) analysis to incorporate internal and external issues of organizations [11] in determining strategies and objectives, (ii) the stakeholders' analysis to identify the needs of interested parties of organizations [3], and (iii) the risk analysis [12]. Furthermore, this planning stage became the focal point of discussion in this publication.

c. Implementation of the transition designs (Do), Evaluation (Check) & Follow-up (Act).

The organizations need to implement the transition designs in order to validate them. After the organizations implement the transition designs, they are then able to evaluate the implementation to ensure the fulfilment of the Standard. Any feedback can be taken to revise the previous designs.

Table 2. The mixed qualitative & quantitative research methods.

No	Research Step	Research Methods
1	Selection of research objects	***Purposive sampling***. There were four organizations voluntarily involved in the research, the selection criteria were the needs for upgrading the QMS and having been certified ISO 9001:2008
2	Gap identification between current QMS and ISO 9001:2015 requirements	***Questionnaire, observation & structured interview***. Gap identification was done using lists of questions, with two categorical answers i.e. confirm or not, supporting by observation to current QMS and secondary data.
3	Gap analysis	***Descriptive statistics***. Tabulated data and numerical analysis.
4	Proposed action plan to close the gap	***Literature review***. Literature search about the new requirements added to ISO 9001:2015 and their practical implications.
5	SWOT analysis, strategies and objectives formulation	***Observation & interview.*** Using the specific concept to identify SWOT, involving the managerial level to validate the SWOT analysis and to formulate relevant strategies and objectives.
6	Stakeholders' analysis	***Focus group discussion.*** Involving the discussion among the stakeholders' parties.
7	Risk identification	***Observation & interview.*** Using the specific concept to identify risk, involving the operational level to registering the risks and determine the risk level.

3. Results and discussion

The framework of PDCA cycle as research methodology was applied into four organizations during 2016, with the characteristics of each organization described in Table 3. All organizations are located in East Java, one is in service industry (organization A) and the rest are manufacturers. Two of the organizations (C and D) are public-listed companies. All organizations have previously been certified by the earlier version of ISO 9001.

Table 3. Four organizations as research objects during 2016.

Research Objects	Location	Industry	Description
Organization A	Surabaya, East Java	Library Unit of Service in a private university	Certified ISO 9001 since 2009.
Organization B	Surabaya, East Java	Thread manufacturer	Certified ISO 9001 since 2003. Certified ISO 22001 since 2013.
Organization C	Sidoarjo, East Java	Glass manufacturer Public-listed company	Certified ISO 9001 since 1996. Certified ISO 14001 since 2005.
Organization D	Surabaya, East Java	Snack manufacturer Public-listed company	Certified ISO 9001 since 2003. Certified ISO 22001 since 2010.

Gap identification analysis is presented in Table 4. The gap analysis showed that the compliance of current QMS of Organization A and B were 71.8% and 79.5%, respectively, i.e. both organizations indicated similar conformity level. These percentages of compliance (%C) could be categorized as high because both organizations had implemented the earlier version of ISO 9001 in which most of the requirements were similar to those in ISO 9001:2015, especially in clauses 5 (Leadership), 7 (Support), 8 (Operations), 9 (Performance Evaluation) and 10 (Improvement). For the latter organizations (C and D), the conformity levels were higher because both organizations were public-listed companies thus had anticipated in advance to upgrade their current QMS especially by preparing the risk assessment. Based on the highest non-conformity percentage of these four organizations, it can be seen that most of the non-compliance occurred on clauses 4 (Organizational Context) and 6 (Planning), presumably because these were newly added requirements in ISO 9001:2015.

Table 4. The compliance of current QMS with ISO 9001:2015 requirements.

Clause of ISO 9001:2015	Organization A		Organization B		Organization C		Organization D	
	%C	%NC	%C	%NC	%C	%NC	%C	%NC
4. Organizational Context	40.0	*60.0*	40.0	60.0	81.6	*18.4*	50.0	*50.0*
5. Leadership	83.3	16.7	71.4	28.6	97.9	2.13	83.3	16.7
6. Planning	20.0	*80.0*	60.0	*40.0*	85.7	*14.3*	66.7	*33.3*
7. Support	95.0	5.0	100.0	0.0	100.0	0.00	95.0	5.0
8. Operation	73.2	26.8	97.8	2.2	100.0	0.00	100.0	0.0
9. Performance Evaluation	100.0	0.0	100.0	0.0	97.2	2.78	100.0	0.0
10. Improvement	90.9	9.1	87.5	12.5	100.0	0.00	75.0	25.0
Total	**71.8**	**28.2**	**79.5**	**20.5**	**97.0**	**3.0**	**91.8**	**8.2**

Note: %C showed the percentage of compliance with the requirements of ISO9001:2015, %NC showed the percentage of non-compliance.

The impacts of the new version of the standard which could be taken as benefits for organizations were that the standard introduced risk-based thinking and helped the organizations address risks and opportunities in a systematic approach; it put more emphasis on leadership commitment especially on strategic planning. The changes in the new version can be divided into three groups:

a. ***Slight changes***. The existing QMS related to quality policy, resource management, management review, internal audit and corrective action has insignificantly changed and can remain the same as the previous one.
b. ***Moderate changes***. The existing QMS related to supply chain management, product or service provision, QMS scope, quality objectives and performance evaluation should be updated to comply with the standard.
c. ***Significant changes***. A list of new requirements, supported by the gap analysis above, should be incorporated into the current QMS from scratch. The description of the new requirements is described as follow [5]:
   - *Clause 4 Context of the Organization, Sub clause 4.1 Understanding the organization and its content*. The requirement states: "The organization shall determine external and internal issues that are relevant to its purpose and its strategic decision and that affect its ability to achieve the intended result(s) of its QMS." **Proposed action plan: transition design #1 (SWOT analysis),**
   - *Clause 4 Context of the Organization, Sub clause 4.2 Understanding the needs and expectation of interested parties*. The requirement states: "The organization shall determine: (a) the interested parties that are relevant to the QMS, (b) the requirement of these interested parties that are relevant to the QMS. The organizational shall monitor and review information about these interested parties and their relevant requirements." **Proposed action plan: transition design #2 (stakeholder analysis),**
   - *Clause 6 Planning, Sub Clause 6.1 Actions to address risks and opportunities*. The requirement states: "When planning for the QMS, the organization shall consider the issues referred to in 4.1 and 4.2 and determine the risks and opportunities." **Proposed action plan: transition design #3 (risk analysis),** and
   - *Clause 7 Support, Sub Clause 7.1.6 Organizational Knowledge*. The requirement states: "The organization shall determine the knowledge necessary for the operation of its processes and

achieve conformity of products and services." Proposed action plan: management of information system, which is not discussed in this paper because the plan will be different and customized to each organization.

Throughout the literature review [3, 8] and brainstorming with experts and practitioners, there are three primary transition designs that should be prepared by organizations to upgrade their QMS from the earlier version to ISO 9001:2015: (1) the identification of internal and external factors of an organization (SWOT analysis), (2) the identification of interested parties of an organization (stakeholder analysis) and (3) the identification of risk (risk analysis). For further discussion, Organization A will be used as the case study to explain the examples of those three transition designs.

3.1. SWOT analysis

This analysis was aimed to identify the internal and external issues of an organization and the strategies were developed based on the SWOT analysis. A strategy should be followed by measurement indicators; therefore, the quality objectives of the current QMS need to be upgraded accordingly. Figure 2 explains how the SWOT analysis was done to determine the strategies and their quality objectives. Internal factors can be analyzed by using the value chain analysis to identify strengths and weaknesses through examining the organization's business processes. External factors can be analyzed by using Porter's Five Forces Model to identify organization's threats and opportunities. In Organization A, by combining its integrated information system (as the strengths) and the increasing usage of information technology (as the opportunities), a strategy as shown in Table 5 was formulated, i.e. to maximize the utilization of the Library by using the support of information technology. Its relevant quality objective was also formulated.

Figure 2. Transition design #1:
From SWOT analysis to organizational strategy and its quality objectives.

3.2. Stakeholder analysis

The objective of this analysis was to identify the interested parties of the organizations and their needs towards the organization. Thus, the organization can determine the business processes that can provide the realization of those needs. The term "interested parties" refers to the stakeholders and after their needs were determined, the organization could analyze whether the current processes were able to accommodate the needs. If not, the organization could create some new processes to fulfill the stakeholders' needs. In Organization A, the stakeholders of Library Unit of Service included the students, lecturers, owner, employees, government, suppliers, other related unit of services and the users from other universities. Table 6 shows an example on how the needs of students as one of the Library's stakeholders were captured in the QMS processes to accommodate the needs.

Table 5. A strategy and its quality objective were developed based-on the SWOT analysis.

SWOT (Organization A)	Strategy	Related Business Process	Quality Objectives	Target	Measurement Method
Strength: It has an integrated information system ***Opportunity:*** The increasing usage of IT	To maximize the utilization of library service by using IT	The provision of Information	The increased visitors of library user interface or online service (***digilib*** and ***elib***)	110%	Count the visitors of ***digilib*** and ***elib.*** The visitors of this period should be 110% for previous period.

Table 6. Transition design #2: From stakeholder's needs to related business processes.

Stakeholders	***Needs***		***Related QMS Processes***
Students	Library should provide good services which supported by good and complete facilities as well as clear procedures.	• Complete collection • Well-provided facilities • Good service quality	• The procurement process • The circulation service of literature material • The Recruitment process

3.3. Risk analysis

One of the major changes in ISO 9001:2015 is the concept of risk-based thinking, which requires the organization to determine the risk from its business processes or strategies. Risk needs to be addressed; it can cause high negative impact on organizational sustainability unless the organization can take control of it. Among the several methods for risk assessment, FMEA (Failure Mode and Effect Analysis) is the most common one that can be used by organizations for risk analysis [12, 13]. FMEA identifies failure potentials and determines the level of the risk by considering three scales of measurement, i.e. the severity (impact of the failure), occurrence (frequency of the failure) and detection (the inability of current control to detect the failure). The scales are weighted with the score of 1-10 or 1-5 (it is flexible as long as the organizations can define the meaning of the value). These three values are then multiplied to determine the RPN (risk priority number) where higher RPN indicates higher risk level. RPN is used to prioritize the risks; if the RPN reaches 25% of the total RPN then those risks should be anticipated [14].

There were two types of risks that were registered during the risk assessment, i.e. strategic and operational risks. Strategic risks were formulated by identifying any failure that could prevent the deployment of strategies. Operational risks were initially formulated by identifying any failure that is always faced by the process doer (internal source) or derived from the customer complaints (external source).The risk register can be updated subsequently during the period of internal audit to add the unidentified failure. Table 7 shows the examples of both strategic and operational risks in Organization A. The FMEA table above used the score of 1-5 for the value of Severity, Occurrence and Detection, with total maximum RPN of $5\times5\times5=125$. Thus, a failure with RPN higher than 25% of the total maximum RPN ($125\times25\% = 31.25$) should be immediately anticipated; the recommended actions should be proposed and implemented accordingly.

4. Conclusion

Organizations are urged to prepare the transition of their Quality Management System (QMS) from ISO 9001:2008 to ISO 9001:2015 before the previous certification expires. For organizations that already implemented the earlier version, there were already high compliance levels of current QMS with the requirements of ISO 9001:2015(with the average of 75%), although they had not done anything yet in term of transition preparation. The organizations could prepare these three major transition designs. Firstly, the organizations can formulate their strategies and relevant quality objectives based on the SWOT analysis. Secondly, they can accommodate the stakeholders' needs by defining relevant QMS business processes. Finally, they also need to register the strategic and operational risks which can be done by using FMEA. The paper showed how these transition designs could be prepared and the examples as the result of implementation in a service organization were also provided. By performing these three major transition designs, the organization is expected to be able to cover the gap and meet the new requirements in ISO 9001:2015.

Table 7. Risk register: Examples of strategic and operational risks.

Strategies or Business Process	Potential Failure Mode	Potential Effect(s) of Failure	Sev	Potential Cause(s) of Failure	Occ	Current Process Controls	Det	RPN	Recommended Action(s)
Strategic Risk									
To maximize the utilization of library service	Overloaded web	Online catalogue was inaccessible	3	Crowded visitors	3	Historical record of web visitors	3	27	-
Operational Risk									
The Circulation Service of literature material	Books had not been returned by the borrowers	Books were listed in online catalogue but not available for rent	3	Limited quantity	5	Periodic checking	5	**75**	**Adding the activity to send reminder to users through email or smartphone notifications before and after the due date.**

5. References

[1] Stevenson T H and Barnes F C 2001 Fourteen years of ISO 9000: impact, criticisms, costs and benefits *Bus. Horizons* **44** pp 45–51

[2] Rao Tummala V M and Tang C L 1996 Strategic quality management, Malcolm Baldrige and European quality award and ISO 9000 certification *Int. J. Qual. & Reliab. Manage.* **13** pp 8–37

[3] Cochran C 2015 *ISO 9001:2015 in Plain English* (Chico: Paton Professional)

[4] Fonseca L M 2015 From quality gurus and TQM to ISO 9001: 2015: a review of several quality paths *Int. J. Qual. Res.* **9** pp 167–180

[5] BSI 2015 *BS EN ISO 9001:2015 Quality Management Systems Requirements* (London: British Standards Institution)

[6] Sari Y and Rosiawan M 2016 The orchestration of tri dharma activities in university to help increasing the competitive advantages of society *J. of Synergy* **1** pp 52–63

[7] Sari Y, Hadiyat M A and Beatrice C 2015 The design and implementation of Lean Quality Management System *National Conf. Proc. of Production System XI and Quality Engineering Management VI* pp B123–B136

[8] Blecken A, Zobel A and Maurantzas E 2010 Development of a lean quality management system: An integrated management system *Adv. Manuf. and Sustainable Logist. (Lect. Notes Bus. Inf.* vol 46) eds Dangelmaier W, Blecken A, Delius R and Klöpfer S (Berlin, Heidelberg: Springer) pp 141–151

[9] Sari Y, Sugiarti Y, Wahyudi R D and Falencia K C 2016 The design and application of ISO 9001:2008 to help micro and small enterprises facing ASEAN Economic Community *National Proc. of Community Services & Corporate Social Responsibility* pp 176–193

[10] Roy R and Ghose D 2016 Changeover from ISO 9001:2008 to the revised standard ISO 9001:2015 *J. Res. Bus. Manage.* **4** pp 89–93

[11] Harvard Business School 2003 *Managing Change and Transition* (Massachusetts: Harvard Business School Publishing Corporation)

[12] Merna T and Al-Thani F 2008 *Corporate Risk Management* (New Jersey: Wiley)

[13] Paciarotti C, Mazzuto G and D'Ettore D 2014 A revised FMEA application to the quality control management *Int. J. Qual. & Reliab. Manage.* **31** pp 788–810

[14] Kaho V R and Susilo L J 2010 *Risk Management Based-on ISO 31000 for Non-Banking Industry* (Jakarta: PPM)

Improving delivery routes using combined heuristic and optimization in a consumer goods distribution company

E Wibisono, A Santoso and M A Sunaryo

Department of Industrial Engineering, University of Surabaya, Raya Kalirungkut, Surabaya 60293, Indonesia

E-mail: ewibisono@staff.ubaya.ac.id

Abstract. XYZ is a distributor of various consumer goods products. The company plans its delivery routes daily and in order to obtain route construction in a short amount of time, it simplifies the process by assigning drivers based on geographic regions. This approach results in inefficient use of vehicles leading to imbalance workloads. In this paper, we propose a combined method involving heuristic and optimization to obtain better solutions in acceptable computation time. The heuristic is based on a time-oriented, nearest neighbor (TONN) to form clusters if the number of locations is higher than a certain value. The optimization part uses a mathematical modeling formulation based on vehicle routing problem that considers heterogeneous vehicles, time windows, and fixed costs (HVRPTWF) and is used to solve routing problem in clusters. A case study using data from one month of the company's operations is analyzed, and data from one day of operations are detailed in this paper. The analysis shows that the proposed method results in 24% cost savings on that month, but it can be as high as 54% in a day.

Keywords: nearest-neighbor heuristic; vehicle routing problem; heterogeneous vehicles; time windows; consumer goods distributor.

1. Introduction

Logistics operations are oriented toward the fulfillment of customer needs from the point of origin to the point of consumption. One mode of operation in this field is the vehicle routing problem (VRP) that has received considerable attention from academics and practitioners, mainly due to its relevant applications that are often encountered in real world. Being a popular routing model in logistics studies, a number of variants have been developed, tested, and applied. These variants follow the characteristics of the problem being studied, e.g. VRP with time windows (VRPTW) [1], VRP with pickups and deliveries [2], or more specific variants such as VRPTW with evolutionary algorithm [3], VRP with split deliveries [4], or VRP with multiple objectives [5]. More recent VRP studies include, e.g. VRPTW with probabilistic travel times [6], green VRP [7], or VRP in maritime logistics [8].

VRP literature grew exponentially with an annual rate of 6.09% between 1956 and 2005, and the period between 1985 and 2006 recorded 918 published VRP articles [9]. This trend continued with 277 articles appeared between 2009 and 2015 as cited in the latest survey [10]. Despite the massive amount of literature described above, *heterogeneous* VRP (HVRP) is rarely studied due to its increased complexity compared to the basic capacitated VRP (CVRP). While reduced complexity allows more in-depth study of the problem on hand, it deviates from reality in industrial applications. In contrast to the number of general VRP articles cited in [9], only 22 articles related to HVRP were published between 1984 and 2007 [11]. In this particular area where analytical approach is hardly efficient, metaheuristics spur as sensible substitute in dealing with complex VRP problems. A few successful examples worth mentioning are record-to-record [12], scatter search [13], genetic algorithm [14], variable neighborhood search [11], multi-start adaptive memory programming [15], iterated local search [16], and tabu search [17].

This paper demonstrates the application of HVRP in a consumer-goods distribution company operating in Surabaya, Indonesia. Being a large city, there are hundreds of wholesalers and retailers in Surabaya served by the company. The complexity of the problem arises both from the number of customers and the heterogeneity of vehicles (in terms of capacity and fixed cost) that the company uses. Time windows are also demanded by the customers. Thus, the model is called HVRP with time windows and fixed costs (HVRPTWF). Routing of distribution is planned on a daily basis therefore computation time has a higher priority than optimality of the solution. To address the problem complexity, an approach integrating the classical time-oriented, nearest neighbor (TONN) heuristic [18] and optimization via mathematical programming is proposed. This integration is a necessity since mathematical programming alone will clearly be impractical as it will suffer from long computation time on complex problems. The proposed model will be compared to the method that the company applies in its daily operations.

Given the above background, the objectives of this paper are twofold:

- To develop an integrated model combining heuristic and optimization in solving routing problem that considers heterogeneous vehicles and time windows.
- To compare the solution from the proposed model with the actual routing obtained from the company's operations.

2. Methodology

We first described in this section the background of the company in our case study. The company XYZ was founded in 1995 as a distributor of consumer goods products serving mainly the Eastern part of Indonesia. The products varied from toiletries (shampoo etc.) to foods (snacks etc.) and the customers also varied from small retailers to giant hypermarkets. XYZ's fleet of vehicles comprised more than 300 trucks/vans and 100 motorcycles, operating from 85 branch offices and 25 warehouses in Indonesia. Orders were received daily from sales force then processed by the planning department to produce delivery plan for the next day. In our study, we focused on Surabaya office (the company's headquarters) and the foods division.

XYZ's customers were classified in three groups: small, large, and super stores. The small stores were served between 8.00 and 13.00, and in Surabaya they were clustered into six areas where each was visited weekly on different days. The large stores were served between 9.00 and 14.00 and clustered into three areas that could be visited every day based on the orders placed. The super stores, given their large demand, had no regular service hours and clusters, and could be served any time within XYZ's working hours. This clustering was the basis of XYZ in making its daily distribution plan. Since drivers were already assigned areas of operations, it was possible that two vehicles depart to two different stores, adjacent on the map but on different distribution areas, without maximizing the vehicle capacity (not fully loaded). Often it was possible for these stores to actually be served by the same vehicle, but the drivers who were used to work in specific areas found it uncomfortable to go to different ones. In addition, the planning department only provided the list of stores to be visited without route and rely on the drivers' experience to plan their own routes every day. All this led to inefficiency in the operations which was visible from the uneven workload of vehicles where some returned after 12 hours but some did after 18 hours.

To solve this problem, an integration model utilizing a time-oriented, nearest-neighbor heuristic and mathematical programming was proposed. The TONN heuristic was used to cluster the number of drop points if they exceeded a certain number that prevented the mathematical model HVRPTWF to be run in an acceptable time. Figure 1 shows the path of solution in the methodology.

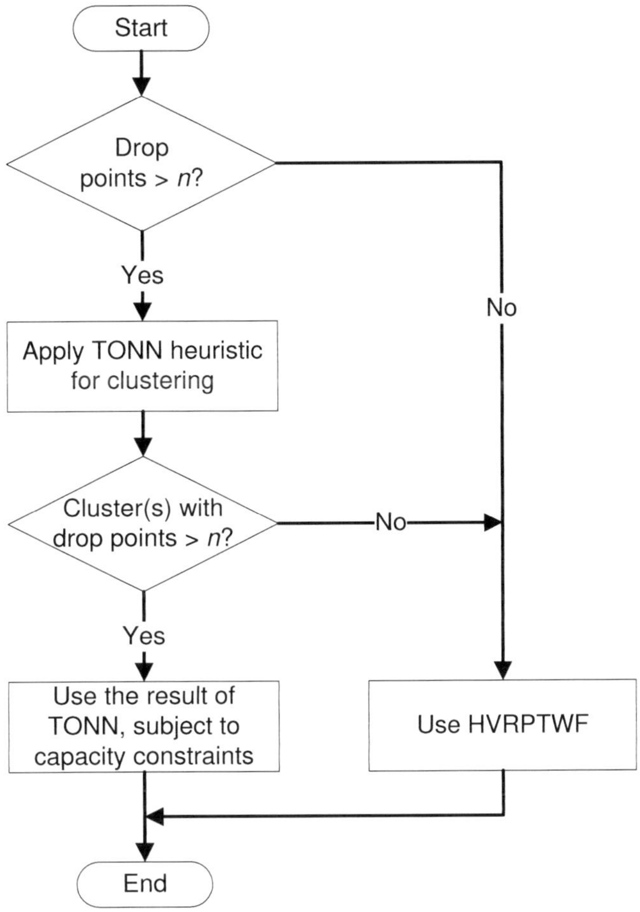

Figure 1. Solution path in proposed methodology.

Figure 1 is explained as follows. First, customers are grouped into drop points because the number of drop points is what affects the routing instead of the number of customers. If there are less than n drop points to be served on a particular day, HVRPTWF is applied. If there are more, TONN heuristic is executed to form the clusters and sequence of visit in each cluster. If, after the heuristic, there are still clusters with more than n drop points, the sequence produced by TONN will be used as the route for vehicles serving that route, subject to capacity constraints. It is possible at this stage to divide a TONN cluster into smaller sub-clusters if vehicle capacity is not sufficient to serve all drop points in that cluster. If the heuristic produces clusters with less than n drop points, HVRPTWF is applied in each of such clusters.

The TONN heuristic is explained with the following example. Suppose $\mathcal{N}$ is the set of all nodes (locations) including node 0 as depot and $\mathcal{C}$ is the set of drop points or $\mathcal{N} \setminus \{0\}$. Given i visited exactly prior to j, s_i and s_j are the arrival times at nodes i and j, respectively, p_i is the service time at node i, and $t_{i,j}$ is the travel time from i to j, then the relationship between s_i and s_j is formulated as $s_i + p_i + t_{i,j} \leq s_j, \forall i,j \in \mathcal{C}$. The lower and upper time windows at node i, L_i and U_i, respectively, bound the arrival time s_i, or $L_i \leq s_i \leq U_i, \forall i \in \mathcal{C}$. Table 1 exhibits distance matrix between all nodes and time windows in each node. The service time p_i is set constant. Note that the distance matrix is symmetrical and also satisfies triangular inequality.

Table 1. Data for Time-Oriented, Nearest-Neighbor heuristic example.

	Distance matrix						Time windows		Service time
	0	1	2	3	4	5	L_i	U_i	p_i
0	-	5	10	13	10	14	-	-	-
1	5	-	7	12	13	17	0	20	2
2	10	7	-	8	14	14	10	30	2
3	13	12	8	-	7	6	10	30	2
4	10	13	14	7	-	5	10	30	2
5	14	17	14	6	5	-	10	30	2

Given the data in Table 1, the first cluster resulting from the heuristic and its sequence from depot is 0→1→2→3→0 and the second cluster and its sequence is 0→4→5→0. The heuristic only solves the clustering problem with respect to time windows and leaves the capacity allocation to the second part in methodology, i.e. mathematical programming with HVRPTWF. In Figure 1, the value of n will be determined after few trial runs of HVRPTWF to see at how many nodes the model becomes impractical with regard to computation time. In the above example, suppose it is found that $n = 2$ (this low figure is only for the sake of this example), then the first cluster will not be optimized with HVRPTWF and the heuristic result will be used with arbitrary vehicle assignment.

For HVRPTWF, the mathematical model follows [19] with more specific elaboration on the lower time window (in [19] the model is applied on maritime logistics case and only the upper time window which reflects due date is used). The sets and equations are explained in the following.

$\mathcal{V}$	Set of vehicles, indexed by v
$\mathcal{A}$	Set of arcs (i, j) denoting a flow from node i to node j
$\mathcal{N}$	Set of all nodes $\mathcal{N} = \{0,1,\ldots,N\}$; 0 is depot
$\mathcal{C}$	Set of droppoints$\mathcal{C} = \mathcal{N} \setminus \{0\}$
f^v	Fixed cost of vehicle v
$c_{i,j}^v$	Variable cost of vehicle$v$ if it goes from node i to node j
$t_{i,j}^v$	Travel time of vehicle$v$ if it goes from node i to node j
C^v	Capacity of vehicle$v$
D_i	Total demand at node i
L_i	Lower time window at node i
U_i	Upper time window at node i
p_i	Service time at node i
M	A large constant
$x_{i,j}^v$	Binary variables for vehicle v in arc (i,j); $x_{i,j}^v = 1$ if the vehicle traverses arc (i,j) and $x_{i,j}^v = 0$ otherwise
s_i^v	Time window for vehicle v at node i

$$\text{Minimize} \sum_{v\in\mathcal{V}} \sum_{i,j\in\mathcal{A}} x_{i,j}^v . c_{i,j}^v + \sum_{v\in\mathcal{V}} \sum_{j\in\mathcal{A}} f^v . x_{0,j}^v \tag{1}$$

Subject to:

$$\sum_{v\in\mathcal{V}} \sum_{i,j\in\mathcal{A}} x_{i,j}^v . C^v \geq D_i \qquad \forall\, i \in \mathcal{C} \tag{2}$$

$$\sum_{i\in\mathcal{C}} D_i \sum_{j\in\mathcal{N}} x_{i,j}^v \leq C^v \qquad \forall\, v \in \mathcal{V} \tag{3}$$

$$\sum_{i\in\mathcal{N}} x_{i,k}^{v} - \sum_{j\in\mathcal{N}} x_{k,j}^{v} = 0 \qquad \forall\, k \in \mathcal{C};\ v \in \mathcal{V} \tag{4}$$

$$x_{i,i}^{v} = 0 \qquad \forall\, i \in \mathcal{N};\ v \in \mathcal{V} \tag{5}$$

$$\sum_{j\in\mathcal{C}} x_{0,j}^{v} \leq 1 \qquad \forall\, v \in \mathcal{V} \tag{6}$$

$$L_i \leq s_i^{v} \leq U_i \qquad \forall\, i \in \mathcal{C};\ v \in \mathcal{V} \tag{7}$$

$$s_i^{v} + t_{i,j}^{v} + p_i - M(1 - x_{i,j}^{v}) \leq s_j^{v} \qquad \forall\, i \in \mathcal{N}; j \in \mathcal{C};\ v \in \mathcal{V} \tag{8}$$

$$x_{i,j}^{v} \in \{0, 1\} \qquad \forall\, i, j \in \mathcal{A};\ v \in \mathcal{V} \tag{9}$$

$$s_0^{v} = 0 \qquad \forall\, v \in \mathcal{V} \tag{10}$$

$$s_i^{v} \geq 0 \qquad \forall\, i \in \mathcal{N};\ v \in \mathcal{V} \tag{11}$$

In the above model, the objective function (1) minimizes total cost that consist both variable and fixed costs of using vehicles. Demand fulfillment is warranted by constraints (2) and vehicle capacity is observed by constraints (3). Constraints (4) balance the incoming and outgoing flows. Constraints (5) are loop prevention and constraints (6) regulate so that a vehicle can take only one trip. Constraints (7) set the lower and upper time windows for a vehicle at a node, while constraints (8) manage the arrival times s_i similar to the explanation in the heuristic section. Sub-tour breaking constraints are not required due to constraints (8). Lastly, constraints (9)–(11) state the nature of decision variables involved:$x_{i,j}^{v}$ are binary and s_i^{v} are continuous, thus the model falls in the category MILP problem.

3. Case study analysis

Our first step in analysis was determining the threshold n to limit the number of drop points that HVRPTWF could still solve efficiently. Using a commercial solver and Intel Core i5 processor running at 2.2 GHz with 4GB RAM on Windows 10, we arrived at the results in Table 2. From this table it can be inferred that n should be set equal to 10. With more than 10 locations, the solver was still unable to find an optimal solution after more than 16 hours.

Table 2. Computation time of HVRPTWF.

n	Average running time	Status
6	1.5 seconds	Global Opt.
7	7.5 seconds	Global Opt.
8	33.5 seconds	Global Opt.
9	21 minutes 54 seconds	Global Opt.
10	1 hour 5 minutes 4 seconds	Global Opt.
11	16 hours 48 minutes 55 seconds	Feasible

Next, a sample of one-month company's operation in the foods division was studied. In this section, we provided an example of one-day distribution and compared the results between the company's method and our proposed methodology combining TONN heuristic and HVRPTWF mathematical programming. The chosen day consisted of 21 nodes (1 depot and 20 drop points) and the 21× 21 distance matrix (in km) was given in Table 3, including the demands (in m^3), service times and time windows(in minutes) at the nodes. The service times were not correlated to the demands, but to the number of items which were not shown here. Data of vehicles are provided in Table 4.

Table 3. Distance matrix (km), demands (m^3), service time and time windows (mins) of locations.

	0	1	2	3	4	5	6	7	8	9	10	11	12	13	14	15	16	17	18	19	20	D_i	p_i	L_i	U_i
0	-	20.1	13.5	14.7	19.0	20.0	18.4	18.3	18.4	18.5	12.7	24.5	21.4	18.7	17.5	19.8	18.8	18.7	16.7	18.3	13.3	-	-	-	-
1		-	6.6	5.4	11.9	11.5	11.4	11.2	4.5	11.4	7.4	19.5	15.2	10.2	10.3	10.3	11.7	11.6	9.5	9.9	9.4	2.08	8.70	0	300
2			-	1.7	15.5	14.0	15.1	14.9	5.2	15.1	4.2	22.0	17.6	12.7	14.0	12.7	15.4	15.3	13.2	12.4	10.5	0.23	25.00	60	360
3				-	13.8	12.7	13.4	13.2	4.6	13.4	2.5	21.0	16.7	11.3	12.3	11.5	13.7	13.6	11.5	11.1	9.8	0.22	25.67	60	360
4					-	3.4	2.5	2.3	14.4	2.5	13.7	10.5	6.5	3.7	1.6	3.1	2.8	2.7	2.4	3.3	5.7	0.07	7.37	0	300
5						-	3.8	4.2	13.7	3.1	13.0	10.6	6.6	3.0	3.8	2.4	4.2	2.9	3.3	2.7	2.7	0.05	4.37	0	300
6							-	1.5	15.1	1.7	14.4	8.6	4.6	4.4	2.4	3.8	0.4	1.8	2.1	4.1	6.0	0.04	2.37	0	300
7								-	14.5	1.1	13.3	10.1	5.7	4.8	1.6	4.2	1.3	1.3	2.5	4.5	4.0	0.04	2.03	0	300
8									-	13.4	5.7	21.3	16.9	12.0	12.9	12.0	14.7	13.3	13.0	11.6	10.8	0.03	17.67	60	360
9										-	13.7	9.1	5.1	3.9	1.1	3.3	1.3	0.2	2.3	3.5	5.2	0.03	2.37	0	300
10											-	21.2	16.9	11.7	12.6	11.7	14.0	13.6	12.3	11.4	8.8	0.02	17.67	60	360
11												-	4.4	11.3	9.2	10.7	8.8	8.9	10.0	10.9	12.4	0.02	2.70	0	300
12													-	7.4	5.7	6.9	4.9	4.9	6.7	7.1	8.2	0.02	2.03	0	300
13	Symmetrical													-	3.2	1.6	4.3	3.8	4.1	1.9	7.0	0.01	3.37	0	300
14															-	4.2	2.0	1.3	2.9	4.5	4.8	0.01	2.03	0	300
15																-	4.2	3.4	5.3	1.9	8.2	0.01	3.03	0	300
16																	-	1.4	2.5	3.7	5.6	0.01	2.03	0	300
17																		-	2.5	3.4	5.4	0.01	2.37	0	300
18																			-	3.4	4.9	0.01	3.37	0	300
19																				-	6.7	0.01	2.37	0	300
20																					-	0.01	4.37	0	300

Table 4. Data of vehicles.

Type	Units	Capacity (m^3)	Cost/day (IDR)	Cost/km (IDR)
A	2	2.4	544,687	812.50
B	1	5.6	502,526	812.50
C	1	9.6	603,725	812.50

Using the company's method, three vehicles were used with the following routings:

- Vehicle A1: 0→2→3→8→10→0
- Vehicle A2: 0→20→6→7→9→16→4→14→17→5→13→15→18→19→12→11→0
- Vehicle C: 0→1→0

The above routing demonstrated inefficient use of vehicle C since, despite its large capacity; it was used to serve one particular store simply because that customer had regularly been served by vehicle C and its driver. This one-on-one relationship between customer and driver familiarized the driver with his region of operations. The driver's knowledge on his region helped on some days and was justified when demand size matched vehicle capacity, but more often led to inefficiency as shown above. The routes constructed by the drivers also relied on their judgment and experience, but was proven suboptimal compared to the output of heuristic and mathematical programming as outlined in the next section.

4. Results and discussion

Applying the algorithm in Figure 1 yielded the following results. First, two clusters (instead of three) were formed. The first cluster consisted of 15 locations and the second cluster consisted of 5 locations. For the first cluster, total demand was 0.34 m^3 so it could be served by vehicle A (time windows had already been taken care of by the heuristic). For the second cluster, HVRPTWF was applied, and it produced a different routing from the actual one by the driver. The routings in both clusters were given below:

- Vehicle A1: 0→20→14→9→17→7→16→6→18→4→15→13→19→5→12→11→0
- Vehicle B: 0→2→3→1→8→10→0

It is worth to mention here that since we knew from the data that vehicle A could not be used due to its lower capacity (2.4 m^3) than the second cluster's demand (2.58 m^3), and vehicle B's daily cost was smaller than vehicle C's, then vehicle B was a logical choice for serving the second cluster. Further, since time windows were no longer an issue as a result from the heuristic application, routing problem of the second cluster was actually reduced to a travelling salesman problem (TSP). This advantage, however, was not apparent on other cases in different days, so HVRPTWF formulation was still required. Besides, with less than 6 locations, HVRPTWF required not more than 1.5 seconds to solve.

Next, we compared the total cost of actual operations and that of our proposed methodology. As shown in Table 5, the proposed method combining heuristic and mathematical programming was superior compared to the drivers' assignment-based method from the company. The proposed method reduced the number of vehicles required that led to cost savings in fixed and variable costs. The cost savings in fixed cost, however, were only on paper since fixed cost was calculated based on depreciation, drivers' salaries, etc., i.e. it would be incurred regardless of vehicle utilization. The savings in variable cost, on the other hand, were tangible. For this day only, the savings in variable cost were 53.60%. For the overall month, the company's method yields IDR 2,406,548 whereas the proposed method did IDR 1,849,893 or equivalent to a 23.56% cost reduction.

Table 5. Cost comparison of two methods.

Vehicle	Actual (IDR)		Proposed (IDR)	
	Fixed Cost	Var. Cost	Fixed Cost	Var. Cost
A1	544,687	31,038	544,687	34,125
A2	544,687	65,894	Not used	
B	Not used		502,526	35,344
C	603,725	32,663	Not used	
Total	**1,693,009**	**129,595**	**1,047,213**	**69,469**

5. Conclusions and remarks for future research

In this paper we have demonstrated an application of combined heuristic and optimization to solve practical routing problems in a distribution company. The company planned its delivery routes daily so a methodology that could arrive at good solution in an acceptable time would have a high practical value. Our proposed approach beat the company's method that was based on drivers' regional assignment by 24% on average but on certain days, such as one shown in the example, the savings could be as high as 54%. The only drawback from implementing this method was to reorient the drivers from their comfort zone serving only their regular customers in a particular region.

The comparison in this paper was only made against the actual routings used by the company. It is possible there are still better approaches, be they heuristics or optimization, which should be aimed as future research agenda.

References

[1] El-Sherbeny N A 2010 Vehicle routing with time windows: An overview of exact, heuristic and metaheuristic methods *J. King Saud University – Science* **22** pp 123–131

[2] Wassan N A and Nagy G 2014 Vehicle routing problems with deliveries and pickups: Modelling issues and meta-heuristics solution approaches *Int. J. Transportation* **2** pp 95–110

[3] Bräysy O, Dullaert W and Gendreau M 2005 Evolutionary algorithm for the vehicle routing problem with time windows *J. Heuristics* **10** pp 587–611

[4] Archetti C and Speranza M G 2012 Vehicle routing problems with split deliveries *Int. T. Oper. Res.* **19** pp 3–22

[5] Josefowiez N, Semet F and Talbi, E-G 2008 Multi-objective vehicle routing problems *Eur. J. Oper. Res.* **189** pp 293–309

[6] Agra A, Christiansen M, Figueiredo R, Hvattum L M, Poss M and Requejo C 2013 The robust vehicle routing problem with time windows *Comput. Oper. Res.* **40** pp 856–866

[7] Lin C, Choy K L, Ho G T S, Chung S H and Lam H Y 2013 Survey of green vehicle routing problem: Past and future trends *Expert Syst. Appl.* **1** pp 1118–38

[8] Wibisono E and Jittamai P 2015 Collaborative capacity sharing in liner shipping operations *Int. J. Logistics Systems and Management* **22** pp 520–539

[9] Eksioglu B, Vural A V and Reisman A 2009 The vehicle routing problem: A taxonomic review *Comput. Ind. Eng.* **57** pp 1472–83

[10] Braekers K, Ramaekers K and Van Nieuwenhuyse I 2016 The vehicle routing problem: State of the art classification and review *Comput. Ind. Eng.* **99** pp 300–313

[11] Imran A, Salhi S and Wassan N A 2009 A variable neighborhood-based heuristic for the heterogeneous fleet vehicle routing problem *Eur. J. Oper. Res.* **197** pp 509–518

[12] Li F, Golden B and Wasil E 2007 A record-to-record travel algorithm for solving the heterogeneous fleet vehicle routing problem *Comput. Oper. Res.* **34** pp 2734–42

[13] Belfiore P and Yoshizaki H T Y 2009 Scatter search for a real-life heterogeneous fleet vehicle routing problem with time windows and split deliveries in Brazil *Eur. J. Oper. Res.* **199** pp 750–758

[14] Prins C 2009 Two memetic algorithms for heterogeneous fleet vehicle routing problems *Eng. Appl. Artif. Intel.* **22** pp 916–928

[15] Li X, Tian P and Aneja Y P 2010 An adaptive memory programming metaheuristic for the heterogeneous fixed fleet vehicle routing problem *Transport. Res. E-Log.* **46** pp 1111–27

[16] Subramanian A, Penna P H V, Uchoa E and Ochi L S 2012 A hybrid algorithm for the Heterogeneous Fleet Vehicle Routing Problem *Eur. J. Oper. Res.* **221** pp 285–295

[17] Jiang J, Ng K M, Poh K L and Teo K M 2014 Vehicle routing problem with a heterogeneous fleet and time windows *Expert Syst. Appl.* **41** pp 3748–60

[18] Solomon M M 1987 Algorithms for the vehicle routing and scheduling problems with time window constraints *Oper. Res.* **35** pp 254–265

[19] Wibisono E 2016 Application of genetic algorithm in containerships network design problem *Proc. 8th Widyatama Int. Seminar on Sustainability* (Bandung: Widyatama University) pp 297–302

The effect of different concentrations of tween-20 combined with rice husk silica on the stability of o/w emulsion: A kinetic study

L Sapei[1], I G Y H Sandy[1], I M K D Suputra[1] and M Ray[2]

[1] Department of Chemical Engineering, University of Surabaya, Raya Kalirungkut, Surabaya 60293 East Java, Indonesia.
[2] POS Bio-Sciences, John and Charlotte Cross, BioSciences Centre, 118 Veterinary Road, Saskatoon, SK, S7N 2R4, Canada

E-mail: lanny.sapei@staff.ubaya.ac.id

Abstract. Emulsion is a thermodynamically unstable system which undergoes destabilization with time. The destabilization kinetics of "food grade" oil-in-water (O/W) emulsions in the presence of both tween-20 and rice husk silica as emulsifiers were studied. Rice husk silica concentration of 2.5% was combined with various concentrations of tween-20 from 0.1 to 1%. Oil phase fraction was 20% relative to the aqueous phase. Emulsification was conducted using a rotor-stator homogenizer at 20,000 rpm. The emulsions tended to destabilize with time. Their destabilization rates were studied using zero order and first order kinetic models. In general, the kinetics of O/W emulsion destabilization followed first order model. Different concentrations of tween-20 combined with rice husk silica influenced the destabilization rate of o/w emulsions. Destabilization rates of emulsions stabilized using mixed emulsifiers of 1% tween-20 and 2.5% silica were ~50 times and ~3 times lower compared to those stabilized using silica alone and tween-20 alone, respectively.

1. Introduction

Emulsion is a colloidal dispersion whereby dispersed phase and dispersion phase are immiscible. In order to form a stable system, emulsifiers have to be added. Emulsifiers could be natural or synthetics and many are found in foods, such as ice cream, whipped cream, margarine, chocolate, and so on. The synthetic ones could cause obesity [1], thus there have been many attempts to limit or reduce their use. Natural emulsifiers have low toxicity level, easily biodegraded, and do not cause obesity in the long term thus very suitable to be used in the food industries replacing the synthetic ones [2]. Amphiphilic polymers such as gum, polysaccharide, saponin, lecithin, lipoprotein, and solid particles such as silica have been generally used as emulsifiers. Emulsifiers used in this experiment was Tween-20 which was regarded as GRAS (generally recognized as safe) within the limited use and acceptable daily intake and amorphous biosilica isolated from rice husk. Rice husks are a natural reservoir for nanostructured silica which contain as much as ~20% silica of the dry weight of the rice husk [3]. Moreover, the silica within rice husks naturally exists in the form of nanoparticles with primary particles of ca. 4.2 nm in diameter [4].

Previous research showed that mainly emulsions were stabilized using amphiphilic polymer or solid particle alone did not persist for a long time due to some advantages and disadvantages. Amphiphilic emulsifiers were easily adsorbed at the oil-water interface stabilizing the system but they were also easily detached from the interface thus destabilization occurred. Silica particles were difficult to be adsorbed in the interface of oil-water, but once being adsorbed; they stayed and stabilized the system [5]. Therefore, mixtures of amphiphilic polymer and solid particles were used to have a synergism effect [6]. Polymeric emulsifier was attached on the surrounding surface of oil-water interface while reducing the surface tension between two phases and solid particle formed the barrier on the oil globule surfaces thus preventing recoalescence which would lead to oil globule separation [6]. Emulsifier mixtures could reduce the rate of o/w emulsion destabilization.

Coalescence is the merging of smaller oil globules into bigger ones. In general, coalescence process which leads to o/w emulsion destabilization followed the first order kinetic model [7]. By investigating the kinetics of emulsion destabilization, the effect of mixed emulsifiers' concentrations on the emulsion stability could be well investigated. This research was expected to give the beneficial contribution to the food industries dealing with development of a more healthful food emulsions for the determination of product shelf-life and storage condition. The aim of this research is to study the stabilization of o/w emulsion using mixture of various concentration of Tween-20 combined with rice husk silica for the formulation of food grade o/w emulsion with kinetics approach.

2. Materials and methods

2.1 Materials

Tween-20 (Merck, Germany); rice husk derived from Mojokerto, East Java, Indonesia; palm oil consisting of 40% saturated fatty acid mainly palmitic acid and 60% unsaturated fatty acid mainly oleic acid with FFA content of less than 0.3% (SunCo, PT. Milkie Oleo Nabati Industri, Bekasi, Indonesia); and demineralized water.

2.2 Preparation of rice husk silica

Clean and dried rice husks were leached out using citric acid 5% in an agitated flask for the metallic impurities removal. Afterwards, they were rinsed several times using demineralized water and then dried in an oven (Memmert, Germany) at 105°C for 2 hours. Leached dried rice husks were then thermally treated in a furnace (Ney VULCAN D-550, Dentsply Ceramco, USA) at 750°C for 5 hours. White ashes consisting of almost pure silica (> 95%) were obtained after the thermal process. The rice husk silica were milled and screened. Biosilica of 200 mesh were used as the emulsifier of o/w emulsions. Biosilica is amorphous and inherently nanostructured with primary spherical particle of several nm in diameter [4].

2.3 Preparation of o/w emulsion stabilized with tween-20 and rice husk silica mixtures.

Aqueous phase was prepared by adding biosilica (2.5%) and Tween-20 (0.1-1%) into the demineralized water and then mixed using a magnetic bar and heated to 50°C for 15 minutes. Oil was poured into a beaker glass and heated up to 50°C for 15 minutes. Oil of 20% was then added into the aqueous phase and emulsified for 5 minutes using a rotor-stator homogenizer (IKA T25 digital ULTRA TURRAX, *Germany*) at 20,000 rpm. The pH of the emulsion was not adjusted. The resulting o/w emulsion was quickly quenched in an ice bath for about 15 seconds. Some emulsions were prepared without any emulsifiers, with Tween 1% only, and with biosilica 2.5% only as controls. The emulsions were transferred into transparent 40 ml glass vials (ID= 25 mm, height= 95 mm) until reaching the height of ~ 4 cm from the bottom and then stored at a room temperature of ~ 28°C for stability tests.

2.4 Determination of emulsion stability

The height of emulsion was measured every 1 minute for the first ten minutes, followed by the measurement in every 5 minutes afterwards until the first one hour, and then the measurement was taken in every 10 minutes for the next one hour. The stable emulsion layer was indicated by a milky appearance with the formation of neither cream nor silica particles sediment. The emulsion stability (%S) was calculated by using equation (1).

$$\%S = \frac{h_t}{h_0} \times 100\% \quad (1)$$

Whereas h_t is emulsion height at a certain time and h_0 is initial emulsion height.

2.5 Determination of kinetic models of destabilization rate of o/w emulsion.

The data used for the kinetic models determination was within 0-120 minutes, since the emulsion stability remained more or less constant after 120 minutes. The destabilization rate of o/w emulsion was fitted using the zero order or first order kinetic models. The most appropriate kinetic order would be chosen based on the best obtained correlation coefficient (R^2) calculated using the least square procedure. The destabilization rate of the emulsion could be written in an equation as follows [8]:

$$-r = -\frac{dS}{dt} = k\,S^a \quad (2)$$

$$S = S_0 - k_0 t \quad (3)$$

$$\ln(S) = \ln(S_0) - k_1 t \quad (4)$$

with $-\frac{dS}{dt}$ the rate of emulsion destabilization, a the order of emulsion destabilization rate, k_0 and k_1 the emulsion destabilization rate constants for the zero order (% stability/ minute) and for the first order (per minute), respectively, t the storage time (minute), the percentage of emulsion stability after time t and S_0 the initial emulsion stability percentage.

3. Results and discussion

Oil-in-water (o/w) emulsion is thermodynamically an unstable system, thus it tends to destabilize with time. However, the kinetic stability of the emulsion could be improved by the addition of emulsifiers. In these experiments, oil-in-water food emulsions were stabilized using combination of 2.5% of biosilica with various concentrations of tween-20 ranging from 0.1-1%. The concentration of 2.5% silica was selected based on our best preliminary results. Upon emulsifiers' addition, the destabilization rate of the emulsion would be slowed down as the emulsion stability was increased. In these experiments, the emulsion destabilization was observed within 2 hours. The initial emulsion at t=0 had a milky appearance and seemed to be fully stable. In general, the formations of both cream layers on the top and silica sediments at the bottom became obvious after 1 hour till 2 hours. The estimated emulsions stabilities stabilized using various mixture of tween-20 and rice husk silica after 60 minutes and 120 minutes of observation could be seen in Table 1.

Table 1. Comparisons of % emulsion stabilities.

Emulsifiers	Emulsion Stability, S (%)	
	t = 60 min	t = 120 min
Without emulsifiers	46.6 %	42.2 %
2.5 % silica	60 %	57.77 %
1 % tween-20	86.67 %	62.22 %
0.1 % tween-20, 2.5 % silica	51.1 %	47.7 %
0.3 % tween-20, 2.5 % silica	82.2 %	71.1 %
0.5 % tween-20, 2.5 % silica	84.4 %	71.1 %
0.7 % tween-20, 2.5 % silica	86.6 %	73.3 %
0.9 % tween-20, 2.5 % silica	88.8 %	71.1 %
1 % tween-20, 2.5 % silica	88.9 %	73.3 %

It could be seen that the emulsion stability was the highest both in short term (60 min) and long term period (120 min) when mixture of 1% tween-20 was combined with 2.5% silica. The addition of tween-20 only could achieve quite high short-term emulsion stability but after 1 hour, the stability dropped. The use of biosilica as a single emulsifier was also not efficient. The use of mixture biosilica and tween helped increase the emulsion stability both in the short-term and also in the long-term period. The polymeric emulsifier such as tween-20 was easily adsorbed at the interface between oil droplets and aqueous phase lowering the surface tension, but then it was also easily leached out from the interface with time. On the other hand, the use of biosilica only was not efficient since the biosilica was not easily adsorbed at the interface but once it stayed at the interface it would act as a barrier and stabilized the system [4]. Thereby, the synergism was achieved when mixtures of biosilica and tween-20 were used. Tween-20 was initially adsorbed at the interface facilitating the adherence of biosilica particles thus protecting the overall droplets and reducing the droplets coalescences.

The emulsion stability was also studied using the zero order and first order kinetics models. The emulsion destabilization rate constants were the slopes of the graphs as could be seen in Fig. 1 and Fig. 2 for the zero order and first order models, respectively. The samples prepared with no emulsifiers and with single emulsifier were excluded from the figures since they showed several slopes. The kinetic rates were calculated by averaging the slopes. The overall kinetics data could be seen in Table 2. In general, emulsion stability was increasing with the increased concentration of tween-20. The tween-20 concentration of 0.1% seemed to be too low and insufficient to stabilize the emulsion.

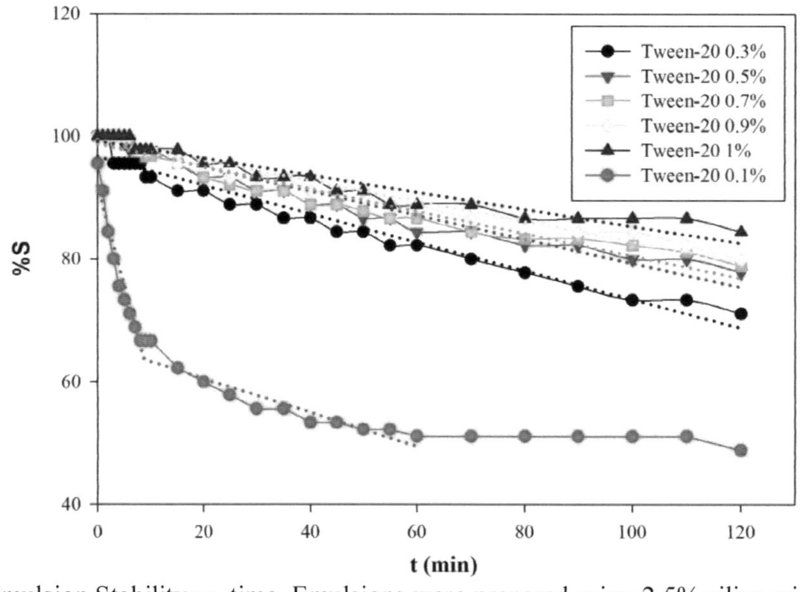

Figure 1. Emulsion Stability *vs*. time. Emulsions were prepared using 2.5% silica with variations of tween-20 concentrations. The dotted line represented the zero order models.

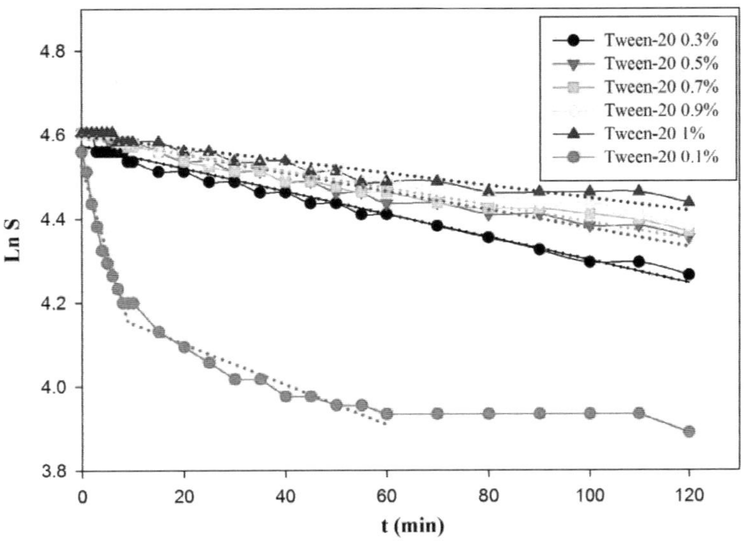

Figure 2. Logarithmic of Emulsion Stability *vs*. time. Emulsions were prepared using 2.5% silica with variations of tween-20 concentrations. The dotted line represented the first order model.

Table 2. Kinetic emulsion destabilizing rates constants and R^2 values according to zero-order and first order kinetic models of o/w emulsions stabilized using 2.5% silica and various Tween-20 concentrations.

Emulsifiers	Orde 0		Orde 1	
	k_0 (% stability/min)	R^2	k_1 (min^{-1})	R^2
Without emulsifiers	4.9621	0.9235	0.0725	0.9434
2.5 % silica	5.7525	0.7749	0.0724	0.7937
1 % tween-20	0.3524	0.9460	0.0040	0.9540
0.1 % tween-20, 2.5 % silica	1.7681	0.9119	0.0231	0.9321
0.3 % tween-20, 2.5 % silica	0.2326	0.9634	0.0027	0.9769
0.5 % tween-20, 2.5 % silica	0.1962	0.9586	0.0022	0.9690
0.7 % tween-20, 2.5 % silica	0.1824	0.9447	0.0020	0.9568
0.9 % tween-20, 2.5 % silica	0.1660	0.9722	0.0018	0.9788
1 % tween-20, 2.5 % silica	0.1390	0.9349	0.0015	0.9422

The kinetics data as could be seen in Table 2 proved that the increasing concentration of tween-20 resulted in the decreasing value of the emulsion destabilization rate constant indicating the higher stability of the resulting o/w emulsions. Moreover, the destabilization kinetic of o/w emulsion more fitted to the first order compared to the zero order based on the R^2 values. The most stable emulsion with the lowest k_1 of 0.0015 min^{-1} was obtained when 2.5% biosilica was combined with 1% tween-20. The destabilization rate constants were ~50 times higher when no emulsifiers or biosilica alone were used, ~3 times higher when tween-20 alone was used and ~ 15 times higher when 0.1% tween-20 was used. The combined 0.1% tween-20 and 2.5% silica stabilized emulsion seemed to be less stable compared to the emulsion stabilized by 1% tween-20 alone. This was due to insufficient amount of tween-20 used. The increasing concentration of tween-20 increased the wettable area responsible for the silica particles adsorption on the interface [9]. This implied that tween-20 concentration mixed with rice husk silica determined the overall emulsion stability. The overall results confirmed the synergism of silica particle and tween-20 in stabilizing the oil droplet dispersions in the aqueous phase. The Pickering emulsion was expected to be more stable in the long-term period especially for food applications. However, the droplets tended to get bigger with time as could be seen in Fig. 3 due to the inherent instability of the emulsion. This could be mainly due to the coalescence of oil droplets due to the desorption of either tween-20 or silica from the interface which followed the first order kinetic model [7].

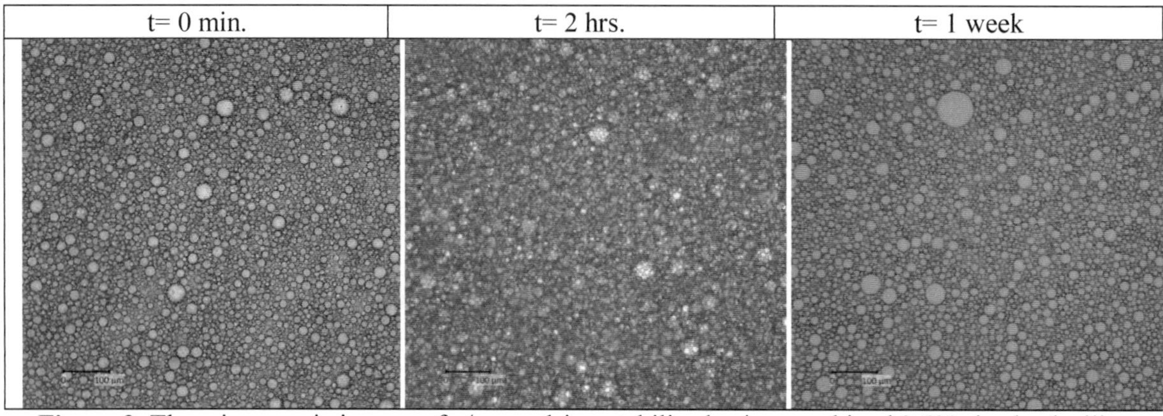

Figure 3. The microscopic images of o/w emulsion stabilized using combined 2.5% rice husk silica and 1% Tween-20 during storage at room temperature (~28°C).

4. Conclusions

The use of combined silica particles and tween-20 as mixed emulsifiers in the oil-in-water emulsion system seemed to be promising. The concentration of tween-20 played a significant role in the overall emulsion stability. The higher the concentration of added tween-20 was, the higher the stability of the emulsion was. The addition of tween-20 enhanced the wettability area for the adsorption of biosilica on the interfaces thus increasing the stability of the resulting Pickering emulsion. The destabilization process of o/w emulsion followed the first order kinetic model as the result of emulsion instability predominantly due to coalescences. The most stable o/w emulsion was obtained using 2.5% rice husk silica and 1% tween-20 with the lowest destabilization rate constant of 0.0015 min^{-1}. The stabilities of o/w emulsions stabilized using mixed silica-tween were higher than 70% after 120 minutes in contrast to those prepared using either biosilica or tween-20 only of which stabilities were lower than 65%. Biosilica seemed promising to be further used in food emulsions as emulsifiers. However lots of optimization processes still have to be conducted in order to obtain o/w food emulsions with much higher stability.

5. References

[1] Simmons A L, Schlezinger J J and Corkey B E 2014 What are we putting in our food that is making us fat? Food additives, contaminants, and other putative contributors to obesity *Current Obesity Reports* **3** pp 273–285

[2] Koglin A, Doetsch V and Bernhard F 2010 Molecular engineering aspects for the production of new and modified biosurfactants *Biosurfactants (Adv. Exp. Med. Biol.* vol 672*)* ed Sen R (New York: Springer) pp 158–169

[3] Liu N, Huo K, McDowell M T, Zhao J and Cui Y 2013 Rice husks as a sustainable source of nanostructured silicon for high performance Li-ion battery anodes *Sci. Rep.* **3** pp 1–7

[4] Wang W, Martin J C, Fan X, Han A, Luo Z and Sun L 2012 Silica nanoparticles and frameworks from rice husk biomass *ACS Appl. Mater. & Inter.* **4** pp 977–981

[5] Pichot R, Spyropoulus F and Norton I T 2010 O/W emulsions stabilised by both low molecular weight surfactants and colloidal particles: The effect of surfactant type and concentration *J. Colloid. Interf. Sci.* **352** pp 128–135
[6] Pichot R, Spyropoulus F and Norton I T 2009 Mixed-emulsifier stabilised emulsions: Investigation of the effect of monoolein and hydrophilic silica particle mixtures on the stability against coalescence *J. Colloid. Interf. Sci.* **329** pp 284–291
[7] Wanli K, Yi L, Baoyan Q, Guangzhi L, Zhenyu Y and Jichun H 2000 Interactions between alkali/surfactant/polymer and their effects on emulsion stability *Colloid. Surface. A* **175** pp 243–247
[8] Levenspiel O 1999 *Chemical Reaction Engineering* 3rd ed (New York: Wiley) pp 47–51
[9] Midmore B R 1998 Synergy between silica and polyoxyethylene surfactants in the formation of o/w emulsions *Colloid. Surface. A* **145** pp 133–143

IOP Conf. Series: Materials Science and Engineering **273** (2017) 012024 doi:10.1088/1757-899X/273/1/012024

Constrained optimization via simulation models for new product innovation

Nugroho A Pujowidianto

R&D Engineer, HP Inc., 138 Depot Road, Singapore 109683

E-mail: npujowidianto@gmail.com

Abstract. We consider the problem of constrained optimization where the decision makers aim to optimize the primary performance measure while constraining the secondary performance measures. This paper provides a brief overview of stochastically constrained optimization via discrete event simulation. Most review papers tend to be methodology-based. This review attempts to be problem-based as decision makers may have already decided on the problem formulation. We consider constrained optimization models as there are usually constraints on secondary performance measures as trade-off in new product development. It starts by laying out different possible methods and the reasons using constrained optimization via simulation models. It is then followed by the review of different simulation optimization approach to address constrained optimization depending on the number of decision variables, the type of constraints, and the risk preferences of the decision makers in handling uncertainties.

1. Introduction

All models are wrong but some are useful [1]. Decision makers use models to assist them in thinking more clearly and weighing various trade-offs that are ubiquitous in this world of limited resources. For example, in developing new printer, we need to consider various key performance indicators or performance measures such as how fast it can print per minute, how many pages it can print given an ink cartridge, how much the cost per page is, how good the output image quality is, how long it can work before it fails, and how noisy it is. These are examples of stochastic performance measures as they depend on the printing usage behavior and the printing content of a printer user. At the same time, there are non-random constraints such as the dimensions or size and the weight of a printer that depends on the targeted user profiles.

There are various models that decision makers can use. Two powerful models are optimization and simulation [2]. Optimization helps the decision makers to list down clearly the decision variables involved, the primary performance measures to be optimized, and the constraints in terms of both decision variables and secondary performance measures. Simulation is useful to account for the external environment such as uncertainties that affect the performance measures.

This paper considers the problem of selecting the best feasible alternatives in the presence of stochastic constraints. In this case, the decision makers would like to optimize the primary performance measure while constraining the secondary performance measures. In some situations, both the primary and secondary performance measures need to be estimated via Discrete Event Simulation as there is no closed form expression to evaluate them. There are two challenges to address the problem. First there is a need to efficiently generate alternatives when the decision space is huge.

Second, multiple simulation replications are needed to estimate the performance measures. As stochastic simulation is computationally intensive, the decision makers have the choice to focus on the worst-case scenario or on average scenario as these correspond to different amount of simulation time.

This paper aims to complement existing surveys on simulation optimization methods by providing a specific overview on the methods for stochastically constrained optimization via simulation. There are many excellent reviews on simulation optimization methods [3-7]. However, most review papers focus on the methods first before breaking down into possible approaches of formulating the problem. This paper takes the other side of the approach, which is to look at the constrained optimization problem first and look at the literatures in addressing it. The outline of this paper is as follows. Section 2 provides possible reasons on why decision makers model problems as constrained optimization together with the different types of possible constraints. Section 3 takes it further by listing the benefits for decision makers to use stochastic model instead of only optimization and when they need to use Discrete Event Simulation instead of other types of simulation model. Section 4 reviews the literatures addressing stochastically constrained optimization while Section 5 concludes this paper.

2. Constrained optimization

Optimization model is one of the tools to make decision. The idea is to find the best possible alternative instead of only aiming for improvement or finding feasible solutions. Each decision maker is given various options on how to model a problem depending on the characteristic of the problem. In the case where all performance measures can be lumped into one performance measure such as cost or profit, we can use a single objective optimization model. In most cases, it is not easy to convert to monetary measures and so finding a good weight for each of the performance measure is difficult. This motivates the use of multi-objective optimization models which can be further divided into either unconstrained multi-objective optimization of which objective is to find the non-dominated solutions or Pareto optimal set or constrained multi-objective optimization. In this paper, we focus on the case where there is only one primary performance measure while the secondary performance measures are constrained. It should be noted that all models including the unconstrained optimization model usually have constraints on the decision variables as resources are limited.

There are at least two reasons for decision makers to model their problem as constrained optimization problems. First it provides clear boundaries and focus on what to optimize and what act as constraints. For example, in the printing context, if the target is office users, one may want to optimize the print speed while constraining other performance measures such as print quality and noise. Secondly, by formulating as constrained optimization, we can identify different types of constraints which may affect the way we make decision.

In new product development, there are three things that customers want: to have a faster, better, and cheaper product [8]. In project management, there is a project management triangle that consists of time, cost, and quality [9]. By formulating as constrained optimization, decision makers are clear which of the three to optimize while constraining the other two. For example, typically cost will be optimized while quality and time are modeled as constraints. This is because there is usually no external regulations on the cost. Quality is demanded by both customers and authorities, while time is crucial as exceeding the time means that the project is not completed.

There are two possible types of penalty or consequences when the constraints are violated: fixed and variable. Fixed penalty refers to pass or fail criteria such as the constraints imposed by authorities. This can be called as hard constraint. Another type of penalty is variable penalty where there are different impacts to the decision makers depending on how much the constraint is exceeded. In the second case, sometimes decision makers may want to convert into the primary performance measure if the weight age of that penalty can be determined.

3. Stochastically constrained optimization

Many models are better than a single model [10]. It is possible to only do optimization without considering uncertainties. For example, if the cost of incorporating simulation models is high, decision makers may opt to use optimization only to find the best alternative. On the other hand, some may only want to use simulation model to evaluate the performance measures of a pre-determined alternative in the presence of uncertainties instead of attempting to find the best alternative. Those are valid options. However, if decision makers want to increase the quality of their decision, using two models such as marrying simulation and optimization to consider the uncertainties are recommended.

There can be many ways to incorporate uncertainties into the model. For example, robust optimization is excellent to consider the uncertainties in the parameters of the formula to evaluate the performance measures. At the same time, there are common cases where the performance measures need to be evaluated using a black-box approach of simulation model. For problems with no time aspect, one can use Monte Carlo simulation which can evaluate the performance of a betting decision given there are various input parameters from different distributions. However, when there is time aspect such as queuing with uncertain arrival and service time; and interaction between the systems we try to model, Discrete Event Simulation is needed. In most cases such as queuing network and printing system, the components are behaving more consistently compared to agents such as human and can be treated as a black box. Otherwise, one may want to use Agent-Based Modeling or combining the Discrete Event Simulation with the concept of agents which can act differently inside the model.

Figure 1 and Table 1 summarize the general decision making model. The simplest case of decision making is when the alternatives are pre-determined, the evaluation effort is minimum such as subjective assessment by a single individual, and there is no uncertainty. When there is no uncertainty, the evaluating and comparing alternatives can be done at the same time. When there is uncertainty in the evaluation and there are no closed-form expressions, the challenge would be on determining the sample size for simulation over different possible scenarios. At the comparing alternatives stage, the risk profile of the decision maker needs to be characterized. A risk-neutral person may prefer central tendency measures such as mean, median, or mode while a risk-averse person may want to include both the central tendency and the spread to model certain percentile level. In addition, the confidence level of the simulation results needs to be specified in determining the sample size of the simulation.

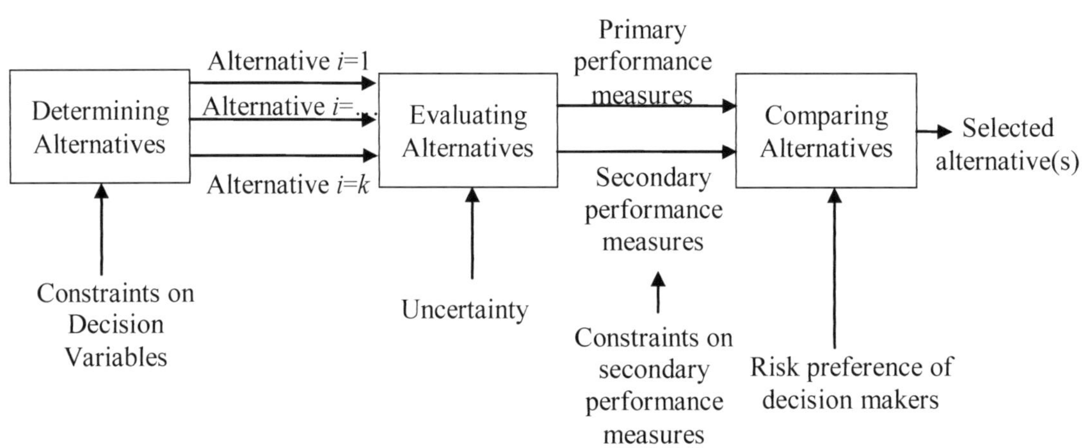

Figure 1. Generalized decision making model.

Table 1. Possible types for each building block.

Determining Alternatives	Evaluating Alternatives	Comparing Alternatives
Pre-determined	Closed form expressions	Weighted Aggregation
Search on Discrete Space	Computer Simulation	Ranking
Continuous Space	Physical Simulation	
	Qualitative Assessment	

4. Literatures on stochastically constrained optimization

4.1. Methods
As both simulation and optimization require significant amount of computing budget, we need to balance the effort between evaluating via simulation and searching for which next sets of alternatives to evaluate [11]. In the case of continuous optimization, Bhatnagar et al. [12] proposed stochastic approximation algorithms when there are stochastic constraints while Kleijnen et al. [13] tackled the problem using response surface method. For the discrete case, there are some Ordinal-Optimization related works [14-16]. In addition, Park and Kim [17] handled the presence of multiple stochastic constraints using a penalty function where the penalty parameter converges to infinity. Luo and Lim [18] used the Lagrangian method where stochastic approximation is applied to the Lagrangian.

When the number of alternatives is finite, it may be possible to evaluate all of the alternatives for at least once that there is no need to spend the computing budget. This is called a constrained ranking and selection problem which decides on the amount of simulation samples to collect for each alternative [19]. For constrained ranking and selection, we have the choice to formulate the problem. This depends on the risk profile of the decision makers. If they want to guarantee the probability of correct selection for the worst case scenario, they can specify the minimum number of simulation samples needed for each alternative [19, 20]. However if they prefer to focus on the general cases, it is more efficient to formulate the simulation budget allocation by either maximizing the probability of correct selection subject to the given budget or to minimize the total simulation budget given the desired probability of correct selection. These two turns out to be dual problems to each other [21]. Another approach would be to consider other criteria such as the expected opportunity cost [22]. In the case of the probability correct selection, asymptotic approach is usually used to get the closed-form solution [23]. Otherwise, we need to use some kind of solver [24]. Another approach for improving the simulation efficiency is to minimize the switching cost in running the simulations between alternatives [25]. It is possible to let certain alternatives to be dormant temporarily to save simulation cost [26]. Another approach is to formulate as a chance-constrained selection problem [27].

4.2. Applications
There are several applications of stochastically constrained optimization via discrete event simulation. First, it can be applied for planning remanufacturing systems [15]. Another application is on designing the optimal water quality monitoring network [28]. Discrete Event Simulation is commonly applied in health care [29]. One example of constrained optimization is for determining the best bed allocation policy [30].

5. Conclusions
This paper aims to provide a brief overview on the stochastically constrained optimization via discrete event simulation. This is done by looking at the context where decision makers want to use constrained optimization model to make decision and the respective literatures which provide various methods and examples of applications.

One possible future research direction is to evaluate the impact of using different models in solving the same problem. This is especially when various models are conflicting in terms of the recommendation given. Another research gap is to build a model where different models can be incorporated. For example, the key performance measures may consist of both objective values and the qualitative assessments from experts. In addition, it is possible to incorporate some kinds of weights using multi-criteria decision making and combine them with the ranking approach in comparing the simulation results.

References
[1] Sterman J D 2002 All models are wrong: reflections on becoming a systems scientist *Syst. Dynam. Rev.* **18** pp 501–531

[2] Fu M C 2002 Optimization for simulation: Theory vs. Practice *INFORMS J. Comput.* **14** pp 192–215
[3] Swisher J R, Jacobson S H and Yucesan E 2003 *ACM Trans. Model. Comput. Simul.* **13** pp 134–154
[4] Rosen S L, Harmonosky C M and Traband M T 2008 Optimization of systems with multiple performance measures via simulation: Survey and recommendations *Comput. Ind. Eng.* **54** pp 327–339
[5] Kleijnen J P C 2014 Simulation-optimization via Kriging and bootstrapping: a survey *J. Simul.* **8** pp 241–250
[6] Jian N and Henderson S G 2015 An Introduction to Simulation Optimization *Proc. Winter Simul. Conf.* pp 1780–94
[7] Amaran S, Sahinidis N V, Sharda B and Bury S J 2016 Simulation optimization: A review of algorithms and applications *Ann. Oper. Res.* **240** pp 351–380
[8] Swink M, Talluri S and Pandejpong T 2006 Faster, better, cheaper: A study of NPD project efficiency and performance tradeoffs *J. Oper. Manag.* **24** pp 542–562
[9] Atkinson R 1999 Project management: cost, time and quality, two best guesses and a phenomenon, its time to accept other success criteria *Int. J. Proj. Manag.* **17** pp 337–342
[10] Page S E 2012 Aggregation in agent-based models of economies *Knowl. Eng. Rev.* **27** pp 151–162
[11] Sun L, Hong L J and Hu Z 2014 Balancing exploitation and exploration in discrete optimization via simulation through a gaussian process-based search *Oper. Res.* **62** pp 1416–38
[12] Bhatnagar S, Hemachandra N and Mishra V K 2011 Stochastic approximation algorithms for constrained optimization via simulation *ACM Trans. Model. Comput. Simul.* **21** 1–22
[13] Kleijnen J P, van Beers W and van Nieuwenhuyse I 2010 Constrained optimization in expensive simulation: Novel approach *Eur. J. Oper. Res.* **202** pp 164–174
[14] Li D, Lee L H and Ho Y C 2002 Constraint ordinal optimization *Inf. Sci* **148** pp 201–220
[15] Song C, Guan X, Zhao Q and Jia Q 2005 Planning remanufacturing systems by constrained ordinal optimization method with feasibility model *Proc. IEEE Conf. Decis. Contr.* pp 4676–81
[16] Guan X, Song C, Ho Y C and Zhao Q 2006 Constrained ordinal optimization a feasibility model based approach *Discrete Event Dyn. S.* **16** pp 279–299
[17] Park C and Kim S H 2015 Penalty function with memory for discrete optimization via simulation with stochastic constraints *Oper. Res.* **63** pp 1195–1212
[18] Luo Y and Lim E 2013 Simulation-based optimization over discrete sets with noisy constraints *IIE Trans.* **45** pp 699–715
[19] Andradottir S and Kim S H 2010 Fully sequential procedures for comparing constrained systems via simulation *Nav. Res. Log.* **57** pp 403–421
[20] Healey C M, Andradottir S and Kim S H 2013 Efficient comparison of constrained systems using dormancy *Eur. J. Oper. Res.* **224** pp 340–352
[21] Chen C H and Yucesan E 2005 An alternative simulation budget allocation scheme for efficient simulation *Int. J. Simul. Proc. Model.* **1** pp 49–57
[22] Pujowidianto N A, Lee L H and Chen C H 2013 Minimizing opportunity cost in selecting the best feasible design *Proc. Winter Simul. Conf.* pp 898–907
[23] Lee L H, Pujowidianto N A, Li L W, Chen C and Yap C M 2012 Approximate simulation budget allocation for selecting the best design in the presence of stochastic constraints *IEEE Trans. Automat. Contr.* **57** pp 2940–45
[24] Pasupathy R, Hunter S R, Pujowidianto N A, Lee L H and Chen C H 2014 Stochastically Constrained Ranking and Selection via SCORE *ACM Trans. Model. Comput. Simul.* **25** 1–26
[25] Healey C M, Andradottir S and Kim S H 2015 A minimal switching procedure for constrained ranking and selection under independent or common random numbers *IIE Trans.* **47** pp 1170–84

[26] Healey C, Andradottir S and Kim S H 2014 Selection procedures for simulations with multiple constraints under independent and correlated sampling *ACM Trans. Model. Comput. Simul.* **24** 1–25
[27] Hong L J, Luo J and Nelson B L 2015 Chance constrained selection of the best *INFORMS J. Comput.* **27** pp 317–334
[28] Park C, Telci I T, Kim S H and Aral M M 2014 Designing an optimal water quality monitoring network for river systems using constrained discrete optimization via simulation *Eng. Optimiz.* **46** pp 107–129
[29] Swisher J R, Jun J B and S H Jacobson 1999 Application of discrete-event simulation in health care clinics: a survey *J. Oper. Res. Soc.* **50** pp 109–123
[30] Pujowidianto N A, Lee L H, Pedrielli G, Chen C H and Li H 2016 Constrained optimizaton for hospital bed allocation via discrete event simulation with nested partitions *Proc. Winter Simul. Conf.* pp 1916–25

Affective design identification on the development of batik convection product

H Prastawa and R Purwaningsih

Department of Industrial Engineering, Faculty of Engineering, Diponegoro University, Indonesia

E-mail: heru.prastawa@undip.ac.id

Abstract. The affective design is increasingly applied to product development in order to meet the desires and preferences of customers. Batik is a traditional Indonesian culture containing historical and cultural values. The development of batik design is one of the efforts to strengthen the identity and superiority of Indonesia's creative industries as well as to preserve batik as the cultural heritage of the nation. Batik product designs offered by the manufacturers do not necessarily correspond with the wishes of consumers, especially the affective values involved. Therefore it is necessary to identify consumer perceptions of convection- based batik product in the form of clothing and fabrics, especially the affective value as the consideration for the designer or manufacturer to develop design alternatives to batik convection products. This research aims to obtain information on consumer affective value, to identify the affective value perception differences among X and Y Generation and to classify affective value in the corresponding cluster of the batik products convection. This study uses Kansei engineering to determine the perception of affective design in the form of Kansei word. Cluster Analysis was used to form clusters that classify affective value of the same class. The results showed that there were 16 pairs of Kansei word which was worth as an affective consumer desire, the 3 indicators that had significant differences among X and Y Generation and 4 clusters with different characteristics.

Keywords: Affective Design; Batik; Kansei Engineering; Cluster Analysis

1. Introduction

The affective design is increasingly applied to the product development in order to meet the desires and preferences of customers. Theoretically, consumer's satisfaction and technical aspects (e.g., functionality, ergonomic and comfort) are equally important in determining the success of product design [1]. Batik is an Indonesian traditional culture containing a historical and cultural value. Batik is not limited to the appearance of beauty that is formed from the harmonious composition of motifs and colors but it also has a spiritual beauty that comes through the decoration and preparation of a pattern that has a philosophical meaning Batik has also been recognized by UNESCO as "Intangible World Heritage" since October 2009. The development of batik is a cultural heritage that begins from Kraton then seeps outside the community of Kraton and eventually forms the industry based on culture and is developed from one generation to the next.

The development of batik design is one of the efforts to strengthen the identity and superiority of Indonesia's creative industries as well as to preserve batik as the cultural heritage of the nation [2]. According to Indrojarwo [3] in his research entitled "Development of Indonesia New Batik Design by Exploration and Exploitation of Recent Context", batik design is related to pattern or patra, motif, variety, color, scale, and composition. The development of batik design can be classified by the type of patra, production process, location, and era. Batik design development has quite an opportunity in the derivation of ancient batik designs into thousands various colors, scale, and proportion. Batik designers also use a voice of consumers regarding the wants of consumers in design. It is important to analyze the affective value of consumers such as feelings or emotions and transform them into an appropriate design in new product development.

Kansei engineering was proposed as a methodology for the affective design of the product in early 1970. This methodology aims to translate human psychological processes, such as feelings and emotions, into elements of appropriate product design, such as size, shape, and characteristic. Kansei engineering was defined as translating technology of a consumer's feeling of the product to the design elements. Kansei engineering uses someone's impression obtained from the environment or other factors using the senses of sight, hearing, feeling, smell and taste to develop the products according to these responses. This method has been used by the automotive manufacturing company, Mazda, the sports car called "Miyata" and has been a best seller in the U.S and Japan [4].

Creative industries should be developed because it has a significant contribution to the Indonesian economy and the center of innovated creation and creativity development. Creative industries also attract creative workers who have a high talent and form creative communities and create many innovations echoed up to the international level [5].

This research aimed to obtain information on consumer affective value, to identify the affective value perception differences among X and Y Generation and to classify affective values in the corresponding cluster of the batik products convection. The research tried to develop batik design from the customer's voice by obtaining information about their wants, especially the affective design in terms of colors, and designs. This information helps designers to design batik convection product.

2. Literature review

2.1. Affective design

Affective is something related with attitudes and values. Affective includes characters and behaviors such as feelings, interests, emotional attitude and values. Some experts say that a person's attitude can be foreseen amendments when someone has high-level cognitive powers. Affective is divided into more details in five levels: receiving, responding, valuing, and organization, characterization by value or complex value.

Emotion/motivation comes from the style of design, function, form, usability and user's experience. User/consumer emotion is a hidden desire. Design can cause some responses like happy, upset, excited, frustrated, and so on. The design that can motivate the user/consumer emotion is called affective design. An example is the signage designed for the Umeda Hospital in Tokyo by the Japanese designer Kenya Hara [6]. Signage is a general term for any type of display graphics that are intended to convey information to the user. The results of study indicated that signage was made of cotton cloth; it provides a comfortable, soft feeling to soothe the patient's mind. Hence, this signage can drive the users into a certain kind of emotion response, and this is said to be *Emotional Design* [7].

2.2. Kansei engineering

The word Kansei means feeling, impression and emotion. Kansei engineering is a method to translate the image customers or consumers' feelings into real design components. Kansei engineering is a technology-oriented method enabling image on how consumers feel united to the design process of a new product. Kansei engineering is defined as the translation technology of consumers feeling (Kansei) about upcoming products into a design element. Kansei emotional structures exist under the attitudes or behaviors of humans. This structure refers to Kansei as a person.

2.3. *Cluster analysis*
Cluster analysis is a method of analyzing the data. Cluster analysis as one of the methods of Data Mining aims to group the data with similar characteristics to the same 'region' and the data with different characteristics to the other 'area', to get a group of objects that have the same values or characteristics. In principle, cluster analysis is a process to reduce the number of large objects into fewer called clusters. This method is used by researchers who do not know a member of a group [8].

3. Methodology
3.1. *Sample*
This research used stratified sampling techniques to explore affective design in batik convection product development. Samples were taken by dividing the population into groups according to the classification of Generation X (The Baby Bust) and Generation Y (The Echo of the Baby Boom). Generation X was born between the years 1965 to 1976 or around the age of 40 – 51 years old and Generation Y was born between the years 1977 to 1997 or around age 19 -39 years old. 50 respondents were selected out of the data from the Department of Cooperatives and SMEs Semarang in March 2016. The respondents are people who understand the convection-based batik product. They are a craftsman or producer of batik, and have a knowledge ability to fill out a questionnaire in accordance with their basic knowledge.

3.2. *Determining Kansei word*
Determination of Kansei word was obtained from the magazine source, the existing literature, users, etc. Kansei word identification was done by observation and small group discussion. The selected word is shown in Table 1.

Table 1. Kansei word.

Kansei ID	Kansei word	Kansei ID	Kansei word
X1	Bright – Gloomy	X9	Formal – Casual
X2	Plain – Gaudy	X10	Unoriginal – Creative
X3	Colorful – Colorless	X11	Heavy – Light
X4	Traditional – Modern	X12	Comfortable – Uncomfortable
X5	Childish – Mature	X13	Thin – Thick
X6	Elegant – Not Elegant	X14	Harsh – Soft
X7	Simple – Complex	X15	Flimsy – Substantial
X8	Cute – Not cute	X16	Stiff – Pliable

3.3. *Determining and designing questionnaire*
The questionnaire of semantic differential showed the feelings of the respondents. Tests using Semantic Differential method was done by giving the value on 7 scales of Kansei word. Respondents were asked to rate batik convection product in accordance with Kansei word that had been described on the questionnaire. The questionnaires also provided a column where respondents could pour desires concerned with batik convection products. Examples of scale on the questionnaire can be seen in Table 2.

Table 2. Scale on questionnaire.

Color	Brig	1	2	3	4	5	6	7	Gloom
	Scor	-	-	-	0	1	2	3	Score

4. Results and discussion
4.1. *Perception of batik convection product*
Customer perception of batik convection product identified by Kansei word trend in 16 indicators is shown in Table 3.

Table 3. Customer perception (CP) of product in close-ended questionnaire.

Affective Dimension	CP	Affective Dimension	CP
Bright	+	Gloomy	
Plain	+	Gaudy	
Colorful	+	Colorless	
Traditional	+	Modern	+
Childish		Mature	+
Simple	+	Complex	
Elegant	+	Not-Elegant	
Cute	+	Not-Cute	
Formal	+	Casual	+
Unoriginal		Creative	+
Heavy		Light	+
Comfortable	+	Uncomfortable	
Thin	+	Thick	
Harsh		Soft	+
Flimsy		Substantial	+
Stiff		Pliable	+

Table 3 shows that customer perception of the affective dimension image can be focused on the bright, soft and colorful colors. Respondents preferred a more traditional design but did not leave a modern touch and also the impression of mature and simple but elegant ones. The design is also not separated from visual cuteness and creativity, not a formal design. Respondents chose the type of material that was light, thin and soft that could cause a sense of comfort. Material also should be a strong and flexible.

4.2. Testing the questionnaire

There were 16 indicators performed on this test. Each indicator was valid if the value of r calculations on Corrected Item-Total Correlation was greater than the value of r table. Data were collected from 50 respondents and a significance level of 5% was used. Obtained r table was 0.279 and the value of r table on each indicator that exceeds 0.279 declared valid and was concluded that the result of questionnaire was valid.

Next test was reliability test to indicate the level of consistency of a questionnaire. Test was done by comparison of the value of Cronbach's Alpha to the value of questionnaire test results. When questionnaire test result had a value greater than 0.6, then this questionnaire was reliable.

4.3. T-test

T-test was used to determine differences in the average of two independent sets of data, which in this study was referred to as Generation X and Generation Y. Results from the SPSS by Independent T-Test showed the value sig (2-tailed). If the value of sig (2-tailed) < 0.05, it means that the indicator has a significant difference from others. The result showed that few indicators had significant differences; plain-gaudy, flimsy-substantial and stiff-pliable. It could be concluded that the perception of respondents Generation X and Y could be distinguished by the aspect of color and statement of material belonging to the batik convection products.

In plain-gaudy indicators between Gen X and Y, the difference was that Gen X respondents tended to prefer a positive statement (gaudy), while Gen Y tended to prefer negative statements (plain). At the flimsy-substantial indicators, Gen X chose positive statements (substantial) with high scores while Gen Y was more likely to have a value that was spread on the positive statement. In the stiff-pliable indicator it was seen that Gen X dominated a high value on the positive statement while Gen Y had a distribution value on the positive statement (pliable).

4.4. Cluster analysis

The clusters used here was a hierarchical cluster analysis by using Ward Method. The number of clusters was determined by ward method at the stage of agglomeration. Agglomeration performed grouping one by one [10]. Output of cluster analysis classified respondent into 4 clusters. The next step was to divide the data into a form of clusters using K-Means method. The obtained results are shown in Table 4. The result of the average of each cluster according to the respondents' perception on batik convection product can be seen in Figure 1.

Figure 1 show that respondents in cluster 1 tended to agree with negative statements of Kansei of color aspect (bright, plain and colorful). In design aspects they chose a traditional design (mature, simple, elegant, cute, formal and unoriginal-creative). In material aspect they preferred a light, comfortable, thin, soft, substantial and pliable material. While respondents in cluster 2 had little in common with cluster 1, who were inclined to agree with negative statements of Kansei but with different average values, mainly the color aspect of bright-gloomy, plain and colorful, prefer traditional design (childish, simple, elegant, cute, formal-casual and creative), and on material they preferred a heavy, comfortable, thin, soft, flimsy-substantial and stiff-pliable.

Table 4. The cluster.

Cluster	Member
1	12
2	6
3	21
4	11
Total	**50**

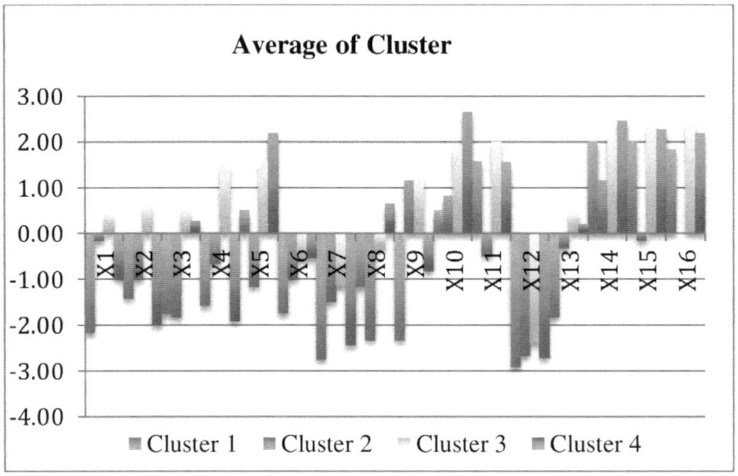

Figure 1. Average of cluster.

Respondents in cluster 3 tended to agree on a positive statement of Kansei on color aspect of gloomy, gaudy, and colorless; on design aspects of modern, mature, simple, elegant, cute, casual and creative; in the aspect of material that was light, comfortable, thick, soft, substantial and pliable. Respondents in cluster 4 looked balanced on both positive and negative statements of Kansei in terms of color aspect of bright, plain and colorless; in design aspects that traditional, mature, simple, elegant, cute-not cute, formal and creative; in the aspect of material that was light, comfortable, thin-thick, soft, substantial and pliable.

5. Conclusion

Based on the analysis and discussion it can be concluded that there are 16 pairs of Kansei word to reflect customer emotion on batik convection product. Overall, respondents tend to choose the bright, plain and colorful in the aspect of color; traditional-modern, mature, simple, elegant, cute, formal-casual and creative on design aspect; and light, comfortable, thin, soft, substantial and pliable in the aspect of materials.

Respondents in Generation X and Generation Y do not pay too much attention to other indicators. The respondents' perception of Generation X and Y can be distinguished by the aspect of color and preference on material of products. Gen X respondents tend to prefer a positive statement (gaudy), while Gen Y tends to prefer negative statements (plain). In flimsy-substantial indicators, Gen X chooses positive statements (substantial) with high scores while Gen Y is more likely to have a value that is spread on the positive statement. In the stiff-pliable indicator it is seen that Gen X dominates a high value on the positive statement while Gen Y has a distribution value on the positive statement (pliable).

Cluster analysis form respondent into four clusters. The two dominant clusters are cluster three and cluster one. Cluster one with twelve respondents, has a high perception in color that is bright, plain and colorful; in design that is traditional, mature, simple, elegant, cute, formal and unoriginal-creative; the material is that light, comfortable, thin, soft, substantial and pliable. Cluster three consisting of 21 respondents has a high perception of the color that gloomy, gaudy and colorless; the design that is modern, mature, simple, elegant, cute, casual and creative; the material that is light, comfortable, thick, soft, substantial and pliable.

6. References

[1] Lokman A M 2010 *Design & emotion: The Kansei engineering methodology* (Selangor: Universiti Teknologi MARA) pp 1–14
[2] Noviana M and Sujoko H 2014 The deployment of Quality Function Deployment (QFD) for developing batik design incorporating the local content of East Kalimantan *J. Teknik Industri* **9** pp 87–92
[3] Indrojarwo B T 2008 Development of Indonesia new batik design by exploration and exploitation of recent context *Research report* (Surabaya: Sepuluh November Institute of Technology)
[4] Nagamachi M 1995 Kansei engineering: A new ergonomic consumer-oriented technology for product development *Int. J. Ind. Ergonom.* **15** pp 3–11
[5] Pangestu M E 2008 *The development of Indonesian creative industry 2025* (Jakarta: Ministry of Trade, Republic of Indonesia)
[6] Mollerup P 2005 *Wayshowing: A Guide to Environmental Signage Principles and Practices* (Baden: Lars Müller Publishers)
[7] Ho A G and Siu K W M 2012 Emotion design, emotional design, emotionalize design: A review on their relationships from a new perspective *Des. J.* **15** pp 9–32
[8] Hirota K and Pedrycz W 1995 D-fuzzy clustering *Pattern Recogn. Lett.* **16** 193–200
[9] Schutte S 2005 *Engineering Emotional Values in Product Design, Kansei Engineering in Development* (Linkoping: Linkoping University)
[10] Santoso S and Tjiptono F 2004 *Market Research: The Concept and Its Application Using SPSS* (Jakarta: PT. Elex Media Komputindo)

Estimating life cycle cost for a product family design: The challenges

T J Suteja[1], A Karim[2], P K D V Yarlagadda[2] and C Yan[2]

[1] Department of Manufacturing Engineering, University of Surabaya, Raya Kalirungkut, Surabaya 60293, Indonesia
[2] Science and Engineering Faculty, Queensland University of Technology, 2 George Street, Brisbane, Australia

E-mail: jayasuteja@staff.ubaya.ac.id

Abstract. A cost estimation system is required to assist in designing a product family. The aim of this paper is to identify the requirements and the problems in estimating the life cycle cost of a product family. Then, this paper also presents the state-of-the-art and the research challenges in developing a life cycle cost estimation system for a product family design. As the conclusion, the life cycle cost estimation process for a product family still needs to face the challenges to determine the end of life strategy of each sub module of a product family, to integrate the end of life strategy to estimate the life cycle cost of a product family, to estimate the life cycle cost of each component level of a product family for design purposes and for different technologies and approaches, to reduce the required time and effort for updating process in estimating the life cycle cost for different structures of different product families, and to transform the available information into the required information in order to estimate the life cycle cost of a product family at the early stage of product development.

1. Introduction

Manufacturing companies are now under pressure to provide a variety of products in a shorter lead-time and in a cost effective way, without sacrificing product quality and environmental considerations, to be competitive in the market. Simpson (as cited in [1]) proposes that an effective means of fulfilling the requirements, to provide a variety of products in a cost effective way, is through a product family design. The implementation of a product family design is not only beneficial at the manufacturing and supply chain stages but also at the end of life stage. In general, a product family is defined as "a group of related products that is derived from a product platform to satisfy a variety of market niches" [2]. Derived from various definitions of a product platform as cited by Simpson et al., the groups of products are related because they share common functions, features, components, technologies, processes, knowledge, people and/or relationships, which form a product platform [2].

In this research, a product family is divided into five levels of components, which are part, product platform, variant, product variant, and product family level. Each component level is related to each other by its market segment. A product family consists of more than one product variant that share common things. Each product variant is developed to satisfy customers from a certain market segment. Each product variant consists of product platform and variant. A product platform consists of several parts that form a common structure. A variant is a part or a collection of parts that differentiates one member of a product family from other members. A part is the lowest level component of a product family. Each part inherits the market segment attributes of the product variant to which it belongs.

The growing implementation of a product family design has led to the need for a product family design support system especially the cost estimation system. However, most of the product family design methods provide a limited support for the cost estimation process [3]. Therefore, the aim of this paper is to identify the requirements and the problems in estimating the cost of a product family. In addition, this paper also presents the state-of-the-art and the research challenges in developing a cost estimation system for a product family design.

The next section describes all the requirements in estimating the cost of a product family. Then, various problems in implementing the product family approach especially in estimating the cost of a product family are explained. In section 4, existing cost estimation systems are reviewed in order to fulfill the requirements and solve the identified problems for estimating the cost of a product family. Next, the requirements and the problems in estimating the cost of a product family and how the existing techniques handle the problems are discussed. Finally, the challenges in estimating the cost for a product family design are concluded in the last section.

2. Cost estimation system requirements for a product family

A cost estimation system is required to assist in estimating the profitability of the product family design, performing an economic evaluation of the technical choices design, and controlling other processes in the product family design. It is also useful in performing the trade-off between the benefit and the complexity in optimizing the product family design. As a product family has five levels of components, the cost estimation of each level of component is required. Therefore, an appropriate cost estimation system is required to estimate the cost of each component level in order to assist in evaluating the design of a product family.

A product family could have various structures. Even different product variants in the same product family could use different structures. The product structure defines how the functions or components of a product are arranged in a hierarchical structure. It describes the relationship among functions or components of a product. Different product family structures will influence the product family cost differently. Therefore, the product family structure must be taken into account in estimating the cost of a product family.

The product family approach has been implemented by various types and sizes of companies. As various types and sizes of companies exist in the world, the cost estimation system must be able to be applied in different types and sizes of companies that implement different technologies and approaches. In addition, as technology is changing rapidly, the cost estimation system must be able to accommodate the changes or it will become unreliable. For that reason, the cost estimation system for a product family must be able to be implemented in different situations and take into account any changes.

The cost estimation system must be able to estimate the life cycle cost of a product family. It involves all the incurred costs during the product design, the production, the product use, the support, and the disposal or recycling stage [4]. The life cycle cost is important to be used as an engineering tool for providing decision support in the design process. Many researchers have proposed a variety of cost estimation techniques. However, the majority of the cost estimation techniques are proposed to estimate the cost incurred at the pre-production and the production stages. The estimation of the post-production costs, especially the end of life cost, is becoming important in performing an evaluation of the product family because the end of life cost now has a significant influence on the total cost of a product. The influence of the end of life cost is becoming significant because the costs of product take back, product recovery, and product disposal are now being imposed on manufacturers as a result of the increasing environmental awareness and the stricter regulations related to the environment. The estimation of the end of life cost is also important in understanding the potential benefits in remanufacturing and refurbishing.

The life cycle cost estimation of a product family should be performed at the early stage of product development. The early stage of product development refers to the stage after product planning and before the product detail design stage. According to Nevins and Whitney (as cited in [5]), it is important to estimate the life cycle cost at the early stage of product development, because at least 70% of the product life cycle cost is committed at this stage. By performing the life cycle cost estimation at this stage, the cost could be adjusted as early as possible and as a result, decisions can be more easily made. In addition, it is more feasible to reduce the cost at the early stage of product development without resulting in many difficulties caused by late modification.

3. Cost estimation problems

Although the implementation of a product family approach could bring various benefits, it could raise the complexity at all stages during the product life cycle, including the estimation of the life cycle cost of a product family. The implementation of a product family approach increases the range of products. The increased range of products increases the number of indirect activities, which benefit more than one product. As a result, it enlarges the portion of the indirect cost to the total cost. The increased portion of the indirect cost creates a significant error in estimating cost for product family by using a traditional volume-based cost estimation technique. The error becomes significant because a volume-based cost estimation technique assumes that the indirect cost is proportional to the volume of the product. In fact, the indirect cost does not always vary with production volumes. The assumption causes under costing for certain types of products and over costing for other products. To reduce the error, the cost estimation system must be able to allocate the indirect cost properly.

In addition, the implementation of a product family approach is more difficult compared to the implementation of multiple products because there are sharing components, processes, resources, etc. between or among members of a product family. The sharing component, process, resource, etc. between or among members could result in difficulty in allocating the cost of each member. It makes the cost estimation process for a product family much more difficult and could generate more errors than for an individual product and for multiple products. For that reason, the cost estimation system must be able to allocate the life cycle cost for each product variant of a product family. As the life cycle cost estimation of each component level is required in evaluating the design of the product family, the cost estimation system must also be able to allocate the life cycle cost of each component level of a product family.

Another problem that arises in estimating the cost of a product family is that different product structures will result in different costs. As the cost estimation system must be able to estimate the life cycle cost of different structures of different product families, the system could require extensive time and effort in estimating the life cycle cost of various structures of a product family. In addition, it is required to store complete information for each product platform, variant, and product variant. To avoid this, the cost estimation system must be easily updated for different structures of different product families.

The cost estimation system must be able to be applied in different types and sizes of companies [6]. It must also be able to accommodate the trend of technology and reflect any change inside the company or it will become unreliable. The implementation of the cost estimation system in different situations could require extensive time and effort in adapting process. Therefore, the cost estimation system for a product family must be easily adapted to reflect any changes.

In addition, the life cycle cost of a product family is influenced by the end of life strategy of the product family. The end of life strategy of a product determines the end of life recovery process that will be conducted after the product is taken back. It is not realistic to determine the end of life strategy on product level because a product consists of more than one component that could have different attributes and end of life strategies. It is also not practicable to determine the end of life strategy on part level because it is usually less economical to disassemble a product into parts. The factors that can be used to determine the end of life of a sub module could be different compared to a product or a part. Various factors are used to determine the end of life strategy of a product that is at the end of its life

[7, 8, 9]. Other researchers have proposed other factors to determine the end of life strategy of a part [10, 11]. For that reason, the cost estimation system must consider various factors that influence the end of life of a sub module in estimating the life cycle cost.

Furthermore, the need to estimate the cost at the early stage of product development results in another problem in estimating the life cycle cost of a product family. As a product family is represented by several attributes, then these attributes influence the life cycle cost of a product family. Not all attributes required to design a product family are available at the early stage of product development. At the conceptual design stage, geometry and material of the product are not yet known. According to Pahl, Beitz, Feldhusen, and Grote, the available product information is identified problems, required functions, working principles and structure [12]. In addition, Zha describes that the available product information at the beginning of the conceptual design is functional requirements and relevant production requirements [13]. Then, during the conceptual design stage, the information is transformed into critical design requirements. Based on the studies by Fixson, Simpson, et al., and by Park and Simpson, the available product family attributes at the early stage of product development are market segment, production volume, product family structure, and product family function [1, 14]. In order to estimate the cost of a product family at the early stage of product development, the cost estimation system must be able to use the market segment, the production volume, the product family structure, and the product family functions as its input.

4. Life cycle cost estimation techniques for a product family

A large number of systems for estimating the cost of a product have been proposed by many researchers. The techniques used by the existing cost estimation systems can be divided into four classifications, which are intuitive, analogy, parametric, and analytic techniques [15].

Intuitive cost estimation techniques use past experience to generate a body of knowledge related to the cost. This knowledge is then used to provide the required information to estimate the cost. These techniques allow the cost estimation of different product families at the early stage of product development. These techniques provide a transparent estimation process which can assist in cost analysis. However, these techniques heavily rely on past data and previous experience to develop the cost estimation system. They have become unreliable because technology is changing rapidly. In addition, they are mostly used to estimate cost of an individual part or product because they require a large amount of information and time to develop the system. Estimating the cost of various product families that consist of a large number and various parts requires much larger amounts of information and time than estimating an individual part.

Analogy cost estimation techniques estimate the cost based on the historical data of products with known cost. The historical data are used to establish a relationship between the cost and the values of certain selected variables. After that, the relationship can be used to estimate the cost of a new product. These techniques are able to estimate the cost using conceptual and not detailed information as their input. In addition, this does not require a lot of data. These techniques are able to estimate the cost of a new product when the extensive database is not available. Therefore, these techniques could provide quick and easy life cycle cost estimation for a product family at the early stage of product development. However, the application of analogy cost estimation techniques is limited to those product families that have the same structure, because they use historical data of a certain product family. Analogy cost estimation techniques function as a black box in cost estimation and do not provide detailed information related to various factors and their influence on the cost. As a result, analogy cost estimation techniques cannot be used to assist in analyzing the cost of a product family and evaluating its design.

Parametric cost estimation techniques use statistical methodology to develop a cost estimation model. Then, the parametric cost model estimates the cost using certain critical parameters of the product. These techniques can be used in estimating the cost estimation of a product family at the early stage of product development. These techniques are able to estimate the life cycle cost of a product family as long as all critical parameters that influence the life cycle cost are identified. However, as they require and use previous data of a certain product family, they are not able to estimate the cost of different product families with different critical parameters. In addition, the accuracy of the technique is very low because they specify a complex product family with only a limited number of parameters. Analytical cost estimation techniques estimate the product cost by summing the cost of all the decompositions of the product. This method requires detailed information about the decomposition of the product to estimate the cost. These techniques require detailed information and are difficult to be implemented at the early stage of product development.

The existing cost estimation techniques have been analyzed in order to estimate the life cycle cost of a product family at the early stage of product development. In order to assist in evaluating a product family, the cost estimation system must be able to estimate the life cycle cost of different structures of a product family, allocate the life cycle cost of each component level of a product family, be implemented at the early stage of product development, be adapted and updated easily. Each cost estimation technique has its own strengths and limitations in estimating the life cycle cost of a product family. The result of the requirement analysis is summarized in Table 1.

Table 1. Analysis results.

Requirements	Intuitive	Analogy	Parametric	Analytic
Able to estimate the life cycle cost of different structures of a product family	Yes	No	No	Yes
Able to allocate the life cycle cost of each component level of a product family	Yes	No	No	Yes
Able to be implemented at the early stage of product development	Yes	Yes	Yes	No
Able to be adapted and updated easily	No	Yes	Yes	Yes

One of the analytic cost estimation techniques, the activity-based costing technique, has been proposed by several researchers to estimate the cost of a product family. The activity-based costing technique is considered as a potential costing method for estimating the cost of a product family. Even though the activity-based costing technique can solve some difficulties in estimating the life cycle cost of a product family, it still has several disadvantages. Although this technique is able to estimate the life cycle cost of different structures of different product families, it requires an extensive time and effort to estimate the life cycle cost of different product families and is difficult and costly to be updated in order to reflect any changes. This technique also has difficulties in evaluating the life cycle cost in different types and sizes of companies because the consumed activities and resources will be different from one company to another company that use different technologies and approaches. Then, an activity-based costing technique is difficult to be implemented at the early stage of product development because detailed information related to consumed activities and resources should be available in using an activity-based costing technique. Last, the original aim of this technique is to estimate the cost of a product family for accounting purposes. It will need an adaptation to suit the purpose as a design support system.

5. Discussion

In this section, the requirements and the problems in estimating the cost of a product family are discussed. In addition, how the existing techniques handle the problems is also discussed.

The end of life strategy has a significant influence in the end of life cost of a product family. The existing studies have determined the end of life strategy on the product or part level. As it is not feasible to determine the end of life strategy on the product or part level, the end of life strategy must be determined on the sub-module level. In addition, the factors that can be used to determine the end of life of a sub-module could be different compared to a product or a part. Considering this fact, there is a need to investigate various ends of life strategies of a sub module, to develop a new method in order to determine the end of life strategy of each sub module of a product family, and to integrate the end of life strategy for estimating the life cycle cost of a product family.

The existing cost estimation techniques have their own difficulties in fulfilling the requirement of estimating the life cycle cost of each component level of a product family. To solve these difficulties, the activity-based costing technique has been proposed as a potential costing method for estimating the cost of a product family. However, this technique still experiences some difficulties in estimating the life cycle cost in different types and sizes of companies that use different technologies and approaches because they consume different activities and resources. In addition, it is not intended for design purpose. Therefore, there is a need to develop a novel technique that can be used to estimate the life cycle cost of each component level of a product family for design purposes and can be adapted easily for different technologies and approaches.

The existing techniques require extensive time and effort to estimate the life cycle cost of different structures of different product families. They need to store complete information for each product platform, variant, product variant, and product family. Therefore, it is important to develop a new method that can reduce the required time and effort for updating process in estimating the life cycle cost for different structures of different product families.

Last, the available product family attributes at the early stage of product development, which is the market segment, the production volume, the structure, and the function, cannot directly be used to estimate the life cycle cost of a product family. How to use these available attributes of a product family to estimate the life cycle cost at the early stage of product development has not been studied yet. In addition, most of the existing systems do not provide detailed information related to various factors and their influence on the cost. As a result, they cannot be used to assist in analyzing the cost of a product family and evaluating its design. For that reason, it is important to find a way to transform the available information into the required information in order to estimate the life cycle cost and evaluate the design of a product family at the early stage of product development

6. Conclusion

A cost estimation system is required to estimate the life cycle cost in order to evaluate the design of different product families at the early stage of product development. The existing techniques do not provide satisfying answers for the identified problems. Therefore, the life cycle cost estimation process for a product family still needs to face and solve the challenges below:

1. A new method is required to determine the end of life strategy of each sub module of a product family and to integrate the end of life strategy for estimating the life cycle cost of a product family,
2. A novel technique needs to be developed to estimate the life cycle cost of each component level of a product family for design purposes and can be adapted easily for different technologies and approaches,
3. A new method is required to reduce the required time and effort for updating process in estimating the life cycle cost for different structures of different product families,
4. It is important to find a way to transform the available information into the required information in order to estimate the life cycle cost of a product family at the early stage of product development.

References

[1] Park J and Simpson T W 2008 Toward an activity-based costing system for product families and product platforms in the early stages of development *Int. J. Prod. Res.* **46** pp 99–130
[2] Simpson T W, Siddique Z and Jiao J 2006 *Product Platform and Product Family Design: Methods and Applications* (New York: Springer) pp 1–15

[3] Ripperda S and Krause D 2017 Cost effects of modular product family structures: methods and quantification of impacts to support decision making *J. Mech. Design* **139** pp 021103–15
[4] Asiedu Y and Gu P 1998 Product life cycle cost analysis: state of the art review *Int. J. Prod. Res.* **36** pp 883–908
[5] Dowlatshahi S 1992 Product design in a concurrent engineering environment: an optimization approach *Int. J. Prod. Res.* **30** pp 1803–18
[6] Tyagi S, Cai X and Yang K 2015 Product life-cycle cost estimation: a focus on the multi-generation manufacturing-based product *Res. Eng. Des.* **26** pp 277–288
[7] Du Y, Cao H, Liu F, Li C and Chen X 2012 An integrated method for evaluating the remanufacturability of used machine tool *J. Clean. Prod.* **20** pp 82–91
[8] Rao R V and Padmanabhan K K 2010 Selection of best product end-of-life scenario using digraph and matrix methods *J. Eng. Design* **21** pp 455–472
[9] Sabharwal S and Garg S 2013 Determining cost effectiveness index of remanufacturing: a graph theoretic approach *Int. J. Prod. Econ.* **144** pp 521–532
[10] Zhang X, Zhang H, Jiang Z and Wang Y 2016 A decision-making approach for end-of-life strategies selection of used parts *Int. J. Adv. Manuf. Technol.* **87** pp 1457–64
[11] Ghazalli Z and Murata A 2012 Evaluation method for product end-of-life selection strategy *Advances in Ergonomics in Manufacturing* eds Trzcielinski S and Karwowski W (CRC Press) chapter 50 pp 449–462
[12] Pahl G, Beitz W, Feldhusen J and Grote K H 2007 *Engineering Design: A Systematic Approach* (London: Springer Verlag)
[13] Zha X F 2005 A web-based advisory system for process and material selection in concurrent product design for a manufacturing environment *Int. J. Adv. Manuf. Technol.* **25** pp 233–243
[14] Fixson S K 2006 *A Roadmap for Product Architecture Costing* (Boston: Springer) pp 305–334
[15] Niazi A, Dai J S, Balabani S and Seneviratne L 2006 Product cost estimation: Technique classification and methodology review *J. Manuf. Sci. Eng.* **128** pp 563–575

An integrative fuzzy Kansei engineering and Kano model for logistics services

M Hartono[1], T K Chuan[2], D N Prayogo[1] and A Santoso[1]

[1] Department of Industrial Engineering, University of Surabaya, Raya Kalirungkut, Surabaya 60293, Indonesia
[2] Department of Industrial & Systems Engineering, National University of Singapore

E-mail: markus@staff.ubaya.ac.id

Abstract. Nowadays, customer emotional needs (known as Kansei) in product and especially in services become a major concern. One of the emerging services is the logistics services. In obtaining a global competitive advantage, logistics services should understand and satisfy their customer affective impressions (Kansei). How to capture, model and analyze the customer emotions has been well structured by Kansei Engineering, equipped with Kano model to strengthen its methodology. However, its methodology lacks of the dynamics of customer perception. More specifically, there is a criticism of perceived scores on user preferences, in both perceived service quality and Kansei response, whether they represent an exact numerical value. Thus, this paper is proposed to discuss an approach of fuzzy Kansei in logistics service experiences. A case study in IT-based logistics services involving 100 subjects has been conducted. Its findings including the service gaps accompanied with prioritized improvement initiatives are discussed.

Keywords: Kansei engineering; fuzzy; logistics services

1. Introduction

Research on Kansei Engineering (KE) is of high interest due to the increase of customer expectation in terms of emotional needs and satisfaction. Its application in services becomes more and more critical since many service companies provide offerings with almost the same quality, price and delivery service. Surely, it makes confusion to customers in deciding which product or service they need to choose and buy. Hence, it needs something to differentiate and superior among all provided criteria. One of the most prominent considerations in making successful transaction is the emotional satisfaction and impression (known as Kansei, in Japanese) [1]. According to Hartono & Raharjo [2], both cognitive and affective satisfaction (known as Kansei and Chisei) are important in service-related encounters. In other words, it can be said that cognitive requirement should be fulfilled first, and then it moves to the realization of Kansei.

KE has been proven as one of the most powerful ergonomics-based products and also service development methodologies, incorporating emotional needs. According to Hartono [3, 4], with regard to service application, it covers but not limited to general KE methodology, SERVQUAL and Kano model embedded [1–3, 5, 6], cultures [3, 6], TRIZ (*TeoriyaResheniyaIzobretatelskikhZadach*) methodology [3, 4], and sustainability approach [4, 7].

In judging whether a particular service is emotionally appeal or not, customer delight is regarded as the most prominent criterion. It refers to customer experience and interaction, rather than just service offering without any interventions. It starts with the gap between what has been expected and perceived by the customer. It is a measure of customer satisfaction based on the service quality. However, quality itself is not enough. Service quality should be strengthened by total customer satisfaction and delight [8]. Recent studies of KE show that Kano model has been successfully integrated into KE methodology, in order to achieve more efficient improvement strategies (see [1] for details). Furthermore, it has been extended by incorporating the issue of sustainability and TRIZ methodology [3, 4]. Its objective is to contribute to the solution for today's issue, yet to maintain the efficiency of proposed methodology. For instance, the use of TRIZ is to provide problem-solving principles to resolve any contradictions. More specifically, according to Hartono [3, 4], TRIZ is used to generate ideas for improvement with the possible lowest contradiction among them.

Inherently, the attention on the reliable and valid perceived service quality and Kansei has been raised. With respect to customer emotional satisfaction, KE has shown its superiority against some similar methodologies. However, it might have missed its capability in modeling exact values of perceived customer impression. A criticism of perceived scores on user preferences has been occurred [9]. In this study, it may refer to both perceived service quality and Kansei responses. In other words, it lacks of the attention on customer dynamics. Hence, this paper discusses an approach of fuzzy Kansei in logistics service quality, accompanied by an empirical study on IT-based supporting logistics services. The Kano model is engaged in order to strengthen KE methodology in focusing more on delighting service attributes (see [1] for details).

This paper consists of 5 main parts. After the introduction section, a short literature review on fuzzy in Kansei and SERVQUAL, and Kano model in logistics services is provided. Afterwards, research methodology and an empirical study on IT-based supporting logistics services, followed by analysis and discussion are presented. The last section will be conclusion and future recommendation.

2. Literature review

2.1. Fuzzy in Kansei and SERVQUAL

According to Hartono & Tan [1], KE in services is defined as the methodology which takes into account the customer emotional needs (known as Kansei) and translates them into service design and development. Since Kansei is a function of design characteristics/attributes [10], the spanning followed by selection and evaluation of service attributes is critical. More specifically, what service attributes affect most to many Kansei will be of high interest, and followed up by either continuous improvement or enhancement.

KE as the backbone of this recent study has been considered to be superior against other similar method. This method provides some remarkable advantages [5], such as ability to translate emotional needs into concrete design parameters, ability to build mathematical model to minimize subjectivity, ability to optimize the intangible properties which are dealing with significant feelings, and to showcase the relationship model of cognitive and affective process.

The generic model of service attributes which is deemed to be the predictor for Kansei refers to SERVQUAL dimensions [1]. In the future Kansei type, it may refer to Kansei quality management. It is defined as KE methodology taking into account customer emotional needs in service design and development to maximize total customer satisfaction. It is expected to serve consistent Kansei at all interaction-based service processes (e.g., reception, offering, and post-purchase) [8]. In achieving that consistency, service quality control is needed. It starts with the identification of service gap, and the magnitude of service satisfaction.

The challenge for Kansei research is its dynamics. How to make Kansei consistent or to judge whether it is still relevant in any particular service encounters over time is so challenging. A study on the evaluation of perceived Kansei and service performance in the steady state has been conducted by Hartono et al. [5]. It showed, for example, some Kansei words were deemed to be important since they had a very little gap between their perceived value in the current and future state. Therefore, in practical point of view, these potential Kansei words should be maintained over time, and more importantly, to what extent they were connected to particular service attributes. Given a very limited resources, this study was hoped to provide practical contribution in terms of prioritized improvement strategies. However, an intensive attention on this research field is less explored. Another concern occurs, that is, a critic on how to get the exact values of Kansei and perceived services. Once the user says a particular attribute performance is good, then a question is raised "How good it is? Is it applied to all actual users? Does it show a score of 4, or 4.5 or 5?" Hence, it comes to the concept of fuzzy taken into account in evaluating both Kansei and service performance.

2.2. Kano model as a catalyst

Kano model has simplified the service or product performance into three main categories, namely, basic/must-be (M), linear/one-dimensional (O), and delighting/attractive (A). Kano's M category is a must, a provision of basic features of product or service in which not give any significant satisfaction once it is improved dramatically [11]. In KE studies, especially in services, Kano is utilized to strengthen the KE methodology by shorten its prioritized improvement steps [1]. Thus, Kano's O and A are regarded as the prominent categories dealt with Kansei. More than the better performance, the higher the customer satisfaction is. It discusses more on delighter, which is beyond expectation. Since most customers have not realized their attractive needs, it is so relevant to engage Kano's a category into Kansei-based service design or development. Schütte [12] mentioned that Kano's attractive feature is closely connected to affect.

Kano model shows its flexibility and capability to be engaged with other quality-based tools/methods for product and service designs. The idea of considering fuzzy mode on Kano categorization, for instance, related to QFD on product management has been developed by Lee et al. [13]. Kano has been treated to be more objective in the course of weighing. Another study by Lopez and Jeronimo [14] proposed a Kano model incorporated fuzzy distances and 2-tuple fuzzy-linguistic model to manage a more efficient and effective logistics services.

With respect to Kansei-based service improvement strategies, Kano serves as a catalyst. Referring to the definition of catalyst, the involvement of Kano in KE methodology increases the efficiency and effectiveness of the formulized improvement strategies. In logistics services, as one of the emerging services nowadays, the contribution of Kano model in KE methodology is expected to be promising.

3. Research methodology

Survey method through personal interview and face-to-face questionnaire were used for data collection, considering the effectiveness of interaction between respondents and the interviewer. Also, this type of method promotes clarification on ambiguous questions and doubts effectively and efficiently. Purposive sampling (known as judgment sampling) was utilized to select subjects or respondents. Those respondents were chosen by the judgment of the researcher [15].

A questionnaire has been prepared and tested through a pilot study. Only one copy of the questionnaires was targeted for one group of participants. One group can be one single respondent or one family. Approximately, it took about 15 minutes to complete one questionnaire [1].

4. Empirical study and discussion

According to Hartono & Tan [1], this study followed the steps as discussed and shown in Figure 1. It started with the choice of logistics services domain. It is, then, followed by the measurement of perceived Kansei and SERVQUAL, and the incorporation of fuzziness as the basic contribution. A membership function of triangular fuzzy number (TFN) will be used. Kano categorization will take place to enhance the efficiency of perceived SERVQUAL which is more focused on Kano's A and O category. Afterwards, the gap between perceived and expected services will be measured, and followed by customer satisfaction score for determining prioritized improvement strategies.

A case study on logistics services of company named "ABC" has been conducted. Those who were experienced services (i.e., there were 23 logistics service items; these were services aiming to deliver goods or documents to particular destinations) from this company at least once in a month within January to December 2016 were targeted and selected as the potential subjects. In total, there were 100 subjects collected (43% male, 57% female; 45% ranged 21-25 years old; 54% working in private sector).

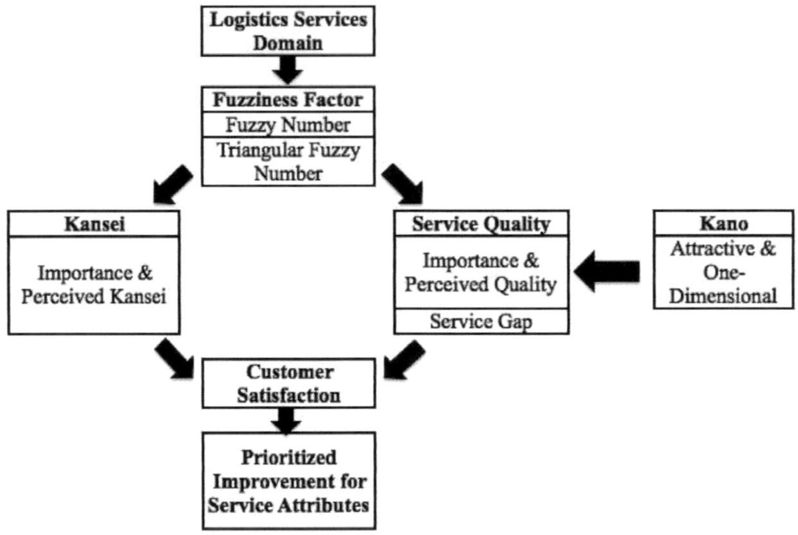

Figure 1. Application framework of fuzzy KE and Kano model in logistics services.

5. Results and discussion

According to Stefano et al. [9], the assessment of customer satisfaction in services is derived from what has been expected and perceived by customer. It is when conventional ordinal scales applied. Due to criticism on the appropriateness of scales used [9], by taking into account exact preference judgment shown in an exact numerical value, fuzziness factor has been applied. Taking a case study on IT-based logistics services, this study used linguistic variables in rating perceived and expected service quality and perceived Kansei scores, consisted of "very low, low, fair, high, very high" for expectation, and "very poor, poor, fair, good, very good" for perception. Afterwards, those linguistic variables were converted into triangular fuzzy numbers (as shown in Table 1).

Table 1. Linguistic variables and triangular fuzzy numbers.

Scale	Linguistic variable (expectation)	Linguistic variable (perception)	Triangular fuzzy numbers
1	Very low	Very poor	{1.19; 1.56; 2.16}
2	Low	Poor	{2.97; 3.86; 4.63}
3	Fair	Fair	{5.66; 6.54; 7.27}
4	High	Good	{8.16; 9.00; 9.67}
5	Very high	Very good	{10.37; 11.20; 11.96}

Considering the fuzzy numbers, the mean of expected and perceived scores of service quality and also the service gap have been calculated, and provided in Table 2. It shows that the attribute "office waiting room" had the highest gap. Afterwards, through linear mathematical modeling of Kansei as the function of perceived logistics attributes with Kano's O and A category, it has been found that (see [1] for the details of weighting process of Kano and Kansei scores into particular service attributes) the attribute "politeness of staffs" showed the greatest importance weight (Table 3). Given a very limited resource in terms of budget, time and effort, this company "ABC" should put more concern on the improvement of the attribute "politeness of staffs". In other words, the attribute of politeness of staffs was deemed to be a very sensitive to the customer emotional satisfaction.

Table 2. Expected and perceived service gap.

Dimension	Label	Logistics service attributes	Expected	Perceived	Gap
Tangible	T_1	Uniform for staffs	9.226	7.277	-1.949
	T_2	Appearance of staffs	8.536	6.800	-1.736
	T_3	ID for staffs	9.588	6.961	-2.627
	T_4	Cleanliness of office counter	9.487	6.916	-2.571
	T_5	Interior of office	8.498	6.302	-2.196
	T_6	**Office waiting room**	9.621	6.787	**-2.834***
Reliability	RL_7	Condition of transportation vehicles	8.892	7.527	-1.365
	RL_8	Promptness of service	10.391	7.966	-2.425
	RL_9	Accuracy of delivery	10.802	9.340	-1.462
	RL_{10}	Lead-time of delivery	10.440	8.261	-2.179
	RL_{11}	Tracking system	9.998	8.450	-1.547
Responsiveness	RP_{12}	Accuracy of tariff	10.011	9.272	-0.738
	RP_{13}	Responsiveness to any problems	10.070	8.069	-2.001
	RP_{14}	Availability of staffs	9.763	8.310	-1.453
	RP_{15}	Completeness of service given by staffs	10.406	9.065	-1.341
	RP_{16}	Clarity of information given by staffs	10.315	8.524	-1.792
Assurance	A_{17}	Security of parking lot	9.976	8.226	-1.749
	A_{18}	Discrepancy of packet/good delivered	10.530	8.829	-1.702
	A_{19}	Readiness of transportation vehicles	9.338	8.210	-1.128
Empathy	E_{20}	Hospitality of staffs	9.868	7.912	-1.956
	E_{21}	Politeness of staffs	9.964	8.389	-1.575
	E_{22}	Proactiveness of staffs	9.560	7.410	-2.150
	E_{23}	Apology raised by staffs	9.506	7.510	-1.996

**it shows the highest service gap*

Table 3. Weighted importance score for logistics service attributes.

Logistics services attributes	Label	|Satisfaction score|*	Kano category and weight		Kansei word and score		Weighted importance score**
ID for staffs	T_3	23.828	O	2	Friendly	6.323	301.3
Cleanliness of office counter	T_4	22.574	O	2	Prompt	6.900	1,063.7
					Secured	9.698	
					Tidy	6.963	
Promptness of service	RL_8	25.224	A	4	Calm	6.301	635.7
Accuracy of	RL_9	15.684	A	4	Prompt	6.900	829.6

Logistics services attributes	Label	|Satisfaction score|*	Kano category and weight		Kansei word and score		Weighted importance score**
					Friendly	6.323	
Clarity of information given by staffs	RP_{16}	18.497	A	4	Calm	6.301	466.2
Security of parking lot	A_{17}	17.556 Article I.	A	4	Secured	9.698	2,621.0 Article II. Article III.
					Happy	6.958	
					Tidy	6.963	
					Comfortable	6.465	
					Accurate	7.240	
Discrepancy of packet/good delivered	A_{18}	17.749	A	4	Comfortable	6.465	458.9
Hospitality of staffs	E_{20}	19.188	A	4	Friendly	6.323	485.3
Politeness of staffs	E_{21}	15.159	A	4	Secured	9.698	**2,626.4*****
					Interesting	5.913	
					Satisfied	7.200	
					Calm	6.301	
					Tidy	6.963	
					Accurate	7.240	
Proactiveness of staffs	E_{22}	20.285	O	2	Secured	9.698	1,181.1
					Interesting	5.913	
					Satisfied	7.200	
					Calm	6.301	
Apology raised by staffs	E_{23}	18.794	A	4	-	-	75.2

|satisfaction score| = importance level x gap
***weighted importance score = |satisfaction score| x Kano weight x number of significant Kansei x Kansei score*
****it shows the highest importance weight*

Following up the critical service attribute of politeness of staffs, this company "ABC" can promote training for staffs regarding personality and teamwork building which always brings value on customer focus.

6. Conclusion, limitation and future recommendation

This study provides a grounded framework of how to fulfill customer emotional needs (known as Kansei) in logistics services by taking into account fuzzy-based KE methodology equipped with Kano model. Practically, given a very limited resource, this study provides guidance to service manager in improving prioritized logistics service attributes with higher degree of certainty.

Due to very limited sample size and only tested in IT-based logistics services, this model of fuzzy-based KE and Kano model should be applied into other logistics services. Involving more samples is also encouraged.

References

[1] Hartono M and Tan K C 2011 How Kano Model Contributes to Kansei Engineering in Services *Ergonomics* **54** pp 987–1004
[2] Hartono M and Raharjo H 2015 Exploring the mediating role of affective and cognitive satisfaction on the effect of service quality on loyalty *Total Qual. Manag. Bus.* **26** pp 971–985
[3] Hartono M 2016 The extended integrated model of Kansei engineering, Kano, and TRIZ incorporating cultural differences into services *Int. J. Technol.* **1** pp 97–104
[4] Hartono M 2016 A Conceptual Integrative Model of Kansei Engineering, Kano and TRIZ Towards Sustainability in Services *Proc. 8th Widyatama Int. Seminar on Sustainability (Bandung)* (Bandung: Widyatama University) pp 303–307
[5] Hartono M 2012 Incorporating service quality tools into Kansei engineering in services: A case study of Indonesian tourists *Procedia Economics and Finance* **4** pp 201–212
[6] Hartono M, Tan K C and Peacock J B 2013 Applying Kansei engineering, the Kano model and QFD to services *Int. J. Services, Economics and Management* **5** pp 256–274
[7] Rasamoelina F, Bouchard C and Aoussat A 2013 Towards a Kansei-based user modeling methodology for eco-design *Int. J. Affective Engineering* **12** pp 337–348
[8] Nagamachi M and Lokman A M 2011 *Innovations of Kansei Engineering* (Boca Raton: CRC Press)
[9] Stefano N M, Casarotto F N, Barichello R and Sohn A P 2015 A fuzzy SERVQUAL based method for evaluation of service quality in the hotel industry *Procedia CIRP* **30** pp 433–438
[10] Nagamachi M 1995 Kansei Engineering: A new ergonomic consumer-oriented technology for product development *Int. J. Ind. Ergonom.* **15** pp 3–11
[11] Kano N, Seraku N and Takahashi F 1984 Attractive quality and must be quality *Quality* **14** pp 39–44
[12] Schütte S 2005 *Engineering Emotional Values in Product Design: Kansei Engineering in Development* (Linköping: Linköping University)
[13] Lee Y-C, Sheu L-C and Tsou Y-G 2008 Quality Function Deployment implementation based on fuzzy Kano model: An application in PLM system *Comput. Ind. Eng.* **55** pp 48–63
[14] Lopez R F and Jeronimo J M R 2012 Managing logistics customer service under certainty: An integrative fuzzy Kano framework *Inform. Sciences* **202** pp 41–57
[15] Black K 2010 *Business Statistics: Contemporary Decision Making* (Danvers: Wiley)

The impact of expatriates directors on the Indonesian company's performance

I M Ronyastra

Department of Industrial Engineering, University of Surabaya, Raya Kalirungkut, Surabaya 60293, Indonesia

E-mail: ronyastra@staff.ubaya.ac.id

Abstract. This research examined the impact of employing expatriates as board of directors (BOD) to the financial performance of Indonesian companies. Using samples from Kompas 100 index in Indonesian Stock Exchange, the research performed analyses on three performance indicators i.e. Return on Asset (ROA), Return on Equity (ROE), and Tobin's Q. Binary variable of whether a company employing expatriate and the proportion of expatriate in the BOD were used as the proxy for the independent variable. The research did not find enough evidence to support the hypothesis that employing expatriate in the BOD would make the financial performance different.

Keywords: Expatriate; Financial Performance; Kompas 100; Organization Design

1. Introduction

As a logical consequence of a free trade era, the competition among countries is getting more complex as the market for products and services expands. Labor selection plays important part in the companies' effort to improve their competitiveness. Furthermore, the companies can seek for candidates both domestic and international which create an expansion in the labor market [1]. Indonesia, which is currently sitting on 16th World's GDP ranking, is exposed to the risk in becoming the destination country for labor force. This condition can be a threat for local labors force whose numbers is predicted to surge as a result of demographic bonus in the next 5 years [2].

This research is interested in examining the impact of employing expatriates on the corporate financial performance which reflected in the accounting performance and market performance. The focus is on expatriates with executive level position i.e. board of director (BOD) because they have the greatest authority to plan and direct the companies' strategic decisions. The relationship between board of director and corporate performance has attracted the interest of many researchers to examine it from many angles. Lindorff and Johnson analyzed the CEO's business education level and firm performance [3], while the CEO's working experience was previously examined by Elsaid, Wang and Davidson [4]. Some other researchers conducted the research using data from Indonesian companies. As done by Darmadi who found out the negative relationship between gender of the top management and the firm performance in Indonesia [5] and by Lindrianasari & Hartono who revealed that decreasing performance could encourage a CEO's turnover [6].

There is an observable increasing trend of hiring expatriates as directors in Indonesian companies from only 10% in 2011 to 13% in 2014. The researcher would like to gather empirical evidence on the outcome of this decision towards the financial performance of the companies.

The remainder of this paper includes literature review and hypothesis on the second section, followed by research methodology on the third section. The data processing and result are discussed in section 4 and closed by some concluding remarks in section 5.

2. Literature review and hypotheses

2.1. Board of Directors (BOD) and Corporate Performance

As the leader in the management team, BOD has the highest responsibility in making strategic decisions which has the purpose to grow the company. On the external point of view, BOD must also be able to acquire investors' trust so that they feel safe to invest in the company. Some researchers tried to figure out factors that can predict the success of an executive in performing his tasks to improve corporate financial performance. As the findings in [3], there seems to be not enough evidence to create a relationship between CEO's business education and the performance for the Australian companies. Veprauskaite & Adams are interested in CEO power which was proxied by several variables including tenure, ownership, remunerations and bonus pay [7]. They found out that CEO power is negatively related to financial performance. While Liu & Jiraporn found out that CEO with higher power can cause bond holders to demand higher yield for it will be more difficult to monitor the executives [8].

Another angle was used by Nourayi & Daroca when treating CEO's compensations as the dependent variable which then led to the conclusion that CEO compensations are significantly explained by the company's size and market-based return [9]. Sun, Wei & Huang found out firm efficiency in terms of revenue and cost are positively associated with the CEO compensations [10]. Personal traits of the CEO also attract interest from researchers as Takeda, et al. tried to explore the stereotyping based on CEO's hair color in the Fortune 500 companies [11]. Pillemer, Graham & Burke examined the relationship among CEO's facial appearance, gender-linked traits and the financial performance [12]. While the gender angle was also taken by Chandani, Mehta & Chandrasekaran [13] and Darmadi [5].

2.2. Measuring the financial performance

Sun, Wei, & Huang in [10] measured the financial performance based on revenue efficiency and cost efficiency which wereestimated through data envelopment analysis method. Chandani, Mehta & Chandrasekaran [13]measuredbank performance using CAMEL Model that combinedcapital adequacy, asset quality, manajemen efficiency, profit, and liquidity.The popular metrics to measure the corporate financial performance are Return on Asset (ROA) and Return on Equity (ROE) which can be obtained from the financial statement. ROA, the capability of the corporate to generate profit from the entire asset they have, is also commonly used as the accounting based performance measurement. Corporate' assets usually come from debt financing and paid in capital. ROE is used to measure the profit generation capability from the amount of capital paid by the corporate' owners.

From time to time, a publicly listed company is being valued by investors in the secondary market. Their valuations are reflected in the fluctuating corporate' stock price in the stock exchange. To include this valuation to the corporate' performance, Darmadi [5] and Veprauskaite & Adams [7] used Tobin's Q that can be considered as the market based financial performance.

2.3. Hypotheses

As the objective of this research was to find the impact of expatriate executives to the companies' performance, a performance comparison between a corporate that employs Indonesian directors only and the corporate that employs at least 1 expatriate as a director must be done. The researcher guesses that there will not be any difference between the two groups. Hence:

H_0: there is a difference in financial performance between the two groups of companies

H_1: there is no difference in financial performance between the two groups of companies.

The research also wanted to find the impact of hiring expatriates as directors toward the corporate performance, so we would also find the relationship between numbers of directors hired and the financial performance:

H_2: there is a positive relationship between proportion of expatriates on the BOD and financial performance.

3. Data and research methodology

3.1. Sample data

This research used sample data from companies listed in Indonesian Stock Exchange (IDX), but the sample was limited to those companies belongs to the Kompas 100 index. The Kompas 100 index consisted of 100 chosen companies based on market capitalization, stock liquidity and fundamental condition. The research observed data were obtained from annual reports during the financial year 2012 until 2014. The index member was updated twice a year. The main source of data for this research was the official IDX website.

3.2. Variables

- Dependent Variables: to measure the company financial performance, this research used three dependent variables. Return on Assets (ROA) and Return on Equity (ROE) were used as proxy for the accounting based financial performance. Data for ROA were collected from corporate's annual financial statements which was calculated as ratio of net income to the total assets book value. While ROE measured the ratio of net income to the total equity of the companies. Tobin's Q was used as proxy for the market based financial performance. Tobin's Q was measured using the formula in Darmadi [5] which subtracting book value of equity from the total assets and adding market value of equity to find the market value of the companies and then divided by the total book value of assets. Darmadi [5], also suggest that the Tobin's Q should be included in the regression model using its natural logarithm form.
- Independent Variables: the hypothesis suggested that the independent variables was the presence of expatriates in the BOD. This research used two proxies for the independent variables. The first proxy was a binary variable that defined the presence of expatriates in the BOD, 1 if there was any expatriate and 0 if there was not. The second proxy was the proportion of the expatriates in the BOD which was the ratio of numbers of expatriate directors to the total count of the BOD members.
- Control Variables: this research used two control variables namely the companies size and the company board size. The companies size was estimated using data from total assets as the proxy while board size was stated in number of BOD members. In the regression model, the control variables were also presented in its natural logarithm form.

3.3. Methodology

- Two sample T-test was used to test the hypothesis of different mean between two groups of companies. The T-test was conducted for all three dependent variables using the binary variable as the separator between samples.
- To learn about the relationship of the expatriates BOD and financial performance, this research used the following regression model:

$$FP = \beta_0 + \beta_1 RExpat + \beta_2 LNAssets + \beta_3 LNBSize + \varepsilon \qquad (1)$$

$$FP = \beta_0 + \beta_1 PExpat + \beta_2 LNAssets + \beta_3 LNBSize + \varepsilon \qquad (2)$$

Where FP was the financial performance, as measured by ROA, ROE, and Tobin's Q, RExpat was the proportion of expatriates in the BOD, PExpat was the binary variables stating the presence of Expat in the BOD, LNAssets was the natural logarithm form of the total assets and LNBSize was the natural logarithm form of the board size.

4. Results and discussion

4.1. Descriptive statistics

The descriptive statistics are presented in table 1. The average total assets for the sample is near 43 trillion rupiah which is significantly higher compared those reported on Darmadi [5] at 6.9 trillion rupiah. This data indicates that the Kompas 100 index members are bigger in terms of average asset size than the rest of the IDX members.

Table 1. Descriptive statistics.

Variables	**Mean**	**Median**	**SD**	**Min**	**Max**
ROA	7%	5%	8%	-12%	42%
ROE	17%	13%	47%	-170%	799%
Tobins Q	2.159	1.514	2.079	0.528	17.935
Total Asset (IDR Bio)	42,744	11,754	111,899	174	855,040
Board Size	5.89	6.00	2.21	1.00	13.00
Presence of Expat	0.376	0.000	0.485	0.000	1.000
Proportion of Expat	11%	0%	18%	0%	100%

Table 2. Correlation results.

Variables	*ROA*	*ROE*	*LN Tobins Q*	*R Expat*	*LN Total Asset*	*LN Board Size*
ROA	1.00					
ROE	0.53	1.00				
LN Tobins Q	0.68	0.37	1.00			
RExpat	0.07	0.09	0.11	1.00		
LN Total Asset	-0.13	-0.06	-0.36	0.01	1.00	
LN Board Size	0.07	0.04	-0.07	0.14	0.60	1.00

4.2. Correlation results

The correlation coefficients for all variables are presented in Table 2. It can be seen that there are positive correlations between the RExpat and the financial performance although the coefficients are very small. Positive correlation also found in LN Total Assets (natural logarithm of total asset) and LN Board Size (natural logarithm of board size), 0.60, indicating that the numbers of BOD members are bigger in the larger companies.

4.3. Two sample T-test results

The F-test result suggests that, at alpha of 0.05, both ROA and ROE have equal variance, but not the LN Tobin's Q. Thus, T-test for ROA and ROE are conducted under the equal variance assumption while LN Tobin's Q under the unequal variance assumption. The detail of the test results are summarized in Table 3.The T-test for ROA revealed that the t Stat is -0.81, which indicates that there is no mean difference between the two sets of companies and we can then reject the null hypothesis of difference in performances' mean. The same conclusion can be inferred for the other financial performance indicators based on the results summary in Table 3.

4.4. Regression results

This research employs a multivariate regression analysis to test the hypothesis H_2. Table 4 and Table 5 summarize the regression of financial result on the proportion and on the presence of expatriates in BOD. It is observed that the P-value for both proportion and presence of expat are not significant and must be removed from the regression model since they do not explain the changes in the response. The R-square suggest that the explanatory variables can explain more variations on Tobin's Q than ROA and ROE. These data provide slight hint that the proportion of expatriates on the BOD has lower impact on the accounting based performances than the impact on the market based performances. The relationship between the variables are seen positive and imply that expatriate directors helped the company to gain trust from the capital market and make the market value of the companies' assets higher. It is an interesting finding that although the mean for the financial results from two sets of samples are not different, the presence and proportion of expatriates on the BOD are positively correlated to the market based financial performance.

Table 3. F-test and T-test results summary.

	ROA		ROE		LN Tobin's Q	
PExpat	***No***	***Yes***	***No***	***Yes***	***No***	***Yes***
Mean	0.0670	0.0741	0.1489	0.2001	0.5290	0.5512
Variance	0.0047	0.0078	0.0244	0.5599	0.3453	0.4240
Observations	209	126	209	126	209	126
Df	208	125	208	125	208	125
F	0.6041		0.0436		0.8143	
P(F<=f) one-tail	0.0007		0.0000		0.0960	
F Critical one-tail	0.7718		0.7718		0.7718	
Variance?	**Equal**		**Equal**		**Unequal**	
Pooled Variance	0.0059		0.2254			
Hyp. Mean Difference	0		0		0	
Df	333		333		243	
t Stat	-0.8148		-0.9556		-0.3131	
P(T<=t) two-tail	0.4158		0.3400		0.7544	
t Critical two-tail	1.9671		1.9671		1.9698	

Table 4. Regression of financial performance on proportion of expatriates in BOD.

Independent Variables	**ROA**			**ROE**			**Tobins' Q**		
	Coef.	S.E	P.Val	Coef.	S.E	P.Val	Coef.	S.E	P.Val
Intercept	0.389	0.089	0.000	1.076	0.562	0.056	5.867	0.664	0.000
R Expat	0.018	0.023	0.429	0.200	0.144	0.165	0.271	0.169	0.110
LN Total Asset	-0.013	0.003	0.000	-0.038	0.021	0.072	-0.197	0.025	0.000
LN Board Size	0.041	0.013	0.001	0.122	0.079	0.122	0.327	0.093	0.001
R Square	0.05241			0.01877			0.01750		
Observation	335			335			335		
Significance F	0.00047			0.09873			0.00000		

Table 5. Regression of financial performance on presence of expatriates in BOD.

Independent Variables	**ROA**			**ROE**			**Tobins' Q**		
	Coef.	S.E	P.Val	Coef.	S.E	P.Val	Coef.	S.E	P.Val
Intercept	0.395	0.089	0.000	1.138	0.561	0.043	5.959	0.663	0.000
P Expat	0.003	0.009	0.431	0.040	0.055	0.468	0.026	0.065	0.689
LN Total Asset	-0.013	0.003	0.000	-0.040	0.021	0.057	-0.200	0.025	0.000

LN Board Size	0.041	0.013	0.001	0.128	0.080	0.109	0.344	0.094	0.000
R Square		0.05106			0.01459			0.16905	
Observation		335			335			335	
Significance F		0.00059			0.18141			0.00000	

These results also suggest that the same conclusion as the result from Darmadi [5] in 2010 regarding the size of the BOD members which has positive association with market based financial performance. The more directors in the BOD seems to be believed can create a more focused work management and wider perspective during the strategic decision making process.

5. Conclusion and further research

Hiring expatriates as board of directors of Indonesian companies did not have observable impact on the accounting-based financial performance, but it may have positive impact on the market-based financial performance. The expatriates are able to gain investors' positive image which is translated into higher valuation of the companies' assets. However, they failed to make a difference in the companies' performance which should have been the main reason of hiring them. It is debatable that the companies' performance is influenced by many factors. Thus, it requires further examination on how the expatriates can contribute more to improve the performance. The research can also be extended to find out the other explanation on the increase in number of expatriates hired as directors.

6. References

[1] Lindorff M and Jonson E P 2013 CEO business education and firm financial performance: A case for humility rather than hubris *Education + Training* **55** pp 461–477

[2] Elsaid E, Wang X and Davidson W N 2011 Does experience matter? CEO successions by former CEOs *Manage. Financ.* **37** pp 915–939

[3] Darmadi S 2013 Do women in top management affect firm performance? Evidence from Indonesia *Corporate Governance: The Int. J. of Business in Society* **13** pp 288–304

[4] Lindrianasari and Hartono J 2012 Antecedent and consequence factors of CEO turnover in Indonesia *Management Research Review* **35** pp 206–224

[5] Veprauskaite E and Adams M 2013 Do powerful chief executives influence the financial performance of UK firms? *The British Accounting Review* **45** pp 229–241

[6] Liu Y and Jiraporn P 2010 The effect of CEO power on bond ratings and yields *J. Empir. Financ.* **17** pp 744–762

[7] Nourayi M M and Daroca F P 2008 CEO compensation, firm performance and operational characteristics *Manage. Financ.* **34** pp 562–584

[8] Sun F, Wei X and Huang W 2013 CEO compensation and firm performance *Review of Accounting and Finance* **12** pp 252–267

[9] Takeda M B, Helms M M, Klintworth P and Sompayrac J 2005 Hair colour stereotyping and CEO selection: can you name any blonde CEOs? *Equal Opportunities Int.* **24** pp 1–13

[10] Pillemer J, Graham E R and Burke D M 2014 The face says it all: CEOs, gender, and predicting corporate performance *Leadership Quart.* **25** pp 855–864

[11] Chandani A, Mehta M and Chandrasekaran K 2014 Impact of gender of leader on the financial performance of the bank: a case of ICICI Bank *Procedia Economics and Finance* **11** pp 459–471

Survival analysis for customer satisfaction: A case study

M A Hadiyat, R D Wahyudi and Y Sari

Department of Industrial Engineering, University of Surabaya, Raya Kalirungkut, Surabaya 60293, Indonesia

E-mail: arbi@staff.ubaya.ac.id

Abstract. Most customer satisfaction surveys are conducted periodically to track their dynamics. One of the goals of this survey was to evaluate the service design by recognizing the trend of satisfaction score. Many researchers recommended in redesigning the service when the satisfaction scores were decreasing, so that the service life cycle could be predicted qualitatively. However, these scores were usually set in Likert scale and had quantitative properties. Thus, they should also be analyzed in quantitative model so that the predicted service life cycle would be done by applying the survival analysis. This paper discussed a starting point for customer satisfaction survival analysis with a case study in healthcare service.

Keywords: satisfaction scores; survival analysis; satisfaction life cycle; service re-design predicting

1. Introduction

Many researchers proposed to conduct the customer satisfaction survey periodically to capture the behavior of this dynamic satisfaction scores [1]. The questionnaire for this survey consists of items which influence customers' satisfaction level. Qualitatively, some researchers decide to redesign the service when there are some indications that the scores are decreasing [2]. It means that there are some limitations in predicting when the redesign process should be conducted. One needs to expert judge the redesign time base on past experiences.

However, quantitative approach in designing questionnaire for measuring customer satisfaction should be considered. Items in questionnaire for this survey are usually scaled in Likert items, so that the scores have some quantitative property [3]. Once the survey is conducted, and satisfaction score is calculated, the quantitative analysis can be done. It gives stronger interpretation in predicting the service life cycle.

As in modeling the reliability of an electric product, the similar survival analysis will be fitted for the customer satisfaction scores. The idea for choosing the product reliability models in this paper is based on [4] and [5] who has successfully modeled the customer lifetime value using survival approach. The satisfaction scores are measured from time to time to capture the dynamic behavior of what customers have perceived. Theoretically, once the new service design is deployed to customers, the satisfaction will increase due to the new service innovations. At some levels, the satisfaction scores reach the maximum level, and then decreased as the perceived service is beginning to be unsatisfied. In this point, a new service design should be made and deployed soon to maintain the customer satisfaction level.

The survival analysis helps the management predict the time when new service design should be deployed by constructing the probability that the scores are less than certain levels of satisfaction. It began by control charting the satisfaction scores over time and capturing its dynamic, fitting the best probability distribution, followed by calculating the probability of satisfaction survival. Smaller survival probability of certain level of satisfaction represents the need of service redesign.

2. Literature review

2.1. Parasuraman's satisfaction scores

Parasuraman [6] mentioned that the satisfaction of customer was the gaps measurement between expected and perceived service received by customers, defined as customer gaps. Moreover, another gap was also measured for mining the causes of customer gaps, and all those integrated gaps were summarized as the well-known method called SERVQUAL which stands for service quality.

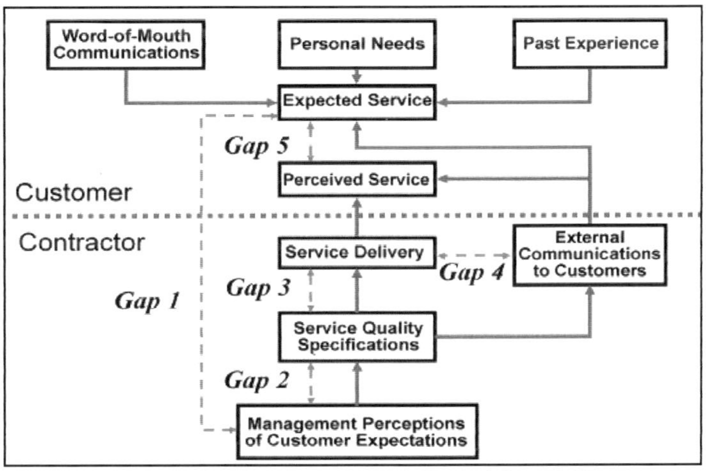

Figure 1. SERVQUAL gaps (adapted from [6]).

Common questionnaires for measuring these customer gaps consist of at least 20 Likert scale questions. All items are generated based on 5 dimensions i.e. tangibles, empathy, reliability, responsiveness, and assurance. The overall satisfaction score of a customer is calculated as the average of each item scores. Scores from all customers measured over time form a dataset to be modeled with survival analysis. In this paper, all satisfaction scores were converted to positive value-low positive mean unsatisfied while high one means satisfied.

2.2. Survival analysis

Similar to the calculation of product reliability, the survival analysis in this paper was based on [7] applied to satisfaction dataset. The steps in survival analysis for the product reliability were as follows:

a. Measure the reliability data *t*(age of component, or satisfaction scores),
b. Fit the dataset with some probability density function (PDF)*f*(*t*), this function could be a product of two or more probability distribution in regards to its dynamically behavior.
c. Use the fitted PDF to calculate the reliability of component or satisfaction. Fitted probability distribution could be calculated using simulated Bayesian approach due to its multimodality.

$$\text{Reliability function } R(t) = P(t > T) = 1 - F(t), \quad (1)$$

$$\text{Where } F(t) = P(t < T) = \int_{-\infty}^{T} f(t)dt$$

d. Predict the probability that a component or satisfaction will decreased at a certain level

The PDF used in this paper referred to the common survival analysis such as Weibull, exponential, gamma, and similar distribution covered in a normal family distribution, multiplied by similar distribution, so that *f*(*t*) could be interpreted.

IOP Conf. Series: Materials Science and Engineering **273** (2017) 012029 doi:10.1088/1757-899X/273/1/012029

3. Research framework and methodology

This paper was a part of preliminary research in modeling customer satisfaction data and predicting the service life cycle. Commonly, the satisfaction level of a customer is related with how long he has been using the service, the longer using the service then the satisfaction level usually decreased, as mentioned in [8]. Meanwhile, since each customer has different satisfaction level variation over time, so there will be two random variables affecting the life-cycle of service, that are the satisfaction level (measured using questionnaire) and how long a customer has been using the service (measured in time). Each of these two variables has different statistical PDF, and both of them should be used to model the survival analysis for prediction the service life-cycle. The main framework included the integration of two probability distribution fitted by Bayesian approach. The satisfaction PDF was called $f(x)$, and how long customers have been using service was denoted by $s(t)$.

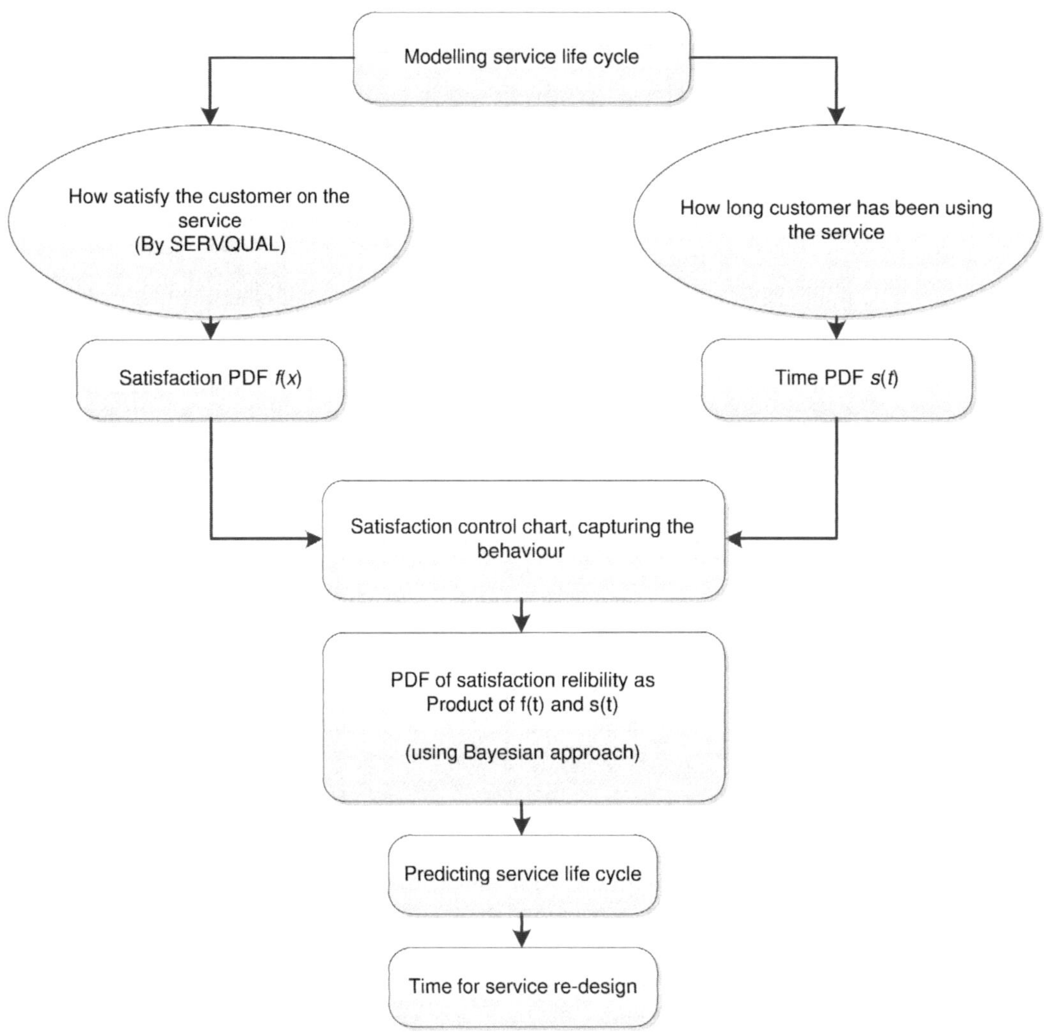

Figure 2. Research framework.

The product PDF as multiplication of $f(x)$ and $s(t)$ was done by simulating those prior distributions so that the posterior PDF was obtained. Mathematically, the posterior PDF (product PDF) for modeling satisfaction reliability survival function is written as follows:

IOP Conf. Series: Materials Science and Engineering **273** (2017) 012029 doi:10.1088/1757-899X/273/1/012029

$$L(xt) = \int f(x)s(t)\,dxdt, \qquad x \geq 0, t \geq 0 \qquad (2)$$

4. Results and discussion

The case study in this paper was taken from [9]. There were 155 data recorded from survey conducted while evaluating hospital patients' perception about its services. The dataset consisted of two variables, patient satisfaction scores (on scale of 1-7) and how long they have been treated and staying in the hospital (in days). First, the behavior of those two variables was captured using multivariate control chart, as in [10] and refers to [11].

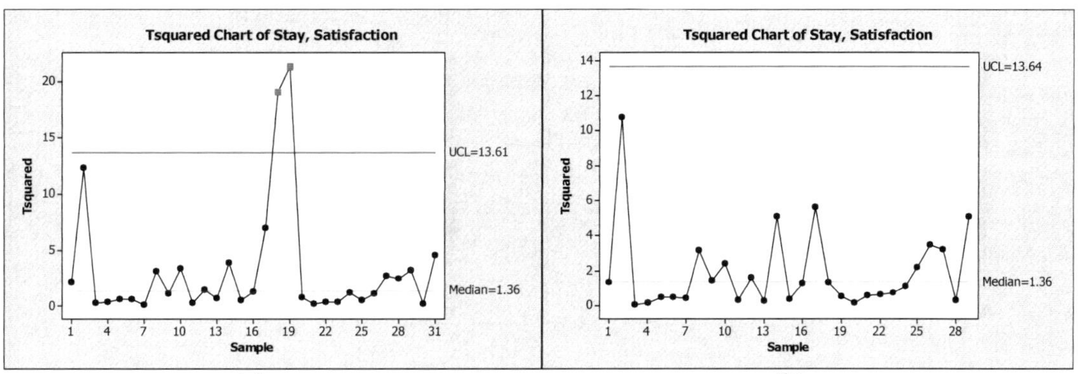

(a) (b)

Figure 3. Multivariate control chart. (a) Before omitting unusual observations, and (b) After removing unusual observations.

Two points were out of control; indicating that some special cases should be investigated at a certain time. If there were some causes, the two points should be removed from analysis. Therefore, the PDF fitting process for those two variables could be held.

Based on parametric survival analysis using MINITAB software, the fitting process for $f(x)$ and $s(t)$ was done. PDF $f(x)$ was well fitted by Weibull distribution, and $s(t)$ was assumed to follow an exponential distribution.

Goodness-of-Fit for Satisfaction

Distribution	Anderson-Darling (adj)	Correlation Coefficient
Weibull	**2.212**	**0.981**
Lognormal	6.213	0.919
Exponential	73.699	*
Loglogistic	6.117	0.920
3-Parameter Lognormal	2.072	0.976
2-Parameter Exponential	57.498	*
3-Parameter Loglogistic	2.706	0.966
Smallest Extreme Value	1.734	0.985
Normal	2.062	0.976
Logistic	2.693	0.966

Figure 4. Minitab output; probability distribution fitting for satisfaction data.

The multiplication process in (2) could not be solved analytically, so Bayesian Markov Chain Monte Carlo [10] simulation was used to fit the PDF $L(xt)$ as the Bayesian posterior distribution. The Bayesian structure for likelihood function $f(t)$, prior distribution $s(t)$, and posterior $L(xt)$ were defined as:

$$L(xt|\alpha,\mu) = \int f(x)s(t)\,dxdt$$
$$f(x|t,\alpha,\mu) \sim weibull(\alpha, s(t))$$
$$s(t) \sim exponential(\mu) \quad (3)$$

The Markov Chain Monte Carlo simulation was conducted using WINBUGS (Bayesian Using Gibss Sampler) software [11].

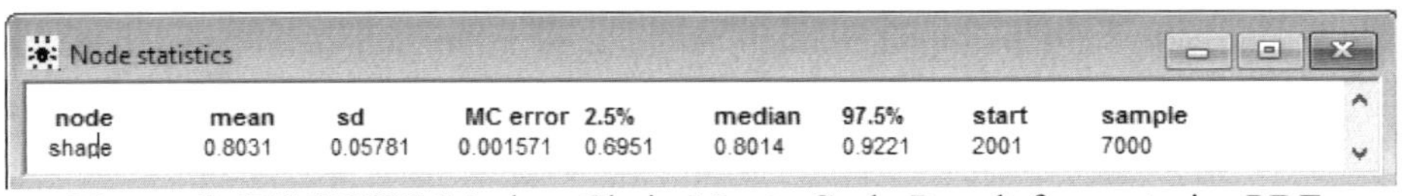
Node statistics

node	mean	sd	MC error	2.5%	median	97.5%	start	sample
shape	0.8031	0.05781	0.001571	0.6951	0.8014	0.9221	2001	7000

Figure 5. WINBUGS Markov Chain Monte Carlo Result for posterior PDF.

Figure 5 shows that the estimated α was 0.803, using a similar way that the estimation of other parameters was obtained. Next, this value would be used to predict the probability of satisfaction survival. Table 1 represents the trends of survival probability.

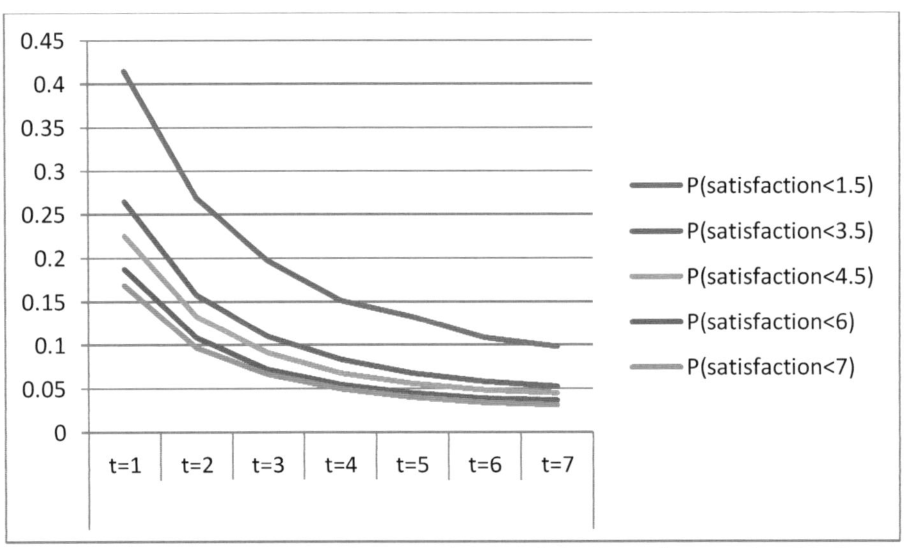

Figure 6. Survival probability over time (satisfaction Likert scale of 1-7).

The survival probability decreased over time; Moreover, the probability at same time t for high satisfaction level gave higher probability than lower ones. Figure 6 describes that until t=7, customer satisfaction becomes higher. On the first day in hospital, patients felt uncomforted because of inadaptable conditions, but after several days the patients perceived more satisfaction and got more complete service.

5. Concluding remarks

This was an initial research for mathematically modeling customer satisfaction due to the service quality of a nonmanufacturing company or public service. The level of satisfaction was dynamic and interesting to study. As shown in the analysis, practically life-cycle of service can be modeled and predicted, and this information is useful in deciding when a service design should be innovate by provider to re-increase the customer satisfaction level. The development of this research leads to the innovation of robust service design.

Acknowledgement
This paper is part of research funded by Directorate General Higher Education, Ministry of Research, Technology and Higher Education of the Republic of Indonesia, year 2017.

References
[1] Hill N, Roche G and Allen R 2007 *Customer Satisfaction: The Customer Experience through the Customer's Eyes* (London: Cogent Publishing) pp 245–250
[2] Lovelock C H, Patterson P and Wirtz J 2015 *Services Marketing* (Melbourne: Pearson Australia) pp 407–410
[3] Audrezet A, Olsen S O and Tudoran A A 2016 The GRID scale: A new tool for measuring service mixed satisfaction *J. Serv. Mark.* **30** pp 29–47
[4] Chen Y, Zhang Hand Zhu P 2009 Study of customer lifetime value model based on survival analysis *Proc. of WRI World Congress on Computer Science and Information Engineering* (Los Angeles: IEEE) pp 266–270
[5] Lu J 2003 Modeling customer lifetime value using survival analysis: An application in the telecommunications industry *Proc. of SAS User Group International Conference 28* (Washington: SAS Institute) pp 120–128
[6] Parasuraman A, Zeithaml V A and Berry L L 1988 SERVQUAL: a multi-item scale for measuring consumer perceptions of the service quality *J. Retailing* **64** pp 12–40.
[7] Ebeling C 2010 *An Introduction to Reliability and Maintainability Engineering* (London: McGraw Hill) pp 20–30
[8] Mittal V, Katrichis J M and Kumar P 2001 Attribute performance and customer satisfaction over time: evidence from two field studies *J. Serv. Mark.* **15** pp 343–356
[9] Trusko B, Pexton C, Harrington J and Gupta P K 2007 *Improving Healthcare Quality and Cost with Six Sigma* (London: FT Press) pp 241–370
[10] Khorsidi H A, Nikfalazar S and Gunawan I 2016 Statistical process control applications on service quality using SERVQUAL and QFD with a case study in trains service *The TQM Journal* **28** pp 195–201
[11] Montgomery D C 2005 *Introduction to Statistical Quality Control* (New York: John Wiley & Sons) pp 487–504
[12] Lunn D, Jackson C, Best N, Thomas A and Spiegelhalter D 2013 *The BUGS Book: A Practical Introduction to Bayesian Analysis* (New York: CRC Press) pp 57–79

Pattern analysis of fraud case in Taiwan, China and Indonesia

A H Kusumo[1], C-F Chi[2] and R S Dewi[3]

[1] Department of Industrial Engineering, University of Surabaya, Raya Kalirungkut, Surabaya 60293, Indonesia
[2] Department of Industrial Management National Taiwan Univ. of Sci. and Technology 43 Keelung Road, Section 4 Taipei, Taiwan 106
[3] Department of Industrial Engineering, Sepuluh Nopember Institute of Technology, Kampus ITS Keputih, Sukolilo, Surabaya, Indonesia, 60111

Email: argohadi@gmail.com, chris@mail.ntust.edu.tw, ratna.sari.dewi80@gmail.com

Abstract. The current study analyzed 125 successful fraud cases happened in Taiwan, China, and Indonesia from 2008 to 2012 published in the English online newspapers. Each of the case report was coded in terms of scam principle, information media (information exchange between fraudsters and victim), money media (media used by fraudsters to obtain unauthorized financial benefit) and other additional information which was judged to be relevant. The Chi-square Automatic Interaction Detector (CHAID) was applied to the coded data of information, scam principle and money media to find a subset of predictors that might derive meaningful classifications. A series of flow diagrams was constructed based on CHAID result to illustrate the flow of information (scam) travelling from information media to money media.

Keywords: coding scheme; flow diagram; CHAID

1. Introduction

Fraud can be associated with injury. One person can injure another either by force or through fraud. The use of force to cause bodily injury is frowned on by most organized societies; using fraud to cause financial injury to another does not always carry the same degree of stigma or punishment [1]. The internet crime complaint (IC3) received more than 250.000 complaints every year associated with the internet crime reaching the amount of one billion in 2015 [2].

A cyber scammer can operate in multiple countries on multiple victims in parallel. A single person or a unit of individuals can realize a large remuneration for the smallest possible investment of time and effort, even when such syndicate might be loosely-coupled and geographically distributed. Whilst there is a constellation of classification schemes available, a consistent classification framework is seemingly nonexistent. Some themes seem to be agreed upon throughout the current literature [3]. Laleh and Azgomi [4] proposed taxonomy of different kinds of frauds including credit card frauds, telecommunication frauds, insurance frauds, internal fraud, computer intrusion, web network fraud and customs frauds. Other taxonomy by Wang, technology-based financial fraud offences, can be divided into two categories: system attacks and non-system attacks [5].However, variance of taxonomy is difficult for comparing sensible transnational or conceiving coordinated operations. To aid in the detection and interception of fraud, a clear and consistent classification scheme needs to be developed.

For classification, the coding scheme is designed to be mutually exclusive individually but complete collectively [6]. A total of 125 successful fraud cases happened in Taiwan, China, and Indonesia from 2008 to 2012 in the English online newspaper were coded and analyzed based on a classification scheme.

Online newspapers as web based computerized data were also being used in several researches. A content analysis of the top 100 circulated daily newspapers in the United States was conducted to provide an understanding as to the current status of these online newspaper sites. A number of categories were examined including newspaper circulation size, updateness, cross media partnership/ownership, and level of convergence [7]. Poria also explored the possible uses of online news discussions that emerge following the publication of news on the Internet [8].

The current study proposed CHAID analysis and Flow Diagram to analyze fraud cases. The CHAID Analysis (Chi Square Automatic Interaction Detection) was applied to the coded data of fraud cases to select the independent (predictor) variable that had the strongest interaction with the dependent variable. Categories of each predictor were merged if they were not significantly different with respect to the dependent variable. The data flow diagram (DFD) was a good process model for such a visualization technique because of its strong hierarchical structure and due to the fact that it was a classic tool for process analysis [9].

A series of flow diagram was constructed based on the information media and money media to illustrate how the fraudsters exchanged information with the victims and how the fraudsters obtained unauthorized financial benefit similar to the prevention of electrical injury [10]. Each of the flow diagrams could be directly linked with feasible prevention strategies by cutting the flow of information or money.

2. Materials and methods

The current study analyzed 125 successful fraud cases happened in Taiwan, China, and Indonesia from 2008 to 2012 in the English online newspaper. This study used the following internet search engines to locate newspaper in each country: Google.com. The case reports were obtained by searching online newspaper with the terms “fraud” and “scam”.

The articles used in this study were chosen based on three criteria as follows [11]: First, Duplicated or reprinted stories eliminated the number of articles included in this study significantly. Second, the case report had to contain information about scam principle, information media and money media. If the critical information were not available in English, it had to be discovered from other sources such as newspapers published in Chinese or Bahasa. Third, articles presented unsuccessful case of fraud or announcement from any institutions were discarded. These kinds of articles were common in the online newspaper. Besides, traditional frauds such as corruption, insurance scam, investment scam, magic/special power ability, trading was also excluded from the analysis.

Using the inclusion criteria above, articles describing specific incidents of fraud and scam could be identified. Articles which did not meet these criteria were eliminated. By only analyzing specific newspaper articles, the report of this issue could be examined. The list of English online newspapers in Taiwan, China and Indonesia is shown in Table 1.

IOP Conf. Series: Materials Science and Engineering **273** (2017) 012030 doi:10.1088/1757-899X/273/1/012030

Table 1. English online newspapers in Taiwan, China, and Indonesia.

Country	Newspaper
Taiwan	China Post; The Taipei Times; Taiwan Economic News; Taiwan Today
China	China.org.cn; China Daily; China News; Economic Observer; Global Times; People's Daily; Shanghai Daily
Indonesia	Inside Indonesia; The Jakarta Globe; The Jakarta Post; The President Post; Tempo Interactive

CHAID and flow diagram were used to identify feasible prevention measures from each distinguishable pattern of flow diagram. The coding scheme was designed to be mutually exclusive individually but complete collectively. The coding schemes were developed based on the flow of scam from fraudsters to victims as shown in Table 2, in terms of scam principle, information media, and money media as variables. However, victim's age and gender were difficult to be defined due to insufficient data in the online newspaper. Besides the ones mentioned in each case, there were many people becoming a victim.

Table 2. Developed coding scheme.

Coding Scheme	Detail
Scam Principle [12].	• Distraction Principle (Advance fee/price/lottery) • Social Compliance Principle • Kindness Principle (Charity) • Time Principle (intimidation) • Unnoticeable Scam
Information media	• Face to Face [13]. • Mobile Phone (Voice and Text Based) [14]. • Internet (Person to person/email and Forum/web) [15]. • ATM fraud [16]. • Other
Money Media	• Financial gain through money transfer (wire transfer, prepaid money card) • Financial gain through stolen information (loosing bank account debit/credit) • Financial gain through direct hit (cash/product)

3. Results

Our current study focused on CHAID analysis of the Scam Principle: Distraction principle, Social compliance principle, Kindness principle, and Time principle (4 out of 7 principles). Other principles from seven principles were not found in data collection. One more principle was added, which was unnoticeable principle such as hacking, manipulated ATM machine, etc.

CHAID divided total population into more distinct groups, based on categories of the ''best'' predictor of a dependent variable (i.e. the information media) and split each of these groups into smaller subgroups based on other predictor variables, e.g. scam principle and money media. Scam principles were found to be significant predictor variables and identified as the ''best'' or most critical predictor ($p < 0.001$) which divided total population in the format of hazard pattern into five different groups (nodes). Money media were found to be the second critical predictor, and it was nested under the levels of the most critical factor, scam principle.

According to the constructed flow diagram as shown in Figure 1, it can be analyzed that the fraudsters can employ more than one media type to deceive their victims. In many cases on online newspaper can be found that the distraction principles which are advance fee/lottery/prize, romance/relationship scan, and job scam employ almost all media type such as mobile phone (voice and text based communication), internet person to person (email), forum/broadcast (website, etc.), magazine/newspaper, etc. In this principle type, while people are distracted by what grabs their interest, fraudsters can do anything to them and they will not notice it [12].

To deceive their victims, the fraudsters use not only one media, but also more than two or even three information media. From one media to another, the fraudsters must use simultaneously the first media before going to another media. In the example of winning lottery story, Fraudster called the victim via voice communication (phone) saying that they won a lottery. Some of the callers who believed in the sweet talk would send money to the suspect, which was requested by him as tax for their premium. Besides contacting the victim by voice communication, fraudsters also use text messages and emails. They also put the bait about lottery in magazine and newspaper. When a victim is interested, they will call the criminal.

From this flow diagram, the information shows that the fraudsters can obtain money from cash/ product (that ended with directly met or "face to face" with the victims) or ask the victims to do a wire transfer from bank or ATM machine (however, the most cases is via ATM) or ask the victim's personal information in a way that the victim must provide credit card number and card security code before their application could be processed. Then they found out that overseas transaction had just been added to his credit card bill.

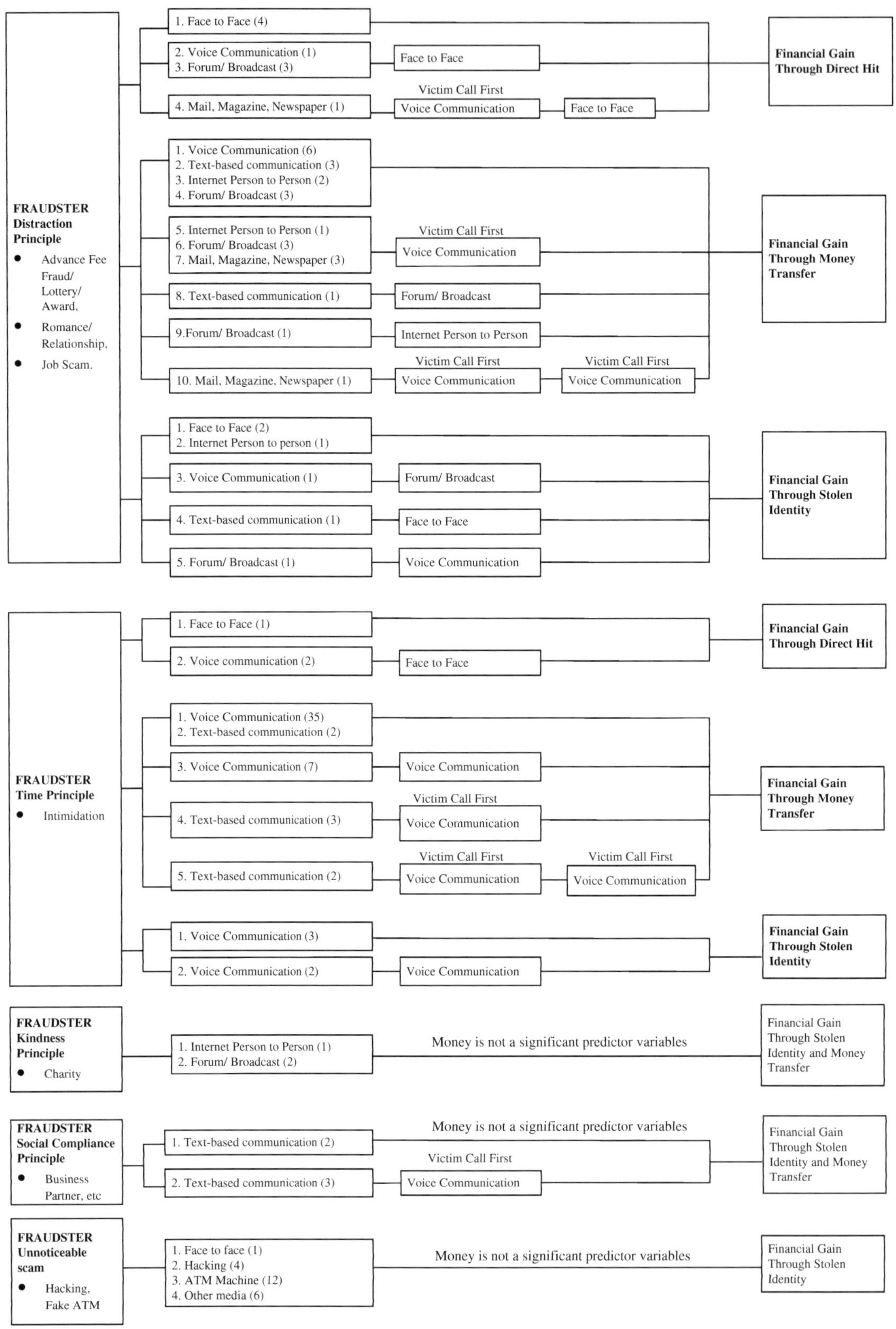

Figure 1. Flow diagram of fraud case.

For time principle (intimidation scam), when under time pressure to make an important choice, people tend to use a different decision strategy. Fraudsters then steer them toward one involving less reasoning [12].Since there was a time pressure, the media used by perpetrators that can be found in this study was only direct communication media, such as mobile phone. Most of the cases, fraudsters randomly called the victims, but there were several cases that fraudsters only sent the text message. Since the victims were panic, they called the fraudsters first. This principle can also happen "face to face". When it happens, fraudsters ask for cash/products. In this flow diagram, besides cash/product, fraudster can also obtain financial benefit from wire transfer and stolen identity from the victims.

In Kindness principle (3 cases), social compliance principle (5 cases), and unnoticeable scam (23 cases), money media were not found to be a significant predictor variable by CHAID. It is possibly caused by the lack of sample data. However, in this condition, it can be analyzed that the media used for charity scam was found only in internet. Fraudster can obtain financial benefit from wire transfer and stolen identity of the victims.

In addition, in social compliance principle, Fraudsters employed mobile phone since it could convince the victims. For unnoticeable scam, for example, the thief performed the scam through a scam called "card copying" or skimmer. Scammers use some technological facilities to steal the card's data when the cardholder uses it at a particular ATM. The data is copied to another card, which is then used to withdraw money from the account. Photographs of the suspects withdrawing money at the ATMs have been obtained and handed over to the police. Beside using skimmer, they do hacking/phishing, make a fake ATM Machine, ask bank employee's help, set up pinhole camera around ATM, collect the information from customer's receipt, and also get customers' information because they have access on it.

4. Discussion

This study collected 125 fraud cases, but the pattern was not clearly seen. We assumed that if the data was collected more than the current study, it would be possible to show the better pattern. This flow diagram cannot be adopted from electrical injury in that each of the flow diagrams can be directly linked with feasible prevention strategies by cutting the flow [10].Since it is dynamic, it means that Fraudster can have the other scenario if one scenario is not successful, e.g. if the fraudsters' demand for wire is not successful, the fraudsters can only ask for the private identity.

We also believed that age, gender and psychological factor of the victim contributed to successful fraud, e.g. there were cases that the youngest victims were in their 50s and lived alone. These people were vulnerable as they were generally less familiar with bank-account transfer practices and rules. To get the money, the fraudsters falsely claimed to be police officers, procurators or bank employees, and then persuade them to transfer money into a certain account due to legal problem or lacking personal information.

Other examples, more than 100 people were taken advantage of in the phone sex scam, most of them were men over the age of 50, divorced, widowed, or inexperienced with relationships. The men falling for such scams was desperately looking for love and was easily exploited by a charming woman who was called by them several times a day and showered them with affection. Victims thought that they have found their "dream girl," and many were willing to do everything for the object of their desire. However, victim's age and gender were difficult to be defined due to insufficient data in the online newspaper.

5. Conclusion

People nowadays face a much higher risk of fraud because they interact with various kinds of information media and money media as a technology to connect them.

This study developed a pattern and a flow diagram based on125 fraud cases reported by online newspaper in Taiwan, China, and Indonesia 2008–2012. Nine patterns of flow diagrams were generated according to information media, scam principle, and money media. The information media had a significant association with scam principle, proven that scam principles were found to be

IOP Conf. Series: Materials Science and Engineering **273** (2017) 012030 doi:10.1088/1757-899X/273/1/012030

significant predictor variables and being identified as the ''best'' or most critical predictor ($p < 0.001$).

Flow diagram has shown that fraudsters can employ more than one information media type to deceive their victims. By using distraction principles, they can employ almost all the information media. However, in time principle, since there is a time pressure, the information media use that can be found in this study is only a direct communication media, such as mobile phone.

References

[1] Singleton T W and Singleton A J 2010 Auditing and forensic accounting *Fraud Principles* vol 4 ed Rivera S (Hoboken: Wiley) chapter 2 p 40

[2] Anonymous 2016 *Internet Crime Report* (Washington, D.C.: U.S. Department of Justice)

[3] Stabek A, Brown S and Watters P A 2009 The case for a consistent cyberscam classification framework (CCCF) *Proc. Symp. and Workshops on Ubiquitous, Autonomic and Trusted Computing (Brisbane)* (Piscataway: IEEE) pp 525–530

[4] Laleh N and Azgomi M A 2009 A taxonomy of frauds and fraud detection techniques *Proc. the 3rd Int. Conf. on Information Systems, Technology and Management, ICISTM (Ghaziabad)* vol 31 (Cham: Springer-Verlag Berlin Heidelberg) pp 256–267

[5] Wang J H, Liao Y L, Tsai T M and Hung G 2006 Technology-based financial frauds in Taiwan: Issues and approaches *Proc. IEEE Int. Conf. on Systems, Man and Cybernetics (Taipei)* vol 2 (Piscataway: IEEE) pp 1120–24

[6] Chi C F, Yang C C and Chen Z L 2009 In-depth accident analysis of electrical fatalities in the construction industry *Int. J. Ind. Ergon.* **39** pp 635–644

[7] duPlessis R C 2003 *A Convergence of Modes: Present Status of Online News Sites A Content Analysis of 100 Online Newspaper Web Sites* (LA: Louisiana State University)

[8] Poria Y and Oppewal H 2003 A new medium for data collection: Online news discussions *Int. J. Contemp. Hosp. M.* **15** pp 232–236

[9] Turetken O and Schuff D 2007 The impact of context-aware fisheye models on understanding business processes: an empirical study of data flow diagrams *Inf. Manage.* **44** pp 40–52

[10] Chi C F, Lin Y Y and Ikhwan M 2012 Flow diagram analysis of electrical fatalities in construction industry *Saf. Sci.* **50** pp 1205–14

[11] Archbold C A, Lytle D, Weatherall C, Romero A and Baumann C 2006 Lawsuits involving the police: A content analysis of newspaper accounts *Int. J. Police Strat. & Mgmt.* **29** pp 625–642

[12] Stajano F and Wilson P 2011 Understanding scam victims: Seven principles for systems security *Commun. ACM* **54** pp 70–75

[13] Nardi B A and Whittaker S 2002 Distributed Work *The Place of Face-to-Face Communication in Distributed Work* (Cambridge: MIT Press) chapter 4 pp 83–110

[14] Schiefer G and Decker M 2008 Taxonomy for mobile terminals - A selective classification scheme *Proc. of the Int. Conf. on e-Business (Porto)* (Setúbal: SciTePress) pp 255–258

[15] Long B and Baecker R 1997 A taxonomy of internet communication tools *Proc. WebNet AACE (Toronto)* (Charlottesville: AACE)

[16] Utakri N 2007 The phantasm of ATM withdrawal *Proc. 5th Australian Information Security Management Conf. (Perth)* (Perth: Edith Cowan University) pp 207–215.

IOP Conf. Series: Materials Science and Engineering **273** (2017) 012031 doi:10.1088/1757-899X/273/1/012031

Outdoor altitude stabilization of QuadRotor based on type-2 fuzzy and fuzzy PID

H Wicaksono, Y G Yusuf, C Kristanto and L Haryanto

Department of Electrical Engineering, University of Surabaya, Raya Kalirungkut, Surabaya 60293, Indonesia

E-mail: hendi@staff.ubaya.ac.id

Abstract. This paper presents a design of altitude stabilization of QuadRotor based on type-2 fuzzy and fuzzy PID. This practical design is implemented outdoor. Barometric and sonar sensor were used in this experiment as an input for the controller YoHe. The throttle signal as a control input was provided by the controller to leveling QuadRotor in particular altitude and known well as altitude stabilization. The parameter of type-2 fuzzy and fuzzy PID was tuned in several heights to get the best control parameter for any height. Type-2 fuzzy produced better result than fuzzy PID but had a slow response in the beginning.

1. Introduction

Nowadays, research about QuadRotor as a lightweight transportation becomes famous. QuadRotor with Vertical Take-Off and Landing (VTOL) ability, make QuadRotor can cooperate with Automated Guided Vehicle (AGV) in Dynamical Cellular Manufacturing System (DCMS) as a part transporter in ways to keep part into some height level storage.

The QuadRotor has four swings with four propellers and also four motors. The QuadRotor with four propellers has three axes: pitch, roll, and yaw by changing the speed of four propellers [1]. There are main components of the QuadRotor. They are X-frame, four propellers, four brushless dc motors, four Electronic Speed Controller (ESC), flight controller, and battery as energy supply. There are several considerations for choosing a specification of components to adjust to the frame. The RX module connects to flight controller to receive signals from TX remote control. Alternatively, if this is an autonomous QuadRotor, the input of the flight controller is from the sensors, such as an ultrasonic sensor, barometric sensor, or camera. That kind of sensors is usually used for measuring an altitude of the QuadRotor.

Currently, there are commercial QuadRotors such as a Phantom and AR-Drone. They have many features on their platform, i.e. localization using GPS data. AR-Drone is used as a platform to investigate the functional of fuzzy logic for trajectory tracking [2]. In this experiment, they used AR Drone because it already has onboard electronics with low price and also provides support for access with mathematic tools. According to [2], AR-Drone makes it simple to compare two fuzzy logic controller scheme performance for position tracking [3]. They investigated the system of fuzzy logic controller for handling AR-Drone moves efficiently.

There are also some researchers who also investigated QuadRotor field from scratch. They used parts of QuadRotor from the market and assembled it into QuadRotor. From our knowledge, there are two studies focusing on developing QuadRotor from scratch. The first is research about QuadRotor attitude. This research focuses on flight controller development because QuadRotor attitude is an essential control for maneuver. Desa designed hybrid control from the fuzzy controller and PID control for attitude stabilization based on Matlab simulation [4]. Rinaldi et al. compared PID and Linear Quadratic Regulation control performance to get the best control of attitude regulation of QuadRotor. Besides conducting the simulation, they also built a physical prototype for an experimental test, but only for one degree of freedom [5].Overall, LQ Regulation is better than traditional control PID because LQ Regulation has a special ability that can directly calibrate the references tracking error and oversight actions. By using QuadRotor test bed one degree of freedom, Wicaksono also conducted and compared T1-fuzzy (type-1 fuzzy) with T2-Fuzzy (type-2 fuzzy) with an objective to perform autonomous hover when the payload has changed before flight or during flight [6]. They concluded that the control of flight controller needs a calibration first if we put some additional payload like camera.

In 2013, Guerrero developed attitude stabilization using nonlinear control with event-triggering function [7]. They also used the real QuadRotor for experimentation. Gonzales proposed a control scheme for altitude stabilization using motor speed feedback [8]. The basic control here is a PD control but with motor speed feedback analysis as the main input of their control scheme idea.

The second focus on QuadRotor field is altitude stabilization. The altitude stabilization is needed for hover act. This feature will help user who handles a QuadRotor by remote control and also for autonomous QuadRotor. There are some sensors as input to measure the altitude of QuadRotor hover mode, such as ultrasonic sensor, barometric sensor, and camera (image sensor). Pah developed the altitude measurement using a camera. For the experiment environment, he simulated a movement QuadRotor by using moveable platform completed with a Livecam 8-bit RGB and the landscape print out in the paper [9]. In 2013, Fatan developed an adaptive neuro-PID for controlling QuadRotor altitude. PID was tuned by an output of the neural network. First parameters contained a starting constant then were changed by an output of the neural network. In this research, he used a simulation which produced the sinusoidal path that reflected the real condition of QuadRotor [10].

Since 2014, author research has focused on QuadRotor altitude control for subsequent studies based on real applications which are already published in several publications. In previous work, the research started from developing the fuzzy logic scheme for QuadRotor altitude controller. An ultrasonic sensor worked from 1 up to 4 meters only. Experiments were conducted with indoor environment and with the real QuadRotor. The best result of optimizing all parts of fuzzy controller scheme is 7 cm of flight oscillation around the desired height 100 cm or about 7% oscillation (local publications). Because of limitation of the ultrasonic sensor (only up to 4 meters), the barometric sensor MS5611-01BA03 is used as a replacement for an input system. The barometric sensor produces a noise together with data measurement. Analysis of sonar and the barometric sensor for altitude hold mode was published in [11].

Type-2 fuzzy can solve a higher nonlinear system. Type-2 fuzzy has a structure to handle uncertainty and an imprecision better than a type-1 fuzzy (traditional fuzzy).There are some schemes for type-2 fuzzy. In this paper, we developed fast geometric type-2 fuzzy. The best result for QuadRotor altitude lock using a barometric sensor is 16% of oscillation [12]. In order to get the best framework for QuadRotor altitude lock especially in the indoor environment, performance comparison between the fuzzy controller and fuzzy-PID was published. The experiments used same QuadRotor, and battery capacity. The best parameters of each controller and also the same situation were identified during testing [13]. The difference between indoor and outdoor environments lies on wind disturbances. In an outdoor setting, we faced soft wind as disturbance. Three controller schemes had a good performance without disruption. Simulation of disturbance was done by moving the left and right extremely in an indoor environment. The benchmarking of type-1 fuzzy, type-2 fuzzy and fuzzy-PID has resulted that there are critical parameters for handling disturbance. First is the response for getting back to the desired altitude. Second are the minimized oscillations after getting back to the desired altitude. The best result is shown from type1 fuzzy for getting back to the desired altitude. The best result of minimized oscillations is produced from type2 fuzzy scheme [14].

Hence, this paper has two main objectives. First, it provides controllers design: type-2 fuzzy and fuzzy PID controller. Second, it provides the comparison of performance of both controllers. The rest of this paper is organized as follows. In Section 2, we provide a short explanation about QuadRotor system configuration. Then we explain controller design starting from type-2 fuzzy and continue to fuzzy PID in Section 3. Our experimental results and discussions are given in Section 4. Finally, the summary is provided in Section 5.

2. QuadRotor system configuration

Our QuadRotor used in this research was built from scratch. We used Whirlwind FY450 as a frame, four brushless motors Sunny Sky X2212 with 980kV, four blades 10 x 4.7 were propellers installed on BLDC motors, and we also used ZTW Spider 30A as an Electronic Speed Controller (ESC). For flight controller, we used low-cost flight controller KK2.0. Due to space limitation, we did not present the detailed wiring and block connection in this paper.

3. Controller design

With the type-2 fuzzy control structure, we guessed and made a hypothesis about the performance of type-2 fuzzy that it will handle uncertainty phenomenon in an outdoor environment better than fuzzy PID, but type-2 fuzzy has more calculation process than fuzzy PID that might make type-2 fuzzy worse than fuzzy PID especially for real-time processing. So, in this paper we designed a type-2 fuzzy and fuzzy PID separately, then tuned until the best performance could be reached, and compared the oscillation error and the responses after getting the disturbances.

We created a YoHe board for realizing our control design which was programmed using C++ language. The YoHe board recognized the height or altitude of QuadRotor time by time using the sonar sensor SRF-05 and the barometric sensors BMP085. The controller board produced a PWM signal as an output to the KK2.0 flight controller. For altitude stabilization purpose, we only controlled the throttle signal. In other words, we manipulated and processed the throttle signal from TX remote control before going to the flight controller. There is a stick/button in TX remote control to activate and deactivated auto mode (altitude stabilization mode).

3.1. Type-2 fuzzy controller

First of all, we introduce the type-2 fuzzy control process as seen in

Figure 1. There are two ways to decide the desired height. It can run manually through the program or get the last height before the auto mode turn ON. So, while auto mode is ON, the controller will keep the altitude of QuadRotor closely possible with the desired height. From many schemes of type-2 fuzzy, fast geometric defuzzification method is suitable for real-time applications without losing many features. A type-1 fuzzy or traditional fuzzy only has a 1-dimentional degree of the membership function. However, type-2 fuzzy has a 2-dimentional degree of the membership function. The clear explanation is seen in Figure 3 and Figure 4.We know the basic of fast geometric defuzzification which divides the membership function areas into five areas. Detailed calculation of type-2 fuzzy defuzzification can be figured from [15], [12]. In Figure 5, we can see one of an output fuzzy set looks.Before deciding a value of each membership functions of type-2 fuzzy, the behavior of QuadRotor learned through controlling QuadRotor using R/C and analysis of the flight data. With our expertise, we decided rule tables and initial values of two Inputs Membership Function (IMFs) and values of an Output Membership Function (OMF). We only focused on both of IMF and OMF in the optimization process. The optimal rules tables is shown in Figure 6 and the final values of Error IMF, Delta Error IMF, and OMF is shown in Figure 7, Figure 8, and Figure 9 respectively.

IOP Conf. Series: Materials Science and Engineering **273** (2017) 012031 doi:10.1088/1757-899X/273/1/012031

3.2. Fuzzy PID controller

PID controller was designed in the previous work. The big difficulty of PID controller is on parameter tuning process, and also one set of PID parameters (Kp, Ki, and Kd) only suits for one or two levels of altitude. Those problems raised the idea to combine PID controller with fuzzy that can have ability for automatic tuning while the altitude changes. The structure of fuzzy PID control process is shown in Figure 2. Fuzzy PID has two IMFs: Error IMF (Figure 10) and Delta Error IMF (Figure 11). With the purpose of fuzzy PID mentioned above, the OMF of this controller has three OMFs: OMF for Delta Kp, Delta Ki, and Delta Kdas seen in Figure 12, Figure 13, and Figure 14.

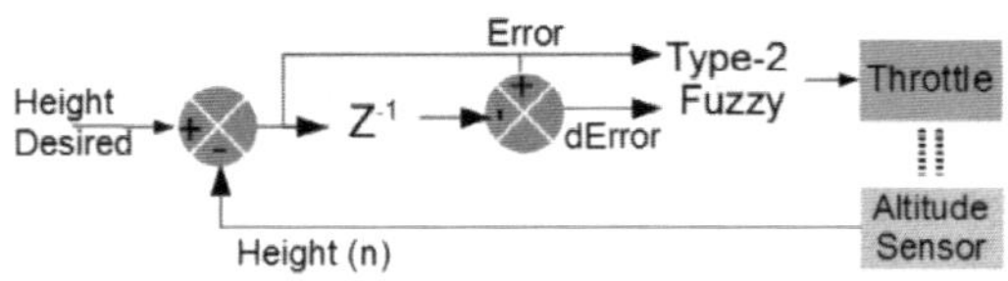

Figure 1. Type-2 Fuzzy Control Process.

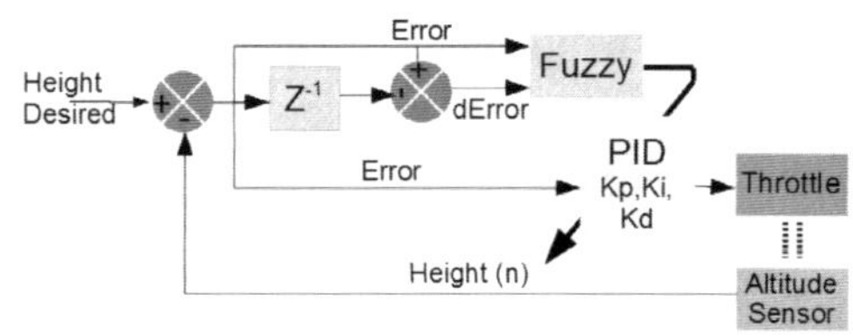

Figure 2. Fuzzy PID Control Process.

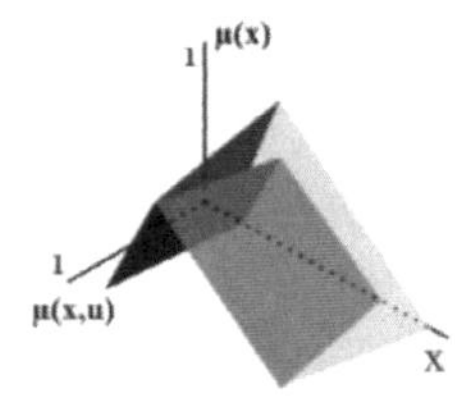

Figure 3. The Degree of Membership Function of Type-2 Fuzzy.

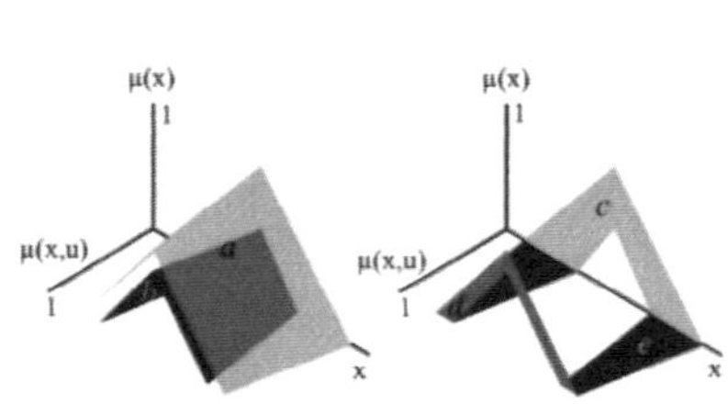

Figure 4. A Method of Fast Geometric Defuzzification.

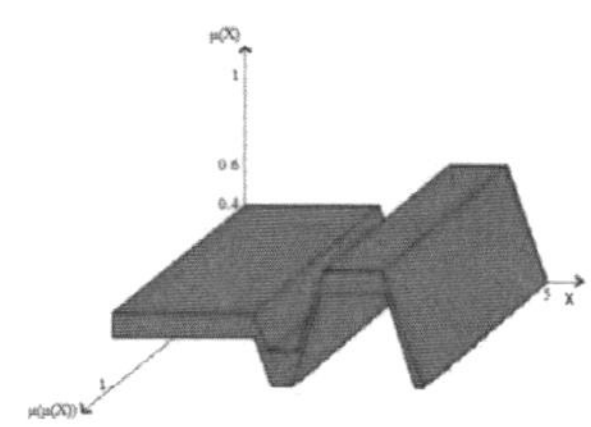

Figure 5. An Example of an Output Fuzzy Sets.

		Delta Error		
		N	Z	P
Error	N	VD	VD	D
	Z	D	S	U
	P	U	VU	VU

N = Negative
Z = Zero
P = Positive
VD = Very Down
D = Down
S = Stay
U = Up
VU = Very Up

Figure 6. Rules Table of type-2 Fuzzy.

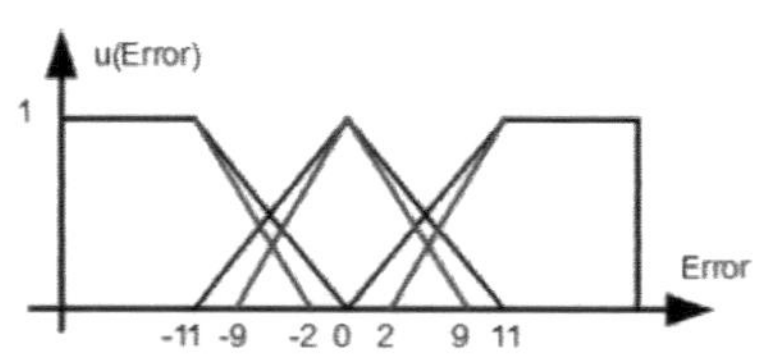

Figure 7. Error IMF of Type-2 Fuzzy.

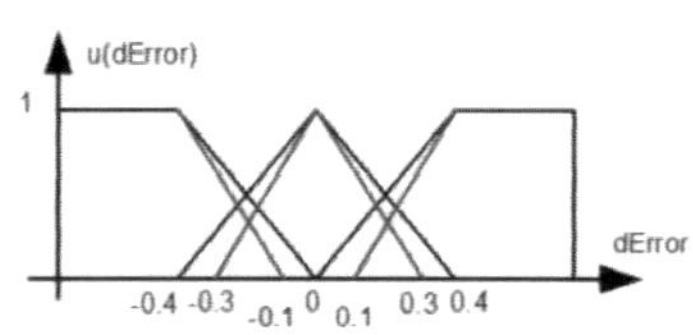

Figure 8. Delta Error IMF of Type-2 Fuzzy.

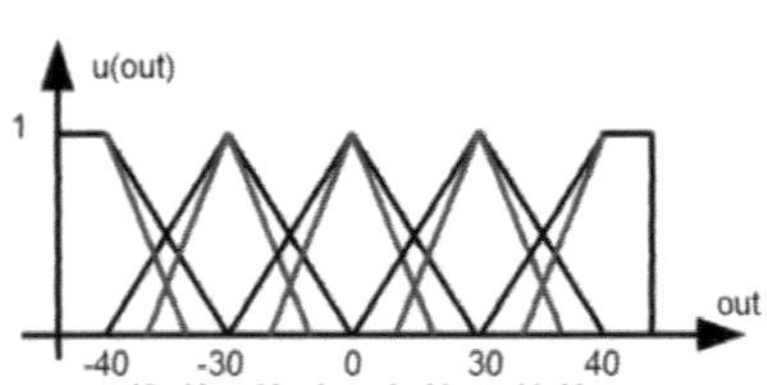

Figure 9. OMF of Type-2 Fuzzy.

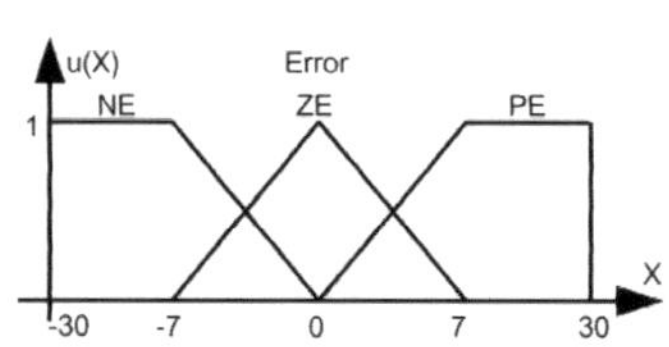

Figure 10. Error IMF of Fuzzy PID.

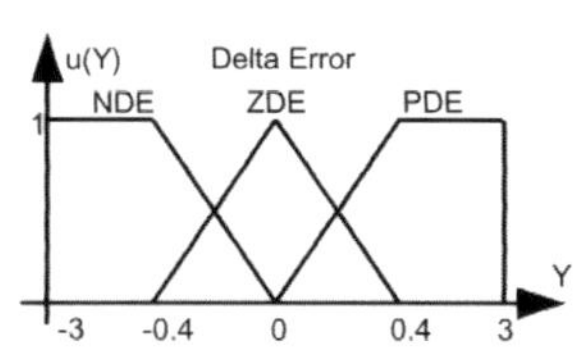

Figure 11. Delta Error IMF of Fuzzy PID.

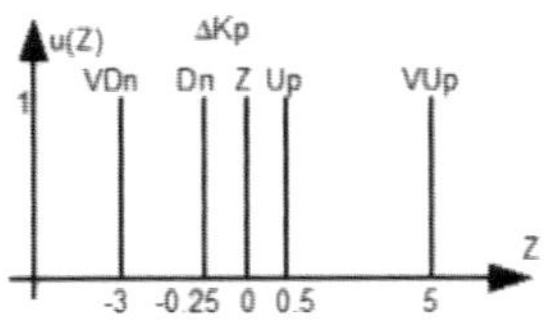

Figure 12. OMF for Delta Kp.

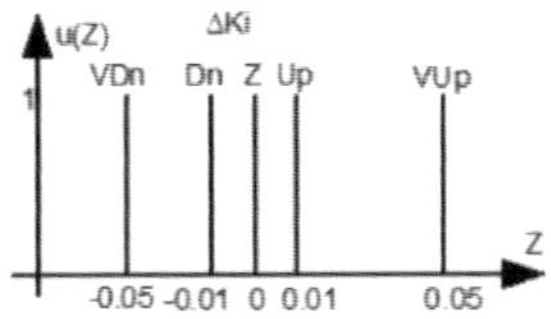

Figure 13. OMF for Delta Ki.

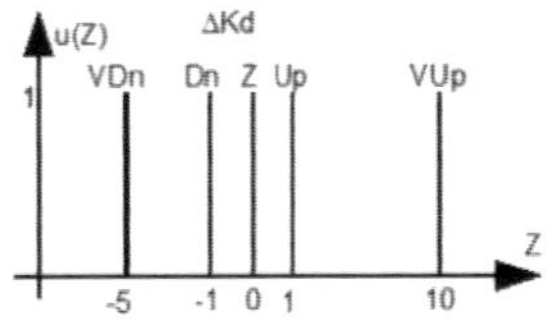

Figure 14. OMF for Delta Kd.

4. Experimental results and discussion

We chose the location for our experiments outdoor (Figure 15) starting from 01:00 pm until 04:00 pm. In these hours, we were able to obtain a medium wind speed. We used QuadRotor as shown in Figure 16. In this paper, because of space limitation, we provided only one experiment for several levels of altitude. Type-2 fuzzy results are illustrated in Figure 17, Figure 19, and Figure 21 for 250 cm, 300 cm, and 350 cm height, respectively. Then results of fuzzy PID are shown in Figure 18, Figure 20, and Figure 22 also for three height levels same with type-2 fuzzy.

Figure 15. The outdoor environment at 02:00 pm.

Figure 16. QuadRotor with Whirlwind FY450 Frame.

Type-2 Fuzzy Results (250, 300, and 350cm).	Fuzzy PID Results (250, 300, and 350cm).

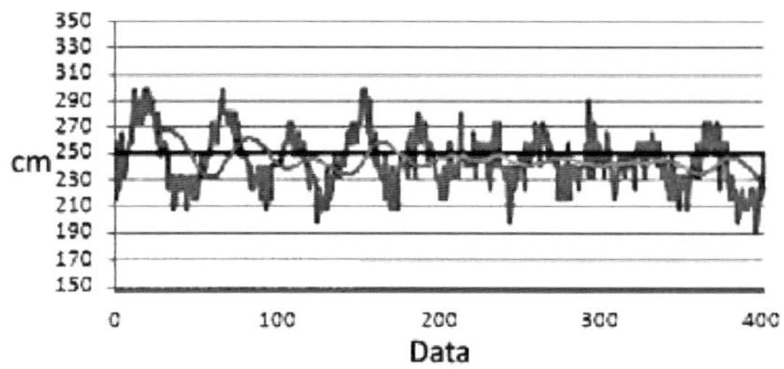

Figure 17. Oscillation = 42.2 cm.

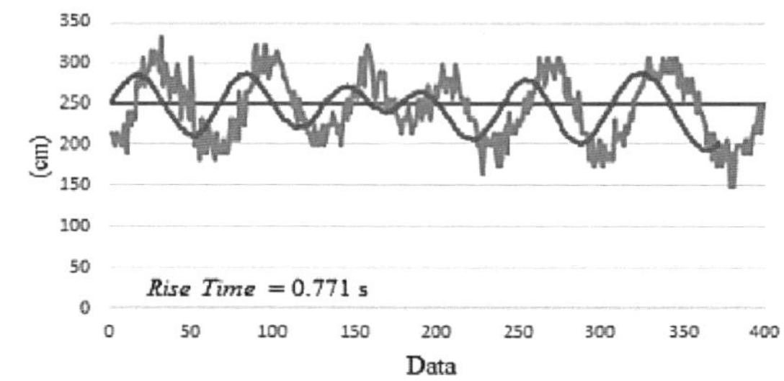

Figure 18. Oscillation = 94 cm.

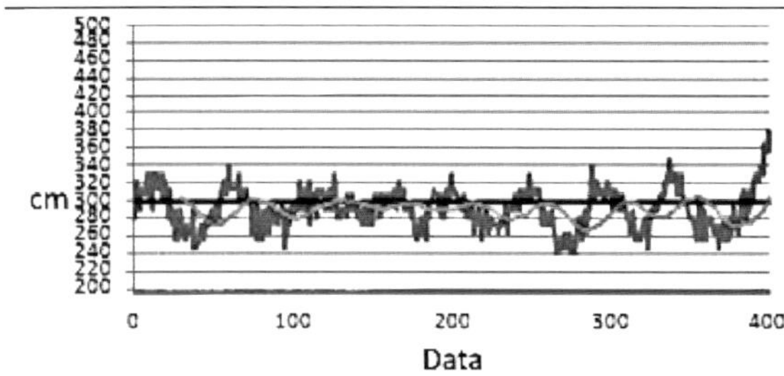

Figure 19. Oscillation = 38.9 cm.

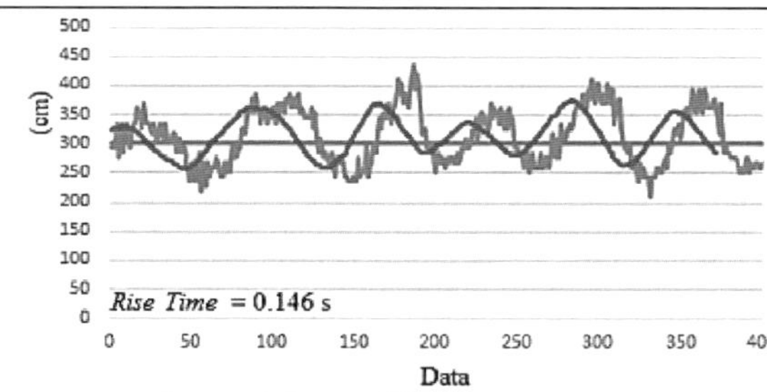

Figure 20. Oscillation = 108 cm.

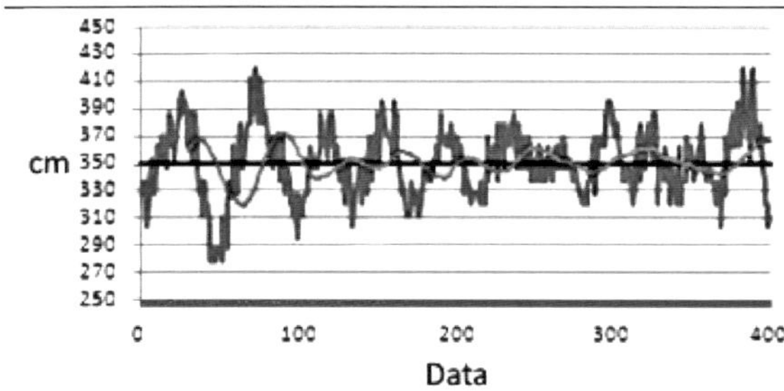

Figure 21. Oscillation = 51.9 cm.

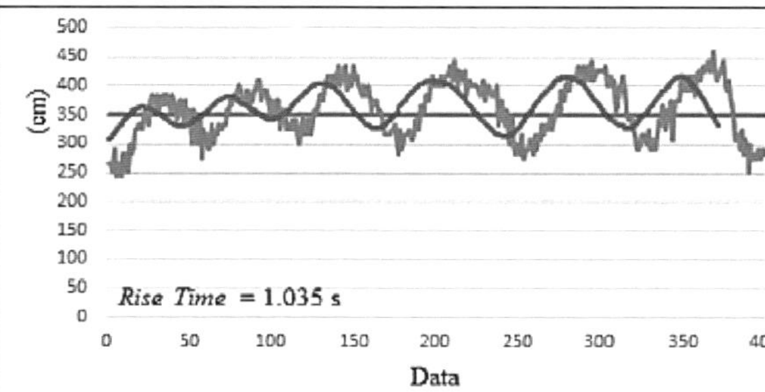

Figure 22. Oscillation = 111 cm.

We obtained oscillation around 40-50 cm for type-2 fuzzy and 90-110 cm for fuzzy PID. If we compare one to another in the same altitude, the oscillation due to fuzzy PID controller is greater almost two times as type-2 fuzzy. If we apply a moving average line for both controllers, we can see that the fuzzy PID controller responses always follow the Barometric sensor. However, type-2 fuzzy could solve the fluctuating values of the measurement.

5. Conclusion

Firstly, type-2 fuzzy and fuzzy PID controller design can be applied on non-commercial QuadRotor in an outdoor environment. Secondly, from several experiments results, we can conclude that type-2 fuzzy can handle well the fluctuating Barometric sensor values, which is in line with our hypothesis. Therefore, type-2 fuzzy which has three-dimensional membership functions is suitable for an outdoor application.

References

[1] Navajas G H T and Prada S R 2014 Building your own quadrotor: A mechatronics system design case study in *2014 III Int. Congress of Engineering Mechatronics and Automation (CIIMA)* **2** pp 1–5
[2] Prayitno A, Indrawati V and Utomo G 2014 Trajectory tracking of AR Drone quadrotor using

fuzzy logic controller *TELKOMNIKA (Telecommunication Comput. Electron. Control)* **12** pp 819–828

[3] Indrawati V, Prayitno A and Utomo G 2015 Comparison of two fuzzy logic controller schemes for position control of AR.Drone in *2015 7th Int. Conference on Information Technology and Electrical Engineering (ICITEE)* pp 360–363

[4] Desa H 2013 Adaptive hybrid control algorithm design for attitude stabilization of quadrotor (UAV) *Arch. Des Sci.* **66** pp 51–64

[5] Rinaldi F, Gargioli A, and Quagliotti F 2013 PID and LQ regulation of a multirotor attitude: Mathematical modelling, simulations and experimental results *J. Intell. Robot. Syst.* **73** pp 33–50

[6] Wicaksono H and Christophorus B 2014 T1-fuzzy vs T2-fuzzy stabilize quadrotor hover with payload position disturbance *Int. J. Appl. Eng. Res.* **9** pp 17883–94

[7] Castellanos J F G, Guzmán J J T, Durand S, Marchand N, Muñoz J U A and V R González-Díaz 2013 Attitude stabilization of a quadrotor by means of event-triggered nonlinear control *J. Intell. Robot. Syst.* **73** pp 123–135

[8] González I, Salazar S, Torres J, Lozano R and Romero H 2013 Real-time attitude stabilization of a mini-UAV quad-rotor using motor speed feedback *J. Intell. Robot. Syst.* **70** pp 93–106

[9] Pah N D and Hermawan H 2015 The development of image-based algorithm to identify altitude change of a quadcopter in *2015 7th International Conference on Information Technology and Electrical Engineering (ICITEE)* pp 76–81

[10] Fatan M, Sefidgari B L and Barenji A V 2013 An adaptive neuro PID for controlling the altitude of quadcopter robot in *2013 18th Int. Conference on Methods & Models in Automation & Robotics (MMAR)* pp 662–665

[11] Yusuf Y G and Agung H W 2016 Optimizing ultrasonic and barometric sensors for quadcopter's altitude-hold using YoHe V1.2 PID and KK V2.0 boards in *2016 International Seminar on Sensors, Instrumentation, Measurement and Metrology (ISSIMM)* 2016 pp 37–42

[12] Wicaksono H, Gunawan Y and Kristanto C 2015 A fast geometric type 2 fuzzy controller using barometric sensor for altitude stabilization quadrotor in *7th International Conference on Information Technology and Electrical Engineering (ICITEE)* **1** pp 520–524

[13] Wicaksono H, Gunawan Y, Yodinata A, and Leonardie 2015 Performance analysis fuzzy-PID versus fuzzy for quadcopter altitude lock system *J. Teknol.* **77** pp 33–38

[14] Wicaksono H, Gunawan Y, and Kristanto C 2016 Altitude lock capability benchmarking: Type 2 fuzzy , type 1 fuzzy , and fuzzy-PID with extreme altitude change as a disturbance *Lect. Notes Electr. Eng.* **365** pp 83–89

[15] Coupland S and John R 2008 A fast geometric method for defuzzification of type-2 fuzzy sets *IEEE Trans. Fuzzy Syst.* **16** pp 929–941

IOP Conf. Series: Materials Science and Engineering **273** (2017) 012032 doi:10.1088/1757-899X/273/1/012032

Investigating the role of Fuzzy as confirmatory tool for service quality assessment (Case study: Comparison of Fuzzy SERVQUAL and SERVQUAL in hotel service evaluation)

R D Wahyudi
Department of Industrial Engineering, University of Surabaya, Raya Kalirungkut, Surabaya 60293, Indonesia

Email: rahman.dwi.wahyudi@staff.ubaya.ac.id

Abstract. The problem was because of some indicators qualitatively assessed had been discussed in engineering field. Whereas, qualitative assessment was presently used in certain occasion including in engineering field, for instance, the assessment of service satisfaction. Probably, understanding of satisfaction definition caused bias if customers had their own definition of satisfactory level of service. Therefore, the use of fuzzy logic in SERVQUAL as service satisfaction measurement tool would probably be useful. This paper aimed to investigate the role of fuzzy in SERVQUAL by comparing result measurement of SERVQUAL and fuzzy SERVQUAL for study case of hotel service evaluation. Based on data processing, initial result showed that there was no significant different between them. Thus, either implementation of fuzzy SERVQUAL in different case or study about the role of fuzzy logic in SERVQUAL would be interesting further discussed topic.

Keywords: Fuzzy logic; SERVQUAL; Fuzzy SERVQUAL; Service Evaluation

1. Introduction

It is recently known that service is increasingly important so that it is even blended with product. Therefore, company pays attention in service improvement strategy as much as in product development strategy. The difference is that product tends to be a tangible stuff, while service tends to be intangible stuff. Consequently, assessing the service performance for customer satisfaction is increasingly difficult because it is linguistic variable and relatively subjective. Meanwhile, assessing the product performance is relatively easier than the service performance because it is able to be conducted using measurable variable, for an instance, durability of product used for product performance.

In term of measuring service performance, customer satisfaction is able to be used as indicator since every customer may have different either standard or definition of satisfaction. Therefore, many strategies should be formulated by company to avoid the misleading because of interpreting the customer satisfaction. For an instance, a company defines that assessing service performance is conducted using Likert scale of 1-5 in which the higher the value is, the higher the satisfaction is. However, the number of 3 may have different interpretation among customers. Some customers interpret that 3 is satisfactory, but some other customers may interpret that 3 is mostly satisfactory.

Probably, company anticipates aforementioned condition by defining that 3 is used for satisfactory situation. However, every customer still has different standard of satisfaction. Certainly, it would be very important in determining poor service needed to be improved. Misleading in interpretation of customer assessment for service performance causes mistake in decision making. Furthermore, it raises some cost to improve probably wrong service.

In many cases, evaluating service performance is conducted by using SERVQUAL as a tool. By using SERVQUAL, company understands the gap derived from the difference between expected service and perceived service. Based on the gap value, company arranges the service improvement priority. The raised problem is that expected and perceived services are linguistic and subjective variables, so that aforementioned problem probably occurs. In term of solving that problem, integrating fuzzy logic into interpretation of expected and perceived service assessment is initiated. Lately, aforementioned idea is called as Fuzzy SERVQUAL.

Fuzzy logic set is commonly used for helping customer giving more objective value by analyzing the ambiguous value. In Fuzzy logic, the ambiguous value is solved by calculating Triangular Fuzzy Numbers (TFN) using centroid method. By its application, Fuzzy has been used for many cases asserting the linguistics variable such as satisfaction assessment, temperature setting, patient recovery assessment in hospital, obsolescence assessment, and so forth. Based on this benefit, investigation about the role of fuzzy in determining SERVQUAL gap is indeed needed. Thus, the purpose of this paper is to compare Fuzzy SERVQUAL and common SERVQUAL in detail.

2. Methodology

In accordance with aforementioned background and purpose, a few steps were conducted. First of all, generating service attribute was needed by doing initial survey to customer and reviewing previous research. Then, questionnaire regarding expected and perceived service was arranged based on customer experiences. In this paper, the raw data used was obtained from previous research held by Gondowidjaja et al. [1]. Based on this raw data, the gap between expected and perceived service for common SERVQUAL could be calculated. Meanwhile, gap for fuzzy SERVQUAL was calculated by integrating fuzzy logic set into expected and perceived service. Commonly, fuzzy methodology is sequentially conducted as linguistic input, fuzzification, fuzzy output, defuzzification and crisp output. All of sequence is implemented for every linguistic input of expected and perceived service data. Thus, there would be two types of gap which would be further compared each other. Comparison result was used for investigating the role of Fuzzy in SERVQUAL. In a flowchart, methodology of this research is described as follows:

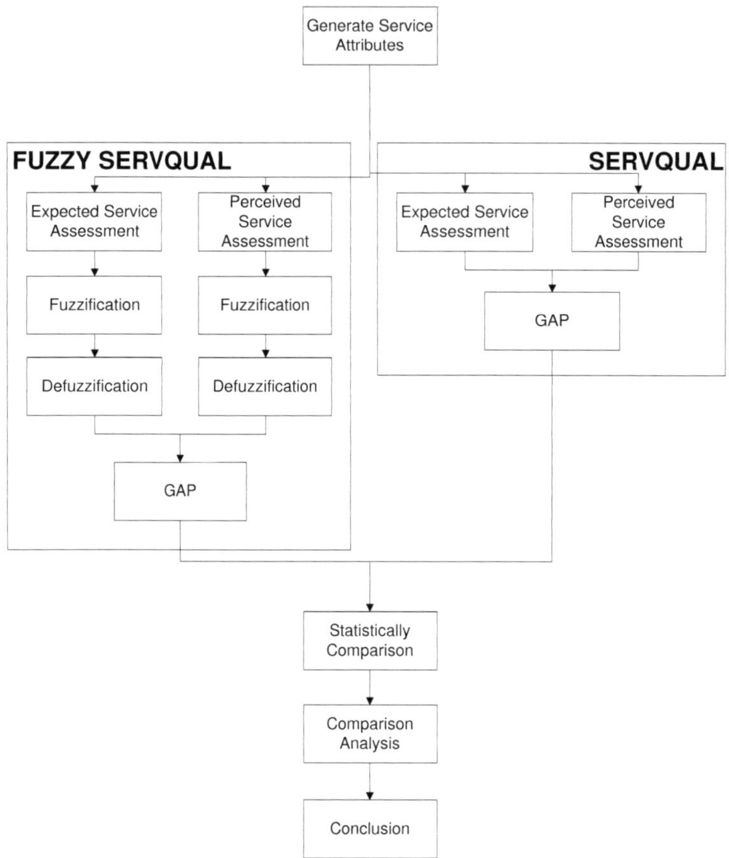

Figure 1. Research flowchart.

3. Literature review

SERVQUAL is a common tool used for measuring a service performance whether the service satisfies customer successfully or not. At first, SERVQUAL is initiated from definition of quality in service itself which many notions defining the definition of quality. One of those notions mentions that quality is a condition when output of process conforms to certain specification. Likewise quality in service, service is qualified when given service conforms to customer satisfaction as certain specification. Then, SERVQUAL as a tool measuring service performance understands that customer satisfaction can be achieved if expected service equals to perceived service. The difference coming from expected service and perceived service is called as Gap 5. Meanwhile, Gap 1-Gap 4 are associated with company's internal process in order to provide excellent service. On the whole process, the occurrence of gaps is presented in figure below.

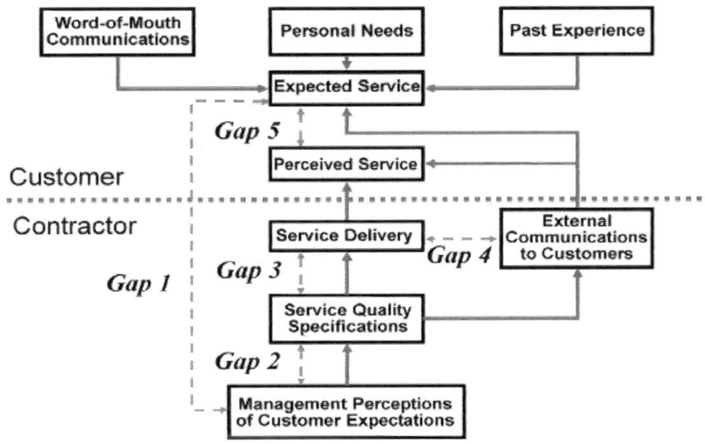

Figure 2. Gaps in SERVQUAL [2].

Certainly, measured gap is related to identified service attributes. Facilitating the data analysis, service attributes are categorized into certain dimension, for instance, there are 5 service dimensions for hotel industry such as tangibility, adequacy in service supply, understanding in service supply, understanding and caring, assurance, convenience [3].

In practice, expected and perceived service assessment is expressed in linguistic variable. Conventionally designed questionnaire frequently uses Likert scale to describe the feeling of respondents. Owing to the fuzziness of human thinking, this approach is inadequate and too simple to rule subject's way and measure complex human thinking and cognition [4].Thus, integration of Fuzzy logic and SERVQUAL will give many benefits. That fuzzy logic is integrated by doing fuzzification and defuzzification into expected and perceived service. Fuzzification of data is carried out on the transformed data by selecting input parameters into the horizontal axis and projecting vertically to the upper boundary of membership function to determine the degree of membership [5].

The efforts to use fuzzy for evaluating customers' satisfaction have been conducted in previous research [6]. It provides a new method of measuring perceived service quality based on triangular fuzzy numbers by replacing perceptions with satisfaction degree as well as expectations with importance degree. Then, inducing general solutions by computing the intersection area between two triangular fuzzy numbers was conducted to determine the worst service attribute. However, inducing general solutions by computing the intersection area between was based on possibility theory so it needed to reinterpret the voice of customers. Other studies tried to fuse fuzzy into several tools to evaluate service such as grey-fuzzy to model the customer expectation [7], fuzzy-AHP [8], fuzzy-AHP-TOPSIS to model the voice of customer [9], and fuzzy linguistics SERVQUAL to recognize the service quality assessment verbally [4]. However, discussion regarding the use of fuzzy in SERVQUAL by comparing common SERVQUAL and fuzzy SERVQUAL directly is important.

4. Data processing

In accordance with the research flowchart, firstly service attributes are generated by doing survey and reviewing the previous researches. Based on the obtained raw data from Gondowidjaja et al. [1], there were 24 service attributes generated. Then, data were processed and Gap 5 was obtained. Gap 5 for SERVQUAL was calculated by subtracting the weighted average of every perceived and expected service attributes. Meanwhile, Gap 5 calculation for Fuzzy SERVQUAL was started by doing fuzzification. In this stage, linguistic variable represented by Likert scale was converted to Triangular Fuzzy Number (TFN). Determining membership set will affect to scalar value in defuzzification stage which is calculated by using its geometric mean.

In this research, the membership sets were dissatisfied, quite satisfied, and satisfied. The value for dissatisfied was 1, 2 and 3; satisfied was 2, 3, and 4; strongly satisfied was3, 4, and 5. Based on the membership set, TFN for every service attribute was obtained. Furthermore, Gap 5 was calculated by subtracting the defuzzification TFN of perceived service and the defuzzification TFN of expected service. The result of Gap 5 calculation is presented as follow:

Table 1. Gap 5 Calculation of SERVQUAL and of Fuzzy SERVQUAL.

		SERVQUAL			FUZZY SERVQUAL								
No	Service Attributes	Expected Service	Perceived Service	Gap 5	Triangular Number			Expexted Service	Triangular Number			Perceived Service	Gap 5
Tangible													
1	Provided facilities (such as: gym center, spa, restaurant, etc) are clean, comfortable and reliable	4.38	3.44	-0.94	3	3.9	4.4	3.7	2.6	3.1	3.8	3.14	-0.56
2	Parking area is large enough	4.48	3.55	-0.93	3	3.9	4.5	3.76	2.7	3.3	3.8	3.22	-0.54
3	Employee Apparel is clean and tidy	4.56	4.72	0.16	3	4	4.6	3.79	2.4	3.7	4.8	3.5	-0.29
4	Food presentattion in restaurant is satisfied	4.6	4.46	-0.14	3	4	4.6	3.79	2.9	3.3	4.5	3.49	-0.3
5	Material related to service (such as soap, shampoo, towel, etc) are proper	4.67	4.7	0.03	3	3.9	4.7	3.81	3	3.7	4.7	3.74	-0.07
6	Room is clean and tidy	4.59	3.74	-0.85	3	4	4.6	3.79	2.7	3.3	4	3.29	-0.5
7	Toilet (both in lobby and room) is clean, quite large and comfortable)	4.49	3.76	-0.73	3	3.9	4.5	3.74	2.7	3.2	4.1	3.25	-0.49
Adequacy in Service Supply													
8	Employee is helpful and available	4.41	4.68	0.27	3	3.9	4.4	3.71	2.9	3.5	4.7	3.62	-0.09
9	Employee is perceptive and deft	4.43	4.42	-0.01	3	3.9	4.4	3.72	3	3.4	4.4	3.57	-0.15
10	Wifi connection is proper and easy to access	4.47	4.55	0.08	3	3.9	4.5	3.72	2.8	3.5	4.6	3.56	-0.16
11	Provided services are conform with promises	4.43	4.54	0.11	3	3.9	4.4	3.73	3	3.6	4.5	3.66	-0.07
Understanding and Caring													
12	Hotel provides flexible service as suitable as customer order	4.36	4.43	0.07	3	3.8	4.4	3.66	2.9	3.3	4.5	3.49	-0.17
13	Employee treats customer kindly	4.41	4.55	0.14	3	3.9	4.4	3.71	2.8	3.6	4.6	3.59	-0.12
14	Employee gives special empathy to customer (such as: greets the by name)	4.61	3.81	-0.8	3	4	4.6	3.8	2.7	3.3	4	3.31	-0.49
15	Employee has knowledge and information regarding to potensial nearby tourism object (shppoing mall, museum)	4.4	3.57	-0.83	3	3.8	4.4	3.68	2.8	3.3	3.8	3.24	-0.44
16	Employe asks for apology is any inconvienence	4.53	4.62	0.09	3	3.9	4.5	3.77	3	3.9	4.6	3.79	0.02
17	Food menu in restaurant is various	4.56	3.81	-0.75	3	3.9	4.6	3.75	2.6	3.3	4.1	3.29	-0.46
Assurance													
18	Hotel provides secure and comfortable environment for customer	4.59	4.68	0.09	3	4	4.6	3.8	3	3.6	4.7	3.7	-0.1
19	Employee has good working knowledge (such as foreign language, professionalism, communication skill)	4.41	3.5	-0.91	3	3.8	4.4	3.7	2.5	3.2	3.9	3.15	-0.55
20	Hotel provides 24 hours of service	4.51	4.63	0.12	3	4	4.5	3.77	2.6	3.2	4.8	3.38	-0.39
Convenience													
21	Reservation is easily done	4.62	3.76	-0.86	3	3.9	4.6	3.79	2.8	3.4	3.9	3.33	-0.46
22	Hotel can be easily accessed	4.6	3.81	-0.79	3	4	4.6	3.79	2.7	3.3	4	3.3	-0.49
23	Information regarding facilities and services is easily gotten	4.66	4.66	0	0	4	4.7	4.32	2.8	3.6	4.7	3.59	-0.73
24	Hotel is able to handle and solve the complaint	4.61	4.68	0.07	3	4	4.6	3.8	2.8	3.4	4.7	3.56	-0.24

Briefly, it is presented that there is different value of Gap 5 between SERVQUAL and FUZZY SERVQUAL. In addition, descending sorting for gap data above results different priority for service improvement. For SERQUAL, the top 5 of the biggest gap needing to be improved are as follows:

1. Provided facilities (such as: gym center, spa, restaurant, etc.) are clean, comfortable and reliable
2. Parking area is large enough
3. Employee has good working knowledge (such as foreign language, professionalism, communication skill)
4. Reservation is easily done
5. Room is clean and tidy

For Fuzzy SERQUAL, the top 5 of the biggest gap needing to be improved are as follows:

1. Information regarding facilities and services is easily gotten
2. Provided facilities (such as: gym center, spa, restaurant, etc.) are clean, comfortable and reliable
3. Employee has good working knowledge (such as foreign language, professionalism, communication skill)
4. Parking area is large enough
5. Room is clean and tidy

Based on the information above, it is known that there are the same four service attributes with different value of gap. However, the aforementioned difference is not necessarily considered as something statistically significant different. Therefore, statistical testing is conducted to state whether there is significant difference among the value of gaps or not. If the result shows that they are the same, then the difference of priority is not significant. Furthermore, if the improving service is conducted with different priority, then the final result is probably not significant different. ANOVA is the tool used for learning whether any significant difference among gaps. The hypothesis test is formulated are built as follows:

H_0: There is no significant different for gap 5
H_1: At least there is significant different for a pair of gap 5
By using Minitab, ANOVA testing results are presented as follows:

One-way ANOVA: S-Gap, F-Gap

Source	DF	SS	MS	F	P
Factor	1	0.006	0.006	0.04	0.834
Error	46	6.028	0.131		
Total	47	6.034			

S = 0.3620 R-Sq = 0.10% R-Sq(adj) = 0.00%

```
                 Individual 95% CIs For Mean Based on Pooled StDev
Level  N   Mean    StDev  ---------+---------+---------+---------+
S-Gap 24 -0.3046  0.4690   (-----------------*-----------------)
F-Gap 24 -0.3267  0.2054  (-----------------*-----------------)
                          ---------+---------+---------+---------+
                              -0.400    -0.320    -0.240    -0.160
```

5. Results and discussion

The ANOVA result above shows that p-value was0.834 which was greater than alpha 0.05, so the decision for this test did not reject H0. Thus, it was concluded that there was no significant difference between gap 5 of SERVQUAL and gap 5 of Fuzzy SERVQUAL. Probably, it was caused by the assessment among customers which was relatively indifferent. This reason was supported by descriptive statistics above titled individual 95% CIs For Mean Based on Pooled ST Dev in which both of data were relatively in the same position. Other possible influencing the result was definition of fuzzy membership which was not really varied and had a large subtraction area, so the conversion value was not strongly bold. However, it was too early to establish hypothesis that the use of fuzzy in SERVQUAL was not significantly powerful. It was necessary to compare and analyze the result for various case studies. Therefore, understanding of nature of faced case study became important.

The insignificant result of ANOVA in this case was influenced by the nature of service in hotel. In this case, the satisfaction assessment for the hotel was easily done with a result which was relatively equal between customers and the data source were domestic customers who were satisfied for service hotel gave. It was supported by survey data showing that the customer's assessment was not significantly varied. In addition, the hotel offered some facilities which were quite complete, well maintained and reliable, so that definition of satisfaction level between customers was relatively the same because they had the same perception about service standard in hotel. The insignificant result was also probably caused by defined Likert scale. Based on obtained data for SERVQUAL, satisfaction level of the customer revolves in Likert scale of 4 to 5 which belongs to the category of satisfied and strongly satisfied in linguistics variable. This range was not significantly wide. One of the consequences was that the value was not significantly sensitive. Likewise, obtained data for fuzzy SERVQUAL revolved in category of strongly satisfied. Thus, the use of fuzzy in SERVQUAL was necessary to be explored based on nature of the case so it would be more powerful. However, the use of fuzzy had already assisted to confirm something ambiguous.

6. Conclusion

In this case, it was obtained that the use of fuzzy in SERVQUAL was not significantly powerful. However, it is not easily stereotyped that the use of fuzzy in SERVQUAL is not significantly powerful for other cases. It is necessary to compare and analyze the result for various case studies. Therefore, understanding of nature of faced case study, data tendency, range of Likert scale and kind of linguistic variable are increasingly important to be considered.

7. References

[1] Gondowidjaja A, Hartono M and Wahyudi R D 2017 *Service Quality Improvement by Using SERVQUAL, KANO and QFD in D'Season Hotel, Surabaya* (Surabaya: Universitas Surabaya)

[2] Parasuraman Z and Berry 1985 A conceptual model of service quality and its implication *J. Mark.* **4** pp 41–50

[3] Akbaba 2006 A measuring service quality in the hotel industry: A study in a business hotel in Turkey (Duzce: Hospitality Management) *Int. J. Hosp. Manag.* **25** pp 170–192

[4] Charles V, Kumar M and Suggu S 2012 *Adapting fuzzy linguistic SERVQUAL model: A comparative analysis of bank service in Malaysia* (Malaysia: Centrum Catolica's Working Paper)

[5] Umoh U A and Nwachukwu E O 2010 Fuzzy rule-based framework for effective control of profitability in a paper recycling plant global *J. Comput. Sci. Technol.* **4** pp 56–67

[6] Chien C J and Tsai H H 2000 Using fuzzy numbers to evaluate perceived service quality *Fuzzy Sets Syst.* **116** pp 289–300

[7] Tseng M L 2009 A causal and effect decision making model of service quality expectation using grey-fuzzy DEMATEL approach *Expert Syst. Appl.* **4** pp 7738–48

[8] Büyüközkan G, Çifçi G and Güleryüz S 2011 Strategic analysis of healthcare service quality using fuzzy AHP methodology *Expert Syst. Appl.* **38** pp 9407–24

[9] Büyüközkan G and Çifçi G 2012 A combined fuzzy AHP and fuzzy TOPSIS based strategic analysis of electronic service quality in healthcare industry *Expert Syst. Appl.* **39** pp 2341–54